IMAGINING PROGRESS

IMAGINING PROGRESS

Science, Faith, and Child Mortality in America

KRISTIN JOHNSON

THE UNIVERSITY OF ALABAMA PRESS
Tuscaloosa

The University of Alabama Press
Tuscaloosa, Alabama 35487-0380
uapress.ua.edu

Copyright © 2024 by the University of Alabama Press
All rights reserved.

Inquiries about reproducing material from this work should be addressed to the University of Alabama Press.

Typeface: Garamond Premier Pro

Cover image: 1895 diphtheria antitoxin bottle; courtesy of the Division of Medicine and Science, National Museum of American History, Smithsonian Institution; all other images stock.adobe.com
Cover design: Lori Lynch

Cataloging-in-Publication data is available from the Library of Congress.
ISBN: 978-0-8173-2201-4 (cloth)
ISBN: 978-0-8173-6149-5 (paper)
E-ISBN: 978-0-8173-9519-3

For my students, as always.

Contents

Acknowledgments ix

Introduction 1

Chapter 1
God's Two Books 7

Chapter 2
The Laws of Nature and of Nature's God 38

Chapter 3
A World Made for Melioration 70

Chapter 4
Nature Groaneth and Travaileth 104

Chapter 5
The Benevolent Arc of History 137

Chapter 6
The Promise and Perils of Salvation 181

Conclusion 223

Notes 229

Bibliography 259

Index 281

Acknowledgments

I am very grateful for the support and encouragement of my fellow contributors to the 2018 History of Science Society (Seattle) session "Anglo-American Science and Liberal Religion" (Matthew Stanley, Bernard Lightman, and Edward Davis), two anonymous reviewers for the University of Alabama Press, and the following friends, colleagues, and family: Keith Bengtsson, Eleanor Bigelow, Bill Brietenbach, John Hedley Brooke, Andrew DeCort, Shannon Dixon, Kena Fox Dobbs, Erik Ellis, Tamra Erickson, James Evans, Amy Fisher, Paul Farber, Mott Greene, Shauna and Chris Hansen, Suzanne Holland, Donna Johnson, Paul Kopperman, Jane Maienschein, Jennifer Neighbors, Elizabeth Nielsen, Bob and Mary Jo Nye, Diane Paul, Leslie Saucedo, Katherine Smith, Addie Taylor, Justin Tiehen, Ariela Tubert, and Peter Wimberger. I would also like to express my earnest thanks to Claire Lewis Evans, Christopher Hellwig, and the rest of the editorial team at the University of Alabama for their support and encouragement. This project was completed with the support of Puget Sound's University Enrichment Committee and the Whiteley Center at Friday Harbor Laboratories. I am very grateful to my students in Science and Religion: Historical Perspectives and History of Medicine for reading draft chapters, and to Thomas and William Ellis-Johnson for being such wonderful and inspiring distractions.

IMAGINING PROGRESS

Introduction

Sometime in 1997 Bill and Melinda Gates read an article by *New York Times* journalist Nicholas Kristof entitled "For Third World, Water Is Still a Deadly Drink." Kristof described how millions of children still die from diseases like diarrhea and pneumonia (both easily treated in wealthier countries). As new parents, Bill and Melinda recalled, the information hit particularly hard: "If there's anything worse than the death of a child, we said to each other, then surely, it's the preventable death of a child." They sent the clipping to Bill Gates Sr. with a note that said, "Dad, maybe we can do something about this," beginning a conversation that eventually led to the Bill & Melinda Gates Foundation.[1]

It isn't hard to imagine why the Gates family assumed child mortality rates could be lowered. After all, eliminating deadly child diseases had been done before. In dozens of countries. For dozens of diseases. In their history of the Foundation, the Gateses included a graph of the steady decline in global infant and child mortality rates over the past thirty years, from 83 under-five deaths per 1,000 births in 1990 to 37 in 2020. Today, graphs like this (and of national mortality rates in places like Iceland, where the under-five mortality rate is 2 per 1,000 births) prove that further progress can be made and more children saved.[2]

Before such extraordinary progress could be achieved, however, it had to be imagined. *Imagining Progress* tells the story of how Americans both explained and responded to child illness and mortality during eras in which no one could appeal to plummeting numbers on a graph as proof that child mortality could be decreased by human effort. It focuses on a time when no one knew for certain that a world could be made in which, in some countries, it would become 170 times less likely that a child would die. Or that the top ten causes of child mortality (smallpox, diphtheria, scarlet fever, measles, influenza, tuberculosis, diarrhea, typhoid fever, pneumonia, and whooping cough) would be removed from bills of mortality of many nations entirely. Or that a world could be created in which, in more countries than not, child loss would become the exception rather than the rule.

By examining how Americans both explained and responded to child mortality from the colonial era to the Scopes trial, this book examines debates over whether and how the world could be improved in the midst of profound evidence that very little could yet be done. Spending time with individuals in the past who had hope but no proof that children could be saved from disease allows us to ask the following questions: How and why did some imagine that child mortality could be reduced, before possessing the evidence of precedent? To what theological and scientific ideas were such hopes linked and why? How did different ideas about the origin of suffering influence what Americans imagined as the best means of saving children? What histories did Americans tell in order to justify particular visions of progress? How did visions of earthly salvation relate to beliefs about eternal salvation? And what limited the ability of different visions of progress to include all of America's children? *Imagining Progress* examines the history of Americans' diverging assumptions about God, nature, and progress at a place where the stakes were at their highest: the bedside of children during eras of high child mortality. Beginning with smallpox inoculation and culminating with the fights over evolution in the 1920s, the book traces the history of debates over whether, how, and for whom a better world could be made.

Each of our guides to this history, like most Americans until the 1880s, experienced the death of at least one son, daughter, or sibling (and sometimes more). Cotton Mather, with whom our story begins, lost thirteen of his fifteen children to smallpox, measles, congenital malformation, and other ailments. Benjamin Franklin lost a son to smallpox. Thomas Jefferson lost at least two daughters and a son in infancy and a daughter to whooping cough. Poet Phillis Wheatley lost three children to unknown illnesses. Transcendentalist Ralph Waldo Emerson lost a son to scarlet fever. Physician James McCune Smith lost six children to cholera and other unknown illnesses. Novelist Harriet Beecher Stowe lost a son to cholera. Physician Josiah Nott lost several children to yellow fever. Abolitionist Frederick Douglass lost a daughter to an unknown illness. And minister Henry Ward Beecher lost two sons to mumps and at least two other children to unknown illnesses.

Even after child mortality rates began to go down during the last quarter of the nineteenth century, the experience of child loss remained common, as our guides from that later time period show. Historian Andrew Dickson White lost a son to typhoid and two infants (a son and a daughter) to unknown ailments. Philanthropist Nathan Straus lost a son and daughter to "diseases of digestion." Civil rights activist Mary Church Terrell lost three infants to unknown illnesses. Sociologist W. E. B. Du Bois lost a son, probably to summer diarrhea. Physicians Abraham and Mary Putnam Jacobi lost a son to diphtheria. Birth control activist Margaret Sanger lost a daughter to polio. Writer Samuel Clemens (pen name: Mark Twain) lost children to scarlet fever and diphtheria. German immigrant Anna Bollinger lost an infant son to congenital malformation. Tulalip cultural leader Harriette Shelton Dover lost a sister to tuberculosis. Other guides, like the feminist Elizabeth Cady Stanton, Darwinian botanist Asa Gray, Omaha physician Susan La Flesche Picotte, eugenicist and physician Harry Haiselden, evangelical preacher Billy Sunday, politician William Jennings Bryan, and lawyer Clarence Darrow did not (to historians' knowledge) lose children, but they watched and commented on the options available to explain and respond to so great and common an affliction.

All of these individuals—physicians, poets, abolitionists, ministers, novelists, biologists, historians, statesmen, social reformers, feminists, lawyers—imagined that children might be saved from early death. In doing so each imagined, often in very different ways, that the world could be improved via human effort. As they imagined, they argued over why such suffering existed in the first place. They debated whether child mortality proved God close, distant, or a figment of the imagination. They argued about the limits of human agency and the lessons of natural law for guiding human action. They quarreled over the best means and how far child mortality might be reduced. They argued over the relation between saving children on earth and other kinds of salvation. And they debated whether visions of earthly progress and salvation should include all of America's children.

In order to understand these stances and debates of the past, we must set aside several things we may know or believe in the present. First, we must forget today's common generalizations regarding the relationships between science, medicine, faith, and religion. This book is about debates, arguments, and quarrels over ideas that concern these things, but it is not a book about conflict *between* them. Historians of science have demonstrated that those who impose on this past a narrative of inevitable conflict (or harmony, for that matter) use definitions of science, medicine, faith, and religion that simply did not exist when our story begins. Using present-day definitions of science, religion, and the boundaries between them might be fine if one is engaged in a philosophical or theological defense of particular ideas in the present, but it is not good history. As historian John Hedley Brooke notes, "Serious scholarship in the history of science has revealed so extraordinarily rich and complex a relationship between science and religion in the past that general theses are difficult to sustain. The real lesson turns out to be the complexity."[3]

Capturing this complexity as historians requires that we set aside our own assumptions about whether and why particular belief systems have legitimacy in the present. The historian's challenging (some would say impossible) task is to witness the past on its own terms rather than through the lens of present-day debates, beliefs, or twenty-twenty hindsight. Religion, for example, is often described by both apologists and critics as possessing power from the purpose and meaning that faith confers on suffering. But as we will see, knowledge claims that most would call "scientific" also derived power from the purpose and meaning they conferred on illness, suffering, and loss, long before science could actually do much to ameliorate such suffering. Similarly, science is often described as possessing power because of what it can do to save lives. But while many of the individuals whose stories are told here had hope, none knew, in the way we know it, that child mortality could be lowered so significantly. Indeed, for most of the time period examined by this book, the stances and debates we are watching must be explained *without* appealing to the ability of science to save children.

Taking just the infant mortality rate (the number of deaths under one year per 1,000 live births), we can see how much progress we need to forget: Today, the US infant mortality rate ranges from 4 to 8 depending on the state. Demographers estimate that for much of the nineteenth century the rate was well over 100. By 1926, when our story concludes, the rate hovered around 74. By the end of the first quarter of the twentieth century, then, some had evidence that progress was possible (as we will see, Americans also knew that these numbers were not dropping at the same rate for everyone), but even that evidence looked very different than it does in the present. Furthermore, parents, ministers, physicians, and scientists did not have access to present-day, retrospective analyses of why those numbers were (inconsistently and slowly) going down. We will witness debates over whether child mortality can be lowered and the best means of doing so. But present-day knowledge of why mortality rates declined will not help us understand the thoughts, hopes, and fears of those who lived in the past. Our task is to understand these stances and debates separately from our own beliefs or evidence of who, in retrospect, was right.[4]

In watching Americans imagine that children could be saved, this book does not argue that such beliefs either began or flourished only within the United States. Those would be absurd and unhistorical claims to make. Rather, *Imagining Progress* traces the particular shapes such beliefs took within colonial America and then, after 1776, the United States. It examines the links between particular visions of progress and changing ideas about God, nature, human agency, the origin of suffering, the shape of history, and the meaning of salvation. And it traces how and why Americans erected, maintained, or undermined boundaries around whose children mattered most as ideas about God, nature, and history changed. Given that we must avoid imposing present-day knowledge and boundaries onto this past, this book combines histories that are often told as separate tales. Ideas about the causes of child mortality, the possibilities of progress, and who should benefit from efforts to save children were debated by physicians, evolutionists, theologians, scientists, pediatricians, public health workers, abolitionists, feminists, preachers, and historians. As a result, this book intertwines the histories of science, medicine, evolution theory, theology, pediatrics, public health, eugenics, and the *history* of the history of science together.

There will be no clear, unambiguous heroes or triumphs in this history, even when mortality rates begin dropping for some communities. We will see new ideas about nature inspire efforts to save the lives of some children. But we will also witness Americans using those same ideas to justify *ignoring* the mortality of some children by declaring such loss to be the inevitable result of natural laws. We will see some Americans imagining radically inclusive visions of progress and others imagining themselves justified in excluding some children from visions of a better earth. And we will see all of these positions defended by appeals to God, nature, or (more often) a powerful combination of the two. Ultimately, this history provides an important reminder that all solutions to the problem of suffering, whether we categorize them as scientific, religious, or a mix of both, are formed within societies that make judgments regarding what constitutes knowledge, the origin of suffering, whose suffering matters, and who gets to say. Those judgments influence not only what can be imagined about the past, present, and future but also the policies pursued and the actions taken or not taken to ameliorate suffering.

This book concludes in the 1920s because by then many of the fault lines of Americans' present-day alternative visions of progress were firmly in place. Will the development of biomedical technology like vaccines save the most lives or is social and economic reform just as, if not more, important? Is belief in God or unbelief the best foundation for faith in progress? What lessons does an evolutionary understanding of human nature have for various visions of progress, if any? Must we resign ourselves to a certain amount of suffering in the world, due to either nature, God, or some complex mix of the two? Is saving children on earth or for heaven more important? Is a choice between these options even demanded? And who can be trusted to answer these questions? The Scopes trial of 1925 was a prominent confrontation between different visions of whether, how, and whose children could and should be saved. Here, that trial is examined as a clash over the extent to which ideas about God, nature, history, human agency, and salvation must change in order to improve the world.

Halting in the 1920s also keeps us within an era in which, although child mortality rates were finally (slowly) moving down for some communities, they had not moved enough to use anachronistic criteria for judging whose vision of progress Americans should have found convincing. We can better understand the hopes and anxieties of those trying to determine the costs and benefits of various visions of progress if we set aside our own knowledge of the triumphs and tragedies that resulted from those stances. That said, the fact that both the triumphs (an overall US infant mortality rate in 2022 of 5.6) and tragedies (Black, Native Hawaiian and other Pacific Islander, Hispanic, and American Indian infants continue to die at higher rates than white infants) of the present have arisen from this past is one reason why studying this history is so important.[5]

Ultimately, tracing this entangled history can help us better understand Americans' present-day, polarized visions of progress. We will be asking several questions about our guides to this past: What did they assume about the origin, meaning, and best response to suffering? How did those assumptions influence what they believed about the possibilities of progress and the limits of human agency? How did the stories they told about the past influence their visions of whether and how progress might be achieved, and for whom? Asking these questions of the past is good practice for asking similar questions of positions taken in the present. It is often tempting to describe stances on science and religion with which we agree as rational and good, and then describe those with which we disagree as ignorant, irrational, and even evil. But accurately mapping why Americans hold such different stances on important issues that involve science and religion requires much more, including careful attention to individuals' values, beliefs, and assumptions about a wide range of things. As we enter this past, it is important to remember two premises of present-day historical scholarship: First, that we can study and understand different positions on science, God, and suffering without adhering to those positions, and second, that developing better maps of the past is a crucial prerequisite of both understanding the present and imagining effective and just amelioration in the future.

Chapter 1

GOD'S TWO BOOKS

> Abigail, William, Katherine, Mary, Joseph, Abigail, Mehetabel, Hannah, Increase, Samuel, Elizabeth, Samuel Nathanael, Jerusha, Eleazer, Martha. Of 15, Dead 9, Living 6.
> —Reverend Cotton Mather's list of his children, only two of whom would, in the end, outlive him, *Diary of Cotton Mather*.

IN THE WINTER OF 1688, the firstborn child of Reverend Cotton Mather died at the age of five months after a series of violent convulsions. Mather described his little girl Abigail as "perhaps one of the comeliest infants that has been seen in the world."[1] Though heartbroken, he reminded himself that such great loss was an affliction felt by many. "A Dead Child," he wrote, "is a sight no more surprising than a broken Pitcher, or a blasted Flower."[2] Few parents escaped such loss, and some felt it over and over again. Mather's grief for Abigail provides a profound reminder of the injustice of the popular belief that high mortality rates desensitized parents to loss because they avoided becoming attached to their children. That myth was apparently first told by Philip Aries in his 1960 book *Centuries of Childhood*. Perhaps its persistence is an indication of how truly unimaginable such a state of existence had become by the mid-twentieth century—so unimaginable that grief was imagined away. Historians have provided extensive evidence that parents in former times felt great anguish at the loss of their children, including accounts of physicians treating women so disturbed by the death of a child that they became physically ill or delirious.[3]

How did settler-colonial parents in the seventeenth century explain the suffering and death of beloved children? How did they find comfort amid both the ever-present potential and common experience of such devastating loss? What actions did they think could be taken in the face of illness? How were particular concepts of God, nature, and history intertwined with their answers? And what alternative explanations of child disease and loss were they intent on replacing? We have relatively few sources with which to answer these questions. However, Mather kept a detailed diary and wrote sermons (with access to a printing press) in which he wrestled with both why so many children died and how parents should respond to such grievous loss. Indeed, Mather stood before his parishioners a day after Abigail died and delivered her funeral sermon. We will begin there.

Right Thoughts in Sad Hours

As Mather sat down to compose his child's funeral sermon, he knew that as a minister he must "write as well as think . . . the things that may be serviceable to the sorrowful." Sitting in the pew before him would be his bereaved eighteen-year-old wife, also named Abigail. (We have no means of knowing what Abigail thought, except to assume, based on her husband's high praise of her character as a Christian, that she believed as he did.) Mather must have supposed he carried out his duty well, for he later published the sermon under the title *Right Thoughts in Sad Hours, Representing the Comfort and the Duties of Good Men under All Their Afflictions; and Particularly, That One, the Untimely Death of Children: In a Sermon Delivered . . . Under a Fresh Experience of That Calamity*. Standing before his parishioners, Mather began: "I am this day visited myself with the sudden death of a dear and only child." Few "earthly anguishes," he said, equaled a child's death: "This Affliction is none the most easy to be born. . . . The Heart of a Parent will have peculiar Passions working in it, and racking of it, at such a time as this." He compared the loss of a child to the tearing off of a limb but reminded his listeners that the death of children was common. One can sense Mather trying to convince himself: he was not the first, and he would not be the last, to feel this terrible affliction.[4]

Mather counseled his listeners to remember Job's words: "Tho' he slay me, yet will I trust in him."[5] The story of Job is Jewish, Christian, and Islamic traditions' most famous tale of God allowing unmerited suffering. In Jewish and Christian versions, when Job is afflicted with terrible suffering, including both painful disease and the loss of his children, his friends offer various explanations. Affliction, offers one, is the result of universal sinfulness and thus is always deserved, whether the individual sufferer is a good man or not. Another friend searches for a more specific origin, insisting Job repent of whatever sin he committed to deserve such punishment. All assume there is some cause of Job's suffering for which he is to blame and for which he must therefore submit, given God is just. In reply, Job maintains his innocence. He rebels against the "merited suffering" theories of his friends. The lesson of Job's "yet will I trust him" is sometimes taken to be that the faithful must renounce any attempt to understand why an infinite God permits what we perceive as evil, for as humans of finite capacity we simply cannot succeed.

Mather seemed to repeat this lesson when he urged his parishioners to remember the words in Isaiah 55:8: "My Thoughts are not as your Thoughts." Only, he then proceeded to give as many suggestions as possible for why God might allow the death of a child. "It is possible that, if thy child had lived," Mather suggested, "it might have made thee the Father of a Fool or . . . the Mother of a Shame," for "it is a very ordinary thing for one Living Child to occasion more trouble than seven Dead ones." Perhaps God sent such affliction because it made us pray more and surely *any* means by which individuals turned toward God and eternal salvation must be a great good. Or perhaps God intended a child's death to direct attention away from earthly things: "To this end," Mather insisted, "the Lord by Afflictions lays Wormwood on the Breasts which you have hung too much upon. He will cause Creatures to be our Grief that they may not be our God." Mather even wondered, while trying to "quiet [his] own tempestuous Rebellious Heart," whether he had been afflicted so that he could counsel his suffering parishioners with more sympathy. For he now truly understood the temptation of believing "all these things are against me." Finally, he proposed that affliction made one more sensible "of the Agonies and Anguishes" felt by Christ for human benefit.[6]

Of course, Mather was wrestling with a problem—namely, the origin and meaning of suffering—with which humans have wrestled for thousands of years. A little over two decades after Mather delivered this sermon, the German philosopher Gottfried Leibniz coined the word *theodicy* (meaning the "justice of God") for efforts to explain why an all-knowing, all-powerful, and all-good God (in whom most settler-colonial Americans believed) permits evil and suffering. Monotheist traditions have developed various theodicies. Some hold that a creation in which free will exists is a greater good but that free will comes with a cost—namely, that humans may act wrongly and in doing so produce evil and suffering for themselves or others. Others emphasize that the world is designed to allow humans to undergo spiritual growth (a great good) that ultimately prepares them for communion with God but that such growth can only happen in a world in which moral choices can be made between good and evil. While still others deny the claim that humans can say for certain that any particular instance of pain or suffering is, in fact, evil.[7]

Child suffering has always raised profound problems for attempts to develop compelling explanations of why a good God allows so much suffering. We will soon witness some, for example, question what possible spiritual and moral benefit could result from a small child coughing up its insides from whooping cough or being strangled to death by diphtheria. Faced by such dilemmas, some, including Mather, transferred the spiritual lesson to the worried parents, who surely (they hoped) prayed more at such hours. Ultimately, and amid variation, Jewish, Christian, and Islamic theodicies insisted that humans must trust that God has a good, redemptive reason for permitting suffering, including that of children. Believers assumed that all suffering has a divine origin, meaning, and purpose. As we will see, very few claimed that, as a result, humans should do nothing to try to alleviate suffering. Mather and his contemporaries supported and used medicine and believed in the duty of charity. But the belief that suffering came from God's hand provided the primary framework within which Jewish, Christian, and Islamic believers explained child illness and loss.

Mather's thinking about suffering and salvation was rooted in the story of the fall of man as recounted in Genesis, the first book of the Hebrew Bible (in Christianity, the Old Testament). There, scripture recorded how God created Adam perfect in the Garden of Eden and gifted Adam the ability to name all of God's creatures. When Adam and Eve disobeyed God by eating from the Tree of Knowledge, evil, toil, sorrow, disease, suffering, and death entered into God's creation as just consequences of their transgression. "Had our First Parents eaten of the *Tree of Life*," Mather wrote, "doubtless a confirm'd state of *Perfect Health*, both in Themselves and their Offspring had been the *Fruit* of it. But our *First Parents* criminally applied themselves to the forbidden *Tree of Knowledge*. This proved a *Tree of Death* . . . and *Sicknesses* are among the *Punishments* of that nefandous Crime."[8] Nor were any of Adam and Eve's children, down to Mather's own babies, exempt from the consequences of that first original sin. Suffering, sin, death—all entered creation as just, vicarious punishment for Adam and Eve's disobedience: "Your Children are born with deadly *wounds* of sin upon their souls," Mather preached. "There is a *Corrupt* Nature in Every Child, in its Infancy: Yea from the *very birth*, they Go *Astray Speaking Lyes*."[9] For Christians, Christ's sacrifice as told in the New Testament constituted vicarious atonement for humanity's sins, thus opening a path toward redemption and eternal salvation, but that path was entirely undeserved and all too easily closed.

As a Christian, Mather believed that the possibility and promise of eternal salvation in heaven provided the ultimate justification of earthly suffering. His belief that the afflicted Christian would be more likely to go to heaven than hell explained, first, why God sent affliction and, second, why one must therefore resign oneself to God's will when a child died. (Although Puritan Christians like

Mather believed in the possibility of infant damnation, historian Peter Gregg Slater notes that ministers tended to speak of specific children as though they were in heaven. The main exceptions were murdered and illegitimate children, perhaps as a deterrent against both infanticide and adultery.)[10] Thus, amid his own "earthly anguish" of having lost a beloved child, Mather tried to remind himself, his bereaved wife, and his parishioners that as a Christian one must not speak of dead children as lost. "Be of good cheer," he urged, "Your Children are in a better place, a better state, than you your selves are yet arrived unto.... Those dear Children are gone from your kind Arms into the sweet Bosom of Jesus, and this is, by far the best of all."[11] This belief that a child was now in heaven could inspire conflicted feelings in parents. In 1660 the British writer Alice Thornton recorded how, after the death of her son William, her daughter admonished her for wishing him back again:

> My dear Naly came to me, then being about 4 years old, and looked very seriously on me, said, "My dear mother, why do you mourn and weep so much for my brother Willy? Do you not think he has gone to heaven? . . . Would you or my father have my brother to live with you, whenas God has taken him to himself to heaven, where he has no sickness, but lives in happiness? Would you have him out of heaven again, where he is in joy in happiness?" . . . At which the child's speech, I did much condemn myself, being instructed by the mouth of one of my own children, and begged that the Lord would give me patience and satisfaction in his gracious goodness, which had put such words into the mouth of so young a child to reprove my immoderate sorrow for him, and begged her life might be spared to me in mercy.[12]

The fact that the call to remember that their children were with Jesus needed constant repeating is evidence of how difficult it was for parents to reconcile themselves to absence (notice Thornton ended her prayer for resignation with a plea for little Naly not to be taken as well). Given pervasive belief in God's close, personal governance and the possibility of eternal salvation, the appropriate response to the loss of a beloved child was widely agreed on: trust in God's will, faith in the possibility of reunion in heaven, and renewed commitment to Christian living. But these were clearly hard things to do. As historian Philippa Koch notes, "There is nothing passive about resignation."[13]

Mather tried to remind his listeners (and himself) that Christians had not been promised respite from tribulations on earth—indeed, quite the opposite. Did not scripture say, "many are the Troubles of the Righteous?" But the Christian also knew (and Mather meant this reminder to be a great comfort) that affliction was not random. Affliction always came from God and always for good reason. Scripture promised that God sends trouble and affliction "only if need be" and that "all things shall work together for good."[14] To doubt either belief was to know better than God how things on earth should be done. And, to Mather, second-guessing God was the devil's province.

Faced with the need to reconcile himself to God's denial of his prayers to spare his children again and again, Mather often preached the folly of limiting the wisdom of God "unto our own way of answering it." "The bravest effort of a true and a strong Faith," he urged, "is, to leave all entirely unto the Lord, and be satisfied with the infinite wisdom of His conduct."[15] Mather had many opportunities to practice what he preached. He lost little Abigail's mother in 1702 to breast cancer, complicated by the terrible symptoms of smallpox. Six of their nine children predeceased her. Mather delivered the funeral sermons for many of those children as well. In private, he recorded his own struggles: "Tis mortal," he scribbled in his diary when writing of one child, "and I will not entangle my affections with it!"[16] Perhaps these exhortations are why historians were once fooled into thinking that colonists did not love their children as much as later generations, but in fact the constant admonitions demonstrate how difficult it was to prevent "entangled" affections.

It is important to note that the primary object of "right thoughts" (faith in heaven) involved hope of much more than a place without suffering. Heaven was also a place where all afflictions would finally *make sense*. In heaven, Mather wrote, "every Labyrinth of providence will be explained, for every one of our Afflictions in this, we shall return a Million Hallelujahs in another and a better world." There, individuals would see "how God has at once Afflicted us and Amended us." Mather was absolutely certain that even the most afflicted would then say, "He hath done all things well."[17] When his wife Abigail died, he carefully recorded her final words after months of dreadful pain: "Heaven, Heaven will make amends for all!" Deeply moved, and perhaps in desperate need of the reminder, Mather placed Abigail's words as the final line of a poem he composed for the nearly one hundred community members who sat at her bedside during her final illness.[18]

Of course, Christian explanations of suffering were neither spoken nor printed on lands devoid of alternatives, as Mather knew very well. Like all Bostonians, he lived and wrote on land originally home to the Massachusetts people. As a minister, he even wrote tracts aimed at convincing Indigenous leaders to replace their heathen beliefs with Christian ones. He printed sermons with Massachusett translation opposite the English text, outlining the key lessons of the Bible: that God created man in his own image, that through Adam and

Eve's disobedience sin and death entered the world, that God had sent his son to give everlasting life to all who believe in him, and that a day of judgment would come. Clearly, Mather believed it his duty to convince "heathens" that their explanations of the world were at best nonsense and at worst marks of sin that deserved eternal damnation.[19]

Further south, in the colony of Georgia, John Wesley, the founder of Methodist Christianity, also preached Christian explanations of suffering with the aim of converting the "heathen aborigines." Lamenting his lack of success, he later wrote that the natives had "no inclination to learn any thing, but least of all Christianity, being full as opinionated of their own parts and wisdom as either modern Chinese or ancient Roman." Wesley once asked a group of Chicasaws if they believed God had made "the sun and the other beloved things." Their leader, Paustoobee, replied, "We cannot tell. Who hath seen?" Asked whether he believed God loved him, Paustoobee answered, "I do not know. I cannot see him." Indeed, upon returning to England, Wesley recorded that he had not "found or heard of any Indians on the Continent of America, who had the least desire of being instructed." (Both Mather and Wesley knew "popish Indians" who had converted to Catholicism, but as Protestants they believed that no better than heathenism.)[20]

Both Mather and Wesley knew that Indigenous Americans had their own explanations of the origin and meaning of great affliction, but neither had any interest in recording those explanations. Historians know that Indigenous explanations of suffering tended to give agency to a greater a range of nonhuman characters. In the nineteenth century, Cherokee elders told, for example, of how "in the old days the beasts, birds, fishes, insects, and plants could all talk, and they and the people lived together in peace and friendship." But when people began to increase and kill animals indiscriminately with new inventions like bows, knives, and hooks, the animals decided to take action. Meeting in councils, the animals created a whole host of diseases, so that man's wicked ways and proliferation would be checked.[21] Christian ministers like Mather and Wesley insisted that the fall of man and the consequent entrance of original sin into creation offered a better explanation of disease than (in the case of the Cherokee, at least) a council of avenging animals. They were also intent on spreading their explanation throughout the land.

The confrontation between these different explanations of suffering and disease occurred as child loss was occurring within Indigenous communities at an apocalyptic scale. European populations grew despite infectious diseases like smallpox and measles, but those same diseases wiped out entire families and communities among those who had called these lands home for millennia.

Again, Indigenous explanations of this extraordinary increase in suffering often imagined a wider cast of characters than Mather's omnipotent God. The Kiowa peoples from the Great Plains, for example, told the story of a trickster hero, Saynday, who once asked Smallpox of its origin: "'I come from far away, across the Eastern Ocean,' Smallpox answered. 'I am one with the white men—they are my people as the Kiowas are yours. Sometimes I travel ahead of them, and sometimes I lurk behind. But I am always their companion and you will find me in their camps and in their houses.' 'What do you do?' Saynday repeated. 'I bring death,' Smallpox replied. 'My breath causes children to wither like young plants in spring snow. I bring destruction. No matter how beautiful a woman is, once she has looked at me she becomes as ugly as death. And to men I bring not death alone, but the destruction of their children and the blighting of their wives. The strongest warriors go down before me. No people who have looked on me will ever be the same.'"[22]

The devastation wrought by diseases like smallpox was so terrible that some Indigenous Americans decided that their own explanations weren't good enough and that ministers like Mather must know something they did not. Historian Julius Rubin has documented how "praying Indians" (the term used for those who converted to Christianity) found in the Gospels a more convincing explanation for their suffering as European populations grew and expanded. In 1652 the Natick "praying Indian" Robin Speene explained an epidemic that took his own children "with great torments in their bowels" as follows: "I see God is angry with me for all sin, and he had afflicted me by the death of three of my children, and I fear God is still angry, because great are my sins." Rubin notes that some were drawn to Christianity's explanation of suffering as ultimately redemptive. After all, the Bible seemed to "account for good and evil by casting events as indicative of God's will, by providing solace in the face of uncertainty, and by positing God as an immanent and active force in the affairs of this world." In making sense of the trauma, disease, and displacement that followed the advance of the English, French, and Spanish, Christianity also prescribed appropriate action: prayer and turning away from sin. Of course, according to both Protestant and Catholic missionaries, avoiding sin required turning away from Indigenous knowledge, beliefs, and customs.[23] In other words, what constituted "right thoughts" when a child died was intimately tied to what one must believe about many things.

At least one individual in Mather's household seems to have found New England ministers' answers to what constituted "right thoughts in sad hours" of little use. Mather lamented the fact that he could inspire neither proper Christian behavior nor correct belief in a young, enslaved African whom his parishioners "gifted" him in 1706 and whom Mather had "putt upon... the Name of *Onesimus*." Although Mather did not connect the two events, hints of Onesimus's rebellion against Mather's authority appeared on the heels of the death of Onesimus's own children. Mather recorded how he tried to take advantage of those deaths to fulfil his own plans for Onesimus's soul. "My Servant burying of his Son," Mather wrote in 1714, "it gives me an Opportunity, to inculcate agreeable Admonitions of Piety upon him." The inculcation seems to have failed, at least according to Mather's standard, for two years later, after the death of a second child, Mather wrote again: "My Servant has newly buried his Son. Lett me make this an Occasion of inculcating the Admonitions of Piety upon him" (by this point Mather had lost nine of his own children).[24]

Onesimus, it seems, had other ideas. A few months later, Mather recorded, "My Servant *Onesimus*, proves wicked, grows useless, Forward, Immorigerous" (meaning "disobedient" and "uncivil"). Indeed, Mather was praying for guidance on how to remove him from the household. Onesimus found the answer: he soon paid the sum required to purchase his replacement. Upon Onesimus's departure, Mather lamented that he had not devoted more effort to saving Onesimus's soul: "Who can tell, what may be done for him, and what a new Creature he may become, if more prayers were employ'd for him!"[25] No record exists of Onesimus "awakening to piety" of Mather's kind or of Onesimus's thoughts on the matter. (Knowing Mather, if he had evidence of such an event, he would have recorded it for posterity.) All we can determine for certain is that these two fathers, confronted with personal experience of the fact that "a Dead Child is a sight no more surprising than a broken Pitcher, or a blasted Flower," must have felt very different things regarding what a parent must believe when a child died.

The Gospel of Medical Care

Cotton Mather believed that, given the deserved prospect of eternal damnation, any means by which God turned individuals' attention toward heaven must ultimately be wise, good, and just, even the death of much-beloved children. Historian Margaret Humphrey Warner notes that no other theme is repeated more often in Mather's various medical writings than that sickness is the rod by which God corrects his offending children.[26] As a result, one's duty in the face of child loss seemed clear, as hard as it was to achieve. "O that every Man would always remember this," Mather lamented in his sermon on the untimely death of children, "the Griefs of thy heart are all ordered by that God by whom the Hairs of thy Head are all numbered. Not so much as the Tongue of a Dog will ever stir against thee, unless managed by the Hand of the Lord."[27] Children were on loan from God, and if taken to Heaven before adulthood, the devout parent must ultimately discipline their hearts into submission. In the words of bereft father, minister, and poet Edward Taylor:

> Grief o're doth flow: and nature fault would find
>   Were not thy Will, my Spell, Charm, Joy, and Gem:
>   That as I said, I say, take, Lord, they're thine.[28]

Faith that God's governance over creation was close, personal, wise, and benevolent was supposed to assure Christians that every affliction, no matter how great, had a divine purpose and meaning, while assuring grieving parents their children were in heaven. But while it might seem that Christianity's emphasis on affliction as heaven-sent rods of correction would lead to a denial of medical care as an impious intervention in God's plan, nothing could be further from the truth. When Mather lost his second wife, newborn twins, and a two-year-old daughter, Jerusha, to a measles epidemic in 1713, Mather recorded his prayers—"I begg'd, I begg'd that such a bitter cup as the Death of that lovely child might pass from me"[29]—*and* his plan to compose a brief tract entitled *Right Management of the Sick under the Distemper of the Measles*. It was only after death that resignation and gratitude for the promise of heaven became the dominant tune, perhaps because at that point one could be sure of what God's will was. Until then Mather tried to sway the course of disease and help others to do so through prayer *and* medicine, fully certain that medical action was also divinely approved.[30]

Clearly Mather did not see efforts to prevent children from dying as interfering in God's will. Just as Christ's words that "the poor will always be with you" accompanied the command to "open your hand wide to your brother" (both were originally from the Torah), belief in the inevitability of sickness went hand in hand with belief in the duty to ameliorate physical suffering. Indeed, Mather was on solid, orthodox ground in believing that hope in eternal salvation and God's omnipotent providence was perfectly reconcilable with both prayer and medicine. Koch notes that "the image of early Americans resigned passively to disease and suffering is, in many ways, a product of the nineteenth century," a story we will trace in chapter 5.[31]

According to most Protestant, Catholic, Jewish, and Islamic traditions, belief in God's close, personal providence did not mean that one should do nothing in the face of illness. In the second century, for example, Clement of Alexandria wrote of medicine as God's gift to humanity, discovered by use of reason (also God's gift). In the fourth century, John Chrysostom also described physicians and medicine as gifts of God.[32] Chrysostom's contemporary, St. Basil the Great, even composed a treatise entitled *Whether Recourse to the Medical Art Is in Keeping with the Practice of Piety* that outlined the prevalent view among Christians: Although the fall of man had entailed a command to Adam and Eve and all their descendants "to return to the earth whence we had been taken" and be "united with the pain ridden flesh doomed to destruction because of sin," God had mercifully provided humanity with "medical art" to "relieve the sick, in some degree at least." "In as much as our body is susceptible to various hurts," Basil wrote, "the medical art has been vouchsafed us by God, who directs our whole life." The Christian must trust that God had a good reason when medicine failed and believe that ultimately the fate of the soul, not the body, was what mattered most, but that did not mean that God's gift of the medical art should be shunned.[33]

Following the Protestant Reformation of the sixteenth century, Protestant Christians maintained this positive stance on medicine and physicians. Lutheran Protestants, for example, could cite Martin Luther's words that "God created medicine and provided us with intelligence to guard and take care of the body so that we can live in good health."[34] Methodist John Wesley wrote a popular list of medical remedies entitled *Primitive Physick: or, An Easy and Natural Method for Curing Most Diseases*, published in 1747 (and in its twenty-fourth American edition by 1880). Indeed, Wesley emphasized that medical knowledge, as the gift of God, helped missionaries demonstrate God's mercy toward a fallen creation.[35] Although differing on many matters, Jewish, Islamic, and Christian physicians emphasized, as one medieval Christian wrote, that "either (medicines) would cure or the patient 'will die for God did not will that they should be healed.'"[36] Cotton Mather echoed this stance in his book *The Angel of Bethesda* when he wrote, "O though afflicted, and under Distemper, go to Physicians, in Obedience to God, who has commanded the Use of Means. But place thy Dependence on God alone to direct and prosper them."[37]

For centuries, then, ministers supported physicians while reminding patients to ultimately look to God for their cure. They often did so by calling upon an ancient distinction (drawn by Greek philosophers) between the first cause (God) and secondary cause (the natural processes, or laws, through which God, as legislator of the universe, normally governed). Mather wrote, for example, in his sermon on the death of children, "It may be one of thy Delights is put into the Ground: O look up, look above Chance, look above all second Causes."[38] Mather was reminding his listeners that, although a physician might explain sickness via secondary causes, all affliction ultimately comes from God. "Disease, whatever be the Next Cause of it," he once wrote, must be viewed as proceeding from God, "the first Cause of all." Clearly, belief in an all-powerful, all-wise, and all-benevolent God, or "first cause," remained the ultimate foundation of Mather's explanation of sickness. But he also believed that God operated through natural processes ("next" or "secondary" causes). In other words, all things come to pass via natural laws governed through, in the words of the Westminster Confession, "the fore-knowledge and decree of God, the first cause."[39] A physician's job was to learn the secondary causes through which God generally governed creation in order to know when one might give, say, a helping hand to the secondary cause through which God healed. Thus, Mather imagined that physicians might, with God's approval, improve knowledge of secondary causes and in doing so learn how to ameliorate suffering.

Mather's confidence was influenced by interpretations of scripture espoused by a famous English statesman and natural philosopher named Francis Bacon. Like most Christians, Bacon explained the entrance of suffering into God's creation as due to the fall of man. That history had made sense of great affliction for ages, from the pain of childbirth to the existence of disease and physical death. Bacon also argued that the story of the Fall explained man's *ignorance* of nature, including medicines. After all, Genesis proclaimed that Adam had once known the names of all animals and plants. To understand what Bacon did next, it is important to remember that the shape of both Jewish and Christian history had always been progressive (rather than cyclical). Tribulation and suffering may have entered the creation at the fall of man, but, for Christians, Christ's sacrifice and atonement for human sin on the cross opened a path of salvation. His promise to return told of a heavenly reign in the future. Bacon interpreted Christ's promises as a signal that humans could participate in the earth's movement toward the second coming by actively recovering knowledge lost at the fall of man. Indeed, he argued that God, in his infinite benevolence, had provided a very specific path by which certain men (and he and his followers did mean, as we will see, just *men*) might recover knowledge and help rebuild the Garden of Eden: Christian men, Bacon argued, must discipline their fallible reason and senses via "artificial enhancements" (like the microscope and telescope) and "new evidentiary canons" (namely, observation and experiment). The resulting increase in knowledge would allow mankind to reproduce the Garden of Eden. Thus, Bacon saw science (or, in seventeenth-century parlance, natural philosophy and natural history) as working in concert with God's purposes. For Bacon, Eve had sinned, not in being curious, but in seeking wisdom by eating from the Tree of Knowledge, rather than the laborious path of finding out for herself via observation and experiment.[40]

Two additional components of Bacon's plan are important to our story. First, Bacon argued that the primary means of producing this knowledge would be the cooperative, organized efforts of learned Christian gentlemen. Second, the knowledge produced would be for "the universal benefit of mankind." To Bacon and his disciples, this divinely approved effort to establish right knowledge thus gave the study of nature a pious and important role in the grand endeavor of recovering what mankind had lost due to the Fall. It also upheld God's benevolence despite the existence of great suffering, outlined very particular methods as the best route toward greater knowledge, and imagined that humanity's ability to reduce suffering would improve over time. Crucially, Bacon imagined that the years prior to Christ's return would be paved with radical new innovations for the "relief of man's estate." Indeed, he imagined that earthly progress via human participation in God's purposes signaled Christ's imminent approach. Eventually, this vision of Christian history would be called *post*millenialist, since Christ's return would follow increasing progress on earth, as opposed to *pre*millenialist Christianity, in which Christ's millennial reign followed *increased suffering*. This distinction will eventually be crucial to our story.

Francis Bacon lived and wrote in England in the first decades of the 1600s. By 1700, Cotton Mather had become the colonies' most ardent proponent of Bacon's vision of the relationship between progress, human agency, and Christian prophecy. In 1702 Mather wrote that Bacon's famous manifestos for the study of nature (like *The Advancement of Learning*, 1605, and *The New Atlantis*, 1627) "show'd em the way to the 'advancement of learning.'"[41] Like Bacon, Mather envisioned mankind taking up the tools provided by God to learn new things and improve the human condition, even as one bowed before the wisdom of God's ways when it was clear that nothing more could be done. Thus, Mather could preach that "the bravest effort of a true and a strong Faith is, to leave all entirely unto the Lord, and be satisfied with the infinite wisdom of His conduct" *and* to emphasize the duty of medical care. Mather cited herbal remedies as evidence that God pitied "us under the sad effects of our offences" and asked, "Can we be any other than charmed with the goodness appearing in it, when we see the plants everywhere starting out of the earth, and hear their courteous invitation, feeble man, I am a remedy, which our gracious Maker has provided for thy feebleness; take me, know me, use me, thou art welcome to all the good that is to be found in me!"[42]

Mather's favorite natural philosophers, naturalists, and physicians agreed (the word *scientists* would not be coined until the 1830s and was not commonly used until the 1870s). The English naturalist John Ray, whom we will meet in the next section, described plants as demonstrating "the illustrious Bounty and Providence of the Almighty and Omniscient Creator, towards his undeserving Creatures." Ray listed plants like the Jesuits' bark tree, the poppy, the rhubarb, the jalap, and others with "uses in Curing Diseases." (The Jesuits' bark tree is the source of quinine, the poppy is the source of opium, and rhubarb and jalap are both strong purgatives. All were obtained from overseas.) This was not necessarily a long list, but Ray was sure "there may be as many more not yet discovered, and which may be reserved on purpose to exercise the Faculties bestowed on Man, to find out what is necessary, convenient, pleasant or profitable to him." After all, God had also made man an adventurous creature with a desire to see unknown lands and bring home "what may be useful and beneficial."[43] The most famous physician of the era, Thomas Sydenham, agreed. In writing of opium, Sydenham (whom Mather greatly admired) exclaimed, "I cannot but break out in praise of the Great God, the giver of all good things, who hath granted to the human race, as a comfort in their afflictions, no medicines of the value of opium."[44]

Calling upon the natural world, including its medicines, to demonstrate the power, wisdom, and goodness of God like this had a name: natural theology. Together, the belief that both natural theology (the study of God based on nature) and revealed theology (the study of God based on scripture) could work in concert was based on a long-standing assumption that God gave humanity two books through which to know him: the Book of Scripture and the Book of Nature. Cotton Mather was one of the colonies' most ardent proponents of this assumption. But to understand what studying the Book of Nature became in the colonies we must cross the Atlantic back to England, to a generation of natural philosophers, naturalists, and physicians who embraced Francis Bacon's vision of a divinely approved route to improving knowledge, an endeavor that, by the 1660s, they were calling the "new sciences."

The New Sciences, the Mechanical Philosophy, and the Wisdom of God

Prior to the seventeenth century, Christian ministers often held that proofs about God's existence and character from scripture were more important than those derived from nature. Indeed, the belief that both man's perception of reality and the creation itself were subject to the corrupting influence of the fall of man meant that knowledge about God obtained from nature might be delusive. This concern ensured that the study of the Book of Nature had long remained subordinate to the Bible as a means of learning about God. By the late seventeenth century, however, wars over what constituted correct Christian belief inspired some to turn increasingly to the Book of Nature as a means of establishing consensus about God. In the wake of the Protestant Reformation, for example, England had descended into civil war from 1642 to 1651. The war culminated in the beheading of King Charles I in 1649 and eleven years during which the church of state (which had been Protestant since the mid-sixteenth century) was abolished. Meanwhile, a chaotic range of sects, each claiming to have discovered the truest interpretation of scripture, spread throughout Britain.

This violent history is crucial background for understanding the priorities of one of the first European scientific societies, the Royal Society of London. The Royal Society was founded the same year that parliamentary monarchy and the (Protestant) Anglican Church (or Church of England) were restored to power in 1660. Having lived through English Civil War, its members were committed to the restoration of social and political order by establishing consensus regarding God's governance. As Protestants (who had repudiated consensus via the pronouncements of a pope), they asked: How can one establish consensus about what God is like, on grounds upon which all reasonable Christian men might agree? With Francis Bacon's writings as models, they decided that agreement could be established via the careful establishment of "matters of fact" about God's other book: nature. No longer could people propose, as some of England's religious sects had done, that intuition, feeling, or individual interpretation of scripture provides accurate knowledge of God. Clearly intuition, feeling, and scriptural interpretation differed among individuals. Royal Society members like Robert Boyle argued that the disciplined experimental and observational study of the Book of Nature, witnessed or replicated by gentlemen (i.e., Christian, European, and upper-class men), provided important means of establishing agreement about the Book of Scripture and God. Based on the pervasive belief that God created nature, consensus regarding natural "facts" (newly defined as statements on which all reasonable men could agree) would, in turn, lead to consensus regarding God's power, wisdom, and knowledge, including God's providential direction of recent British history.[45]

To differentiate their methods and assumptions from alternative ways of studying nature in both the past and present, the Royal Society members (including Isaac Newton, Robert Boyle, and John Ray, often touted as the founders of modern physics, chemistry, and ecology, respectively) declared that they pursued "new sciences." Two things were supposed to differentiate their work from the "ancients": First, they conceptualized the natural world as composed of "inert matter in motion," governed by natural laws. In doing so, they divested matter of agency or "self-motion" in favor of a cosmos entirely under God's governance via divinely created natural laws. (Previous systems of physics imagined that four elements—namely, air, earth, fire, and water—*sought* their "natural place.") Second, and in contrast to knowledge systems that imagined nature as analogous to organisms (fire rising, for example, because, like organic beings, it "seeks" its natural place), Royal Society members imagined that natural things could best be understood as analogous to machines.[46] Together, these assumptions became known as the "mechanical philosophy."

It can be tempting to assume that these new sciences and the mechanical philosophy became powerful because they led to progress in medicine and the amelioration of the human condition. After all, the mechanization of nature and experimental methods adopted by proponents of the new sciences would (two centuries later) place extraordinary, lifesaving techniques in the hands of physicians. However, as historian Steven Shapin notes, seventeenth-century philosophers and physicians did not mechanize nature because doing so improved medical therapeutics. No lives were saved in the seventeenth century, eighteenth century, or much of the nineteenth century for that matter, by imagining the human body as analogous to a machine. Old techniques and therapies like bloodletting were given new, mechanistic interpretations, but treatments mostly remained the same. Rather, proponents of the new sciences mechanized nature because doing so proved useful in theological debates over God's governance. In France, the Catholic priest Marin Mersenne adopted a mechanical view of nature on the grounds it would allow men to distinguish between natural law and miracles (once one knew how matter in motion worked via natural laws, one could discern when natural law was not involved). Also in France, René Descartes mechanized all of creation in order to protect the special status of human beings (humans alone, he argued, possessed an immaterial soul, evident in their unique ability to speak).[47]

The members of the Royal Society explicitly drew upon the mechanization of nature to demonstrate the existence and character of God, establish consensus about God by appealing to man's reason, and defend the piety of the new sciences. To understand how and why these men applied a mechanistic view of nature to theology, it is important to understand that profound dilemmas lurked within the mechanical philosophy, especially given Royal Society members' interest in defending Christianity. Most versions of the mechanical philosophy relied on atomistic concepts of matter (i.e., they assumed matter is made up of indivisible units whose behavior determines all we see, hear, touch, and feel). But for centuries atomism had been linked with heathen, ancient Greek philosophers who declared the world arose from the random interactions of atoms rather than the benevolent design of an attentive creator. Indeed, critics of the Royal Society accused members of holding "atheistical" theories of matter. The dilemma was heightened by the ways in which mechanical philosophers called upon the old distinction between the first cause of phenomena (God) and the secondary causes through which God governed the world (natural laws). For mechanical philosophers, the first cause created the "machine" (e.g., the solar system or a dog), which operated according to secondary causes (the universal law of gravitation in the case of the solar system; the processes of digestion, reproduction, and generation in the case of the dog). Critics argued, however, that the Royal Society's emphasis on secondary causes might make the first cause irrelevant.

Historian John Hedley Brooke explains the dilemma as follows: "What role could be left for God to play in a universe that ran like clockwork? Would one have to side with those who became known as 'deists,' who restricted that role to the initial creation of a law-bound system, and who attacked Christian conceptions of a subsequent revelation? Would God's special providence, His watchful concern for the lives of individuals, not be jeopardized if all events were ultimately reducible to mechanical laws?" Did, in other words, the mechanical philosophy seal the argument for belief in the close, personal God of the theist, or did it at best justify belief in the more distant God of the deist?[48] In a culture in which most believed that theological, social, and political order depended on belief in a close, personal God, answering this question in a manner accessible to both commoners and the king was vital. Newton tried to do so by emphasizing that the perfect placement of the planets and their continued stability could only be the result of God's close, providential care and attention to his creation. But Newton's version of what would become known as the "design argument" depended on complex mathematics and thus could not be easily delivered from the pulpit or even in court.[49]

In 1691 a book appeared that placed demonstrations of God's close, personal attention to creation within reach of even the most unlearned parishioner. Composed by British naturalist John Ray, the book was entitled *The Wisdom of God Manifested in the Works of the Creation*. For 249 pages, Ray argued that the purposeful parts of animals and plants demonstrated the existence of an all-wise, all-powerful, all-benevolent God. In Ray's hands, the comparison of animals and plants to machines bolstered (actually quite old) "design arguments" for the existence and character of God. After all, human experience proved that the purposeful parts of machines (from clocks to watermills) had to be designed by rational mechanics to serve their particular uses. No random interaction of matter had ever produced a clock, much less the extraordinary fit of form to function in a dog's foot. Furthermore, Ray argued that God's extraordinary and benevolent attention to the wise design of his creatures proved the truths of Christianity, the existence of a soul, and the promise of heaven.[50]

Ray reveled in detailed descriptions of animal and human anatomy, all aimed at demonstrating "the exact fitness of the Parts of the Bodies of Animals to every one's Nature and Manner of Living." He carefully described how woodpeckers "have a Tongue which they can shoot forth to a very great length, ending in a sharp stiff bony Rib, dented on each side; and at Pleasure thrust it into the Holes, Clefts, and Crannies of Trees" to draw out insects. "More over," he continued, "they have short, but very strong Legs and their Toes stand two forwards, two backwards ... very convenient for the climbing of Trees, to which also conduces the stiffness of the Feathers of their Tails, and their bending downward, whereby they are fitted to serve as a Prop for them to lean upon, and bear up the Bodies." By accumulating examples like this, Ray wrote that he aimed to "make out in particulars" what scripture asserted in general concerning the works of God—namely, "In Wisdom hast though made them all"—by demonstrating how his works "are all very wisely contriv'd and adapted to Ends both particular and general."[51] (To Ray, "adapted" meant created by God to serve a particular function.)

Ray clearly believed that careful attention to both scripture and the purposeful design evident in nature would prevent the study of secondary causes from dispensing with the first cause. Furthermore, he believed God's intricate design of the tiniest part of his creation (the barbed tongue of the woodpecker was a favorite example) proved God's close attention to human lives. Nature thus justified confidence in scripture's message that a purpose existed in every affliction. One might not be able to discern the purpose or meaning of each particular incident of suffering, but that inability must not undermine faith that such meaning and purpose existed, even amid the greatest of afflictions. Ray wrote with reverence of "the Adapting all the Parts of Animals to their several Uses" as a demonstration of not only God's existence but also God's principal attributes, including infinite power, wisdom, and goodness. All human experience with machines, he argued, from water mills to cathedral clocks, proved that function could arise only from design:

> There is no greater, at least no more palpable and convincing Argument of the Existence of a Deity than the admirable Art and Wisdom that discovers itself in the make and constitution, the order and disposition, the ends and uses of all the parts and members of this stately fabrick of Heaven and Earth. For if in the works of Art, as for example; a curious Edifice or Machine, counsel, design, and direction to an end appearing in the whole frame and in all the several pieces of it, do necessarily infer the being and operation of some intelligent Architect or Engineer, why shall not also in the Works of Nature, that Grandeur and Magnificence, that excellent contrivance for Beauty, Order, Use, &c. which is observable in them, wherein they do as much transcend the

Effects of human Art as infinite Power and Wisdom exceeds finite, infer the existence and efficiency of an Omnipotent and All-wise Creator?[52]

For Ray, the study of human anatomy provided the ultimate, rational demonstration of God's existence and character. Writing of the human body, including William Harvey's account of the pulmonary circulatory system, Ray wrote, "Now to imagine that such a Machine composed of so many Parts, to the right Form, Order and Motion whereof such an infinite number of Intentions are required, could be made without the Contrivance of some wise Agent, must needs be irrational in the highest degree."[53]

It is important to note that the alternative explanation—namely, that different purposeful parts (and thus different species) arose via some purely mechanistic, law-bound process like evolution—was not taken seriously for generations. To propose that a part fitting one function could change into a different kind of part fitting a different function made no sense to most naturalists and anatomists well into the nineteenth century. Such change would require that matter act of its own accord, yet how could a clock rearrange its parts into a different kind of clock? Of course, the idea that God created species "as they are" fit with traditional interpretations of scripture (God created each species "after their kind"), but it also fit with observation (no one, after all, had witnessed one species change into another species). This is why Ray wrote with such confidence that atheist interpretations of the mechanical philosophy were not just impious, but irrational.[54]

Ray's book helped make both natural history and the study of human anatomy popular and effective fortresses against claims that the mechanical philosophy led to atheism or deism. Twelve editions of *The Wisdom of God* were published by 1759. When later naturalists, from German anatomists to French abbots, updated Ray's work throughout the eighteenth and nineteenth centuries, they added new natural history and anatomical facts, but the argument remained essentially the same: the purposeful parts and behaviors of animals and plants, and the extraordinary design of the human body, demonstrate God's wisdom, power, and goodness. Indeed, in an age enamored of the intellectual triumphs of Copernicus, Galileo, and Newton, natural history and anatomy became one of the primary means through which both theologians and laymen could say rational, empirical, meaningful, and devout things about God. As a result, British naturalists, ministers, and laymen believed that the study of plants, animals, and human anatomy supported scriptural messages about both "right thoughts" and "dutiful action" amid suffering—namely, faith that a divine purpose existed behind every affliction and that heaven would indeed make amends for all. In this world, the new sciences were on God's side, and the Book of Nature served as a perfectly pious topic of Christians' attention.

To return across the Atlantic: Mather became colonial New England's most ardent proponent of both Ray's work and the new sciences. Mather, too, believed that the careful study of the "Book of Creatures" (nature) provided knowledge of God's power, wisdom, and goodness. And he meant to convince others. In 1721 he published what would later be described as the first book of science published on New England shores: *The Christian Philosopher*. The book included 1,530 lines (of 11,062) taken directly from Ray's *The Wisdom of God*.[55] Mather described Ray as employed by God to make discoveries, then called upon the authority of a fourth-century church father as evidence that studying nature was a pious thing to do: "Chrysostom, I remember, mentions a twofold book of God: the book of the creatures, and the book of the scriptures. God, having taught us first of all by his works, did it afterwards, by his Words. We will now for a while read the former of these books; 'twill help us in reading the latter. They will admirably assist one another."[56]

Mather's book included a detailed survey of the latest anatomical discoveries, all harnessed into proofs that the human body is "a machine of a most astonishing workmanship and contrivance." "It cannot be without admiration looked upon," he wrote, "that all the bones, and all the muscles, and all the vessels of the body, should be so contrived, so adapted and compacted, for their several motions and uses; all according to the strictest rules of mathematics!" Speaking of the windpipe, Mather noted how, as "a continual respiration is necessary for the support of our lives . . . lest, when we swallow, our meat or drink should fall in to do mischief there, it hath a strong valve, an epiglottis, to cover it when we swallow." Writing of the stomach's ability to digest food, he wrote of God's great wisdom in creating a menstruum that could corrode flesh but not the stomach itself. Mather's citation of vomiting as evidence of God's astonishing craftsmanship is worth reading as a representative excerpt:

> The provision made in the body of man to ward off evils, is very admirable. The secretions made by the glands, whereof Cockburn, Keil, Moreland, and others, give us accounts, are such as cannot be considered without amazement. How many parts of the body stand ready to do what belongs to faithful sentinels. The principal and more essential instruments of life and sense, how well are they barricaded? Of how many parts are we supplied with pairs, to make up for a defect which may happen in any of them? . . . Dr. Sloane admires the contrivance of our blood, which on some occasions, as soon as any thing destructive to the constitution comes into it, immediately by an

intestine commotion endeavours to thrust it forth, and so it is not only freed from the new guest, but sometimes what likewise might long have lain lurking there. What emunctories has the body, and what surprising passages, to carry off mischiefs, which we foolishly bring upon ourselves! and how astonishing the methods and efforts of nature to set all things to rights.[57]

For his sources Mather drew on anatomists both famous and obscure, including Andreas Vesalius (1514–1564), Thomas Willis (1621–1675), Thomas Wharton (1614–1673), Nehemiah Grew (1641–1712), and Ludovicus Bilsius (1624–70). These anatomists combined natural theology with the study of the human body (*anatomia theologica*) to provide both theological and theoretical rationales for dissection. As John Hedley Brooke notes, "How could it be wrong, the argument ran, to practice dissection, if divine craftsmanship was thereby more fully revealed?" Thus, the mechanical philosophy both inspired and justified a practice pursued consistently in no other medical tradition. Meanwhile, emphasizing purpose in nature meant naturalists and anatomists searched for the use of every anatomical fact. Harvey, for example, imagined the blood might circulate after observing valves in veins, for, as he wrote to Boyle, "so provident a cause as nature had not so placed so many valves without design."[58]

No doubt it was easy to see wisdom, power, and goodness reflected in the circulation of blood, the universal law of gravitation, complex organs like the eye, the design of the seasons, a productive landscape, a healthy human body, the birth of an infant, and a whole host of seemingly benevolent provisions. But what of heart failure, kidney stones, blindness, famines, accidents, and the dozens of ailments that cut off infants' lives in their first few days or hours? What of smallpox and "monstrous" births? What of breasts, perfectly designed to suckle, not bringing forth milk? What of the mother, perfectly designed to nurture the new babe, dying amid violent convulsions from childbed fever? What of the fact, according to Mather, that disease had carried off "*nine parts of ten*, (yea, 'tis said, *nineteen* of *twenty*)" of the men, women, and children who lived in the "New World" prior to the arrival of Europeans?[59]

Clearly, neither Ray nor Mather were ignorant of the great suffering caused by disease. Their writings show that both knew the purpose and goodness of many parts of the human body by virtue of what happened when those parts broke. They wrote of the protective purpose of tooth enamel, for example, by describing how teeth sometimes became so worn "that the very nerve lies bare, and for mere pain they can be used no more."[60] But the existence of pain and suffering by no means vitiated the design argument. Indeed, the whole point of natural theology was that the marks of design proved God's close benevolence, wisdom, and goodness despite the obvious existence of great suffering and affliction. For Ray and Mather, the extraordinary design evident in the Book of Nature (when properly interpreted) bolstered Christians' belief that a divine, benevolent purpose existed behind the greatest of suffering.

Natural theology and the design argument would be the primary means through which American naturalists, natural philosophers, and anatomists made sense of what they were doing, and of the world, well into the nineteenth century. Meanwhile, it is crucial to note that they linked a particular explanation of the origin of species (that species are independently created by God) to Christian explanations of the existence of suffering, including the loss of beloved children.

Although, centuries later, all this talk of God's design would be set aside as outside the methodological boundaries of science, the mechanistic view of nature would remain. Allopathic medicine, for example, is still based on the assumptions that, first, nature can be studied by imagining organic beings as "matter in motion, governed by natural laws," and second, that nature is best studied by breaking a system down (or "reducing" it) into parts and then examining how those parts work. Today, this mechanistic, reductionist approach is what distinguishes biomedicine from more "holistic," nonmechanistic traditions.

Scholars have shown that mechanizing nature pushed alternative medical systems outside the boundaries of what eventually became known as "modern science." Language about a great creator resonated with Indigenous concepts to a point, but mechanizing the world, drawing sharp boundaries between the natural and supernatural, and removing spirit from the nonhuman world, did not. Indigenous knowledge systems, for example, neither drew the same line between the material and immaterial nor reserved moral agency solely for human beings. As we have seen, Cherokee histories attributed the origin of sickness and disease to a council of animals who wished to chasten human beings for having expanded beyond their proper bounds. Those same histories explained healing medicines as first appearing not at the behest of God but at the behest of plants, who took pity on suffering, disease-afflicted humans.[61] By contrast, proponents of the new sciences insisted on sharp boundaries between human beings (in possession of the divinely conferred ability, via an immaterial soul, to both reason and make moral decisions) and animals (which, they held, possessed none of these things).

Having established certain rules, the Royal Society and its disciples declared Indigenous (and any other) knowledge systems irrational, superstitious, and, ultimately, unscientific. Some Christian ministers, including most famously John Wesley, conceded that the first Americans possessed "generally infallible" cures for many illnesses, but Wesley attributed that knowledge to both trial and error and Indigenous reliance "on the healing powers of nature, the author of which was God." Wesley could then dispense with the worldviews and traditions within which Indigenous healers explained such things as heathen superstitions. The idea that animals and plants had moral agency or influence in the world had no place in the new sciences, according to which all motion, activity, and purpose must ultimately come from God.[62]

Hope in New Things

Mather was clearly captivated by the new sciences slowly arriving via ships crossing the Atlantic from Europe. He embraced a new ordering of the cosmos (Copernicus, 1543), new anatomy (Vesalius, 1543), new stars and moons (Galileo, 1610), and a new map of the heart (Harvey, 1628). He read these new discoveries through the lens of a progressive, Christian history, lauding new technologies (or "arts"), from pendulum watches to the printing press, as evidence of scriptural promises that the earth would be redeemed from the corrupting results of the fall of man. Each invention or discovery, Mather argued, constituted a well-timed revelation "wherein the superintendance of the glorious Creator and Governor of the world" chose to make things "locked up from human understanding... understood by the children of men." He asked his readers to imagine, "if the mathematics, which have in the two last centuries had such wonderful improvements, do for two hundred years more improve in proportion to the former, who can tell what mankind may come to!" He was certain that extraordinary discoveries were coming to pass and that those discoveries provided clear demonstrations of God's close, benevolent, wise, and powerful governance.[63]

Notice that Mather's hope of progress did not depend on the new sciences either having improved medicine or ameliorated human suffering yet. He had no evidence that the new sciences were either curing disease or driving mortality rates down. Harvey's account of the circulatory system need not (and did not) lead to any therapeutic discoveries to be cited as evidence of a divinely approved progress in the recovery of lost knowledge. Even bloodletting (which might conceivably have been called into question by the idea that a limited quantity of blood circulates in the body) continued, albeit with new, mechanistic justifications.

Mather's faith in progress arose from his particular interpretation of Christian history rather than evidence that suffering had been ameliorated via the new sciences. In the 1720s the extent to which Mather was willing to call upon this faith to justify extraordinary new attempts to prevent child mortality astonished his fellow Bostonians. For in 1721, the same year Mather's *The Christian Philosopher* appeared in Boston bookshops, Mather became a proponent of a radical new (to Europeans) means of saving children: inoculation against smallpox.

While in retrospect it might be tempting to see Mather's campaign as part of the steady advance of the new sciences that allowed mechanistic science (and what is now called biomedicine) to ultimately eradicate this dreaded disease in 1980, the story of inoculation in the New England colonies is much more complicated. First, smallpox inoculation initially appeared within nonmechanistic healing traditions. Second, Mather had no idea (according to the mechanistic criteria established by the new sciences) why inoculation worked. What he did have was a progressive vision of history that inspired him to understand new discoveries as part of a benevolent, divine plan preparing the earth for the second coming of Christ. That vision—combined with both his activist, mission-driven demand to (what he called) "do good" and the threat of losing more children to a dreadful disease—inspired Mather's faith that inoculation was an additional revelation of God's benevolent governance.

Mather knew this terrible disease all too well. At the time smallpox was one of the era's most efficient killers of children and adults alike. The mortality rate could be anywhere from 15 to 30 percent and every case, mortal or not, entailed tremendous suffering. He had lost his first wife, Abigail, to its deadly course some years earlier. When the epidemic of 1721 began, Mather prayed for guidance on whether to send his children out of town and for help in submitting "if these dear Children must lose their lives." These were two prayers he had no doubt made during previous epidemics. But now Mather added a third prayer: Should he inoculate his children against this terrible disease?[64]

Mather first heard of inoculation from his "servant," Onesimus, whom we met earlier. In contrast to those who denied Africans' capacity for rational thought, Mather described Onesimus as a young man "of a promising Aspect and Temper" and best "govern'd and managed" by appeal to his reason. Indeed, he had "resolved with the Help of the Lord, that [he] would use the best Endeavours to make him a Servant of Christ." He was also willing to imagine that God might have revealed extraordinary knowledge to Onesimus's people. In 1716, the same year Onesimus purchased his freedom, Mather wrote a letter to a physician in London describing Onesimus's testimony regarding the practice of giving individuals smallpox to preserve them from the disease: "I had from a servant of my own, an account of its being practiced in Africa. Enquiring of my Negro-Man Onesimus, who is a pretty intelligent fellow, whether he ever had the smallpox, he answered Yes and No; and then told me that he had undergone an operation, which had given him something of the smallpox, and would forever preserve him from it, adding, That it was often used among the *Guramantese* & whoever had the courage to use it was forever free from the fear of the contagion. He described the operation to me, and showed me in his arm the scar."[65]

This is the only account we have of Onesimus's words. "Guramantese" was Mather's spelling of *Coromantee*, a strong hint that Onesimus was from the Gold Coast of West Africa (present-day Ghana). If Mather's designation was correct, Onesimus was most likely raised within the Obeah tradition of healing. When ill, an individual went to an obaya, who mixed human materials and natural elements into sacred powders. A god called Sopona both gave and healed smallpox, but Sopona could be influenced by human effort. Although a "heathen custom," inoculation mapped onto Mather's own view of the relationship between divine and human agency. It fit with his vision of how medicine could be combined with prayer, how human agency interacted with God's, and how ministers might have important things to say about healing. Indeed, Mather found within Onesimus's words the potential for a new revelation of what God might approve and even demand.[66]

Five years after Onesimus showed Mather his scar, in the midst of the latest smallpox epidemic, Mather recorded his internal debate whether to inoculate his son, Sammy. "What shall I do? what shall I do, with regard unto Sammy?" he lamented in his diary.[67] He had already lost so many children. Did he have enough trust in Onesimus's experience, African knowledge, and God to try so strange a procedure on his own son? The struggle was brief. Mather decided that inoculation was one of many wonders, from the discovery of the circulation of blood to the universal law of gravitation, signaling the truth of scripture, God's goodness, and Christ's imminent return. He decided to inoculate Sammy. He then joined a physician named Zabdiel Boylston in an effort to inoculate all of Boston (we know less about Boylston's motivations, theological or otherwise). With Mather defending him in print, Boylston inoculated 248 people that summer, including his own thirteen-year-old son.[68]

Not everyone approved of what Boylston was doing or what Mather was preaching. The London minister Edmund Massey insisted that only God had power to inflict disease, that God always did so with good reason, and that therefore inoculation constituted a dangerous and sinful rebellion against God's governance. "Let the Atheist, Scoffer, the Heathen and the Unbeliever, disclaim dependence upon Providence and dispute the Wisdom of God's Government," Massy exclaimed, "Let them Inoculate, and be Inoculated, whose Hope is only in, and for this Life!" Massey's critique would often be cited in the nineteenth century as the worst kind of theological opposition to scientific progress. But it is important to note that Massey was not against the use of medicine. He, too, wrote of physicians as instruments "in the Hand of Providence, to restore Health, and to prolong Life" who worked "by Virtue of a wonderful Insight into the Nature of the Mineral and Vegetable World." Smallpox inoculation, however, did not entail the use of a perfectly designed herb passed to the hand of man by a benevolent creator to ameliorate the suffering of the sick. Inoculation entailed the *production* of disease by transferring it from a sick to a healthy individual.[69]

Almost every physician in Boston decided against inoculation as well. While using plants as medicines fit within long-standing medical traditions, taking the pus from a smallpox pustule and transferring it to a healthy individual had no precedence in European practice and no justification in theory. Furthermore, given that quarantine served as the primary means of containing epidemics, Mather and Boylston's campaign to spread the disease faster than the (already devastatingly rapid) "common way" flew in the face of existing attempts to halt the disease's spread. Nor did inoculation make sense according to the Royal Society's emphasis on mechanistic explanations and the study of secondary cause as the best route toward progress: What, exactly, was the mechanism or secondary cause through which inoculation might work, for example? (Germ theory, virology, and immunology were nearly a hundred and fifty years in the future; in the meantime, theoretical frameworks were proposed, including by Mather, but they were all highly contested and, in retrospect, all wrong.)

Tales of inoculation were also arriving from China and the Ottoman Empire, but the fact those precedents came from "heathens" carried little weight with most Boston physicians.[70] Onesimus's testimony and the experience of non-Europeans did not, to many European observers, count as evidence that the practice worked. Almost all concluded, at least initially, that intentionally exposing a child to pus from smallpox sores was not only intuitively, but *in fact*, dangerous. (According to today's risk calculations, they were right: a safer method using cowpox—vaccination, rather than variolation—was not developed until the 1790s by Edward Jenner, and even that came with much higher risk than today's medicine would countenance. Meanwhile, because inoculated—versus vaccinated—individuals *were* contagious, the opposition was, it turns out, wise to urge caution.) For many Bostonians, Mather's faith in Onesimus's word, African experience, and God simply wasn't enough to justify scraping deadly material into the skin of children.

A physician named William Douglass led the opposition. He was the only man in the colonies with a medical degree (most physicians learned their trade by apprenticeship). He also despised Puritan ministers' influence in Boston. Mather's prominent role in the Salem Witch Trials a generation earlier placed him well outside the bounds of rationalism in Douglass's eyes. Mather's involvement in the witch trials allowed opponents to point out that he had been deadly wrong before. (By this point in time, even ministers involved in the trials were second-guessing the verdicts. One of the judges, Samuel Sewall, had by this point publicly begged the pardon of both God and men for having sent twenty innocent people to their death. When four of his own children died in five years, he believed it a just divine punishment for his role.)[71] As for Onesimus's testimony, it carried no weight with Douglass, who dismissed African experience and knowledge on the grounds he was certain "there is not a race of men on earth more false."[72]

The controversy over Mather and Boylston's campaign was drawn out and acrimonious. At one point someone threw a grenade into Mather's house with a note that read, "Cotton Mather, You Dog damn you: I'll inoculate you with this, with a pox to you." Mather, for his part, continued to blame opposition on the devil, who, "alarmed lest lives be saved," had taken possession of the people. One of those supposedly devil-possessed individuals was Benjamin Franklin's brother. James Franklin had founded the *New England Courant* that summer, a satirical newspaper known for ridiculing Boston's Puritan ministers, especially Mather and his father, Increase Mather. Criticizing Mather for meddling in medicine was right up the *Courant*'s alley. The paper cited Mather's role in the Salem Witch Trials as evidence that he was not to be trusted to think either rationally or rightly. It even accused Mather of hypocrisy for abandoning orthodox religious principles like resignation and trust in God's ways. (Mather's defenders noted how strange it was to be lectured about God's will by so sacrilegious a newspaper.) Ultimately, the conflict between Mather and the *New England Courant* was as much about the Mathers' political and theological influence in Boston as inoculation per se. Mather and the Franklin brothers agreed on the possibility of progress. Clearly, they did not agree on who should direct that progress.[73]

Mather did not back down in the face of opposition. Eventually, as we will see, the lower mortality rates of those Boylston inoculated compared to those who took smallpox "in the common way" convinced skeptics, including Douglass, that inoculation might save children from this dreaded disease. But Mather's faith in inoculation before anyone had those numbers depended on his interpretation of the Bible. He read his own place in history, for example, as part of a providential, progressive movement signaled in the Book of Revelation, a movement in which human action could work in concert with God's purposes.

Of course, Mather's particular vision of Christianity was not necessary to the practice of inoculation, as Onesimus's knowledge and Turkish, Chinese, and African medial traditions prove. But his campaign emphasizes how Christian views of providence inspired some to imagine that radical action to prevent suffering, disease, and even death, might constitute God's will as well. Yes, resignation was a virtue, but certain kinds of suffering need not be submitted to entirely. Illness could sometimes be reduced or avoided through action, with due attention to the glory of God, who had designed the world in such a way that human efforts to increasingly ameliorate suffering on earth might succeed.

Of course, belief in the possibility of discovering new remedies could be a hard bargain when no remedy had been found, or a known remedy was applied too late. One can sense Mather's awareness that he might, as a sinful creature for whom the fall of man obscured right knowledge of nature, choose wrongly in deciding whether to inoculate Sammy. On the other hand, as we will see, distancing God's hand from suffering by attributing affliction to secondary causes clearly had its appeal. Even Mather, who insisted God's presence could be found in every affliction, was, at least once, tempted to distance God's hand from the death of his children. We have examined how Mather described the body of man as "a Machine of the most Astonishing Workmanship and Contrivance!" And yet by the time Mather wrote these lines, he had plenty of experience with the fact that sometimes God's little machines failed completely. In late March 1693 Mather recorded the loss of an infant named Joseph in his diary: "God gave to my wife, a safe Deliverance of a Son. It was a child of a most comely and hearty Look, and all my Friends entertained his Birth, with very singular Expressions of Satisfaction. But the Child was attended with a very strange Disaster; for it had such an obstruction in the Bowels, as utterly hindered the Passage of its Ordure from it. We used all the Methods that could be devised for its Ease; but nothing we did, could save the Child from Death. It languished, in its Agonies, till Saturday, April 1. about 10 h P.M. and so died, unbaptised."[74]

Five years after he composed the sermon on little Abigail's death, Mather recorded how, in the midst of this new trial, God enabled him to bear this infant's loss with an unexpected measure of resignation to his holy will. Once again, he delivered a sermon the following day, this time on faith, patience, and thankfulness, choosing as his text Job 2:10 (in which, amid his afflictions, Job's wife urged him to curse God. Job replied: "Thou speakest as one of the foolish women speaketh. What? shall we receive good at the hand of God, and shall we not receive evil?").[75] And yet, within a few days of Joseph's death, Mather grasped at an alternative explanation that distanced God's hand from so great an affliction. He wondered whether someone, or some*thing* else, had been involved:

> When the Body of the Child was opened, we found, that the lower End of the Rectum Intestinum, instead of being Musculous, as it should have been, was Membranous, and altogether closed up. I had great Reason to suspect a Witchcraft, in this praeternatural Accident; because my wife, a few weeks before her Deliverance, was affrighted with a horrible Spectre, in our Porch, which Fright caused her Bowells to turn within her; and the Spectres which both before and after, tormented a young Woman in our Neighborhood, brag'd of their giving my Wife that Fright, in the hopes, they said, of doing Mischief unto her Infant at least, if not unto the Mother: and besides all this, the Child was no sooner born, but a suspected Woman sent unto my Father, a Letter full of railing against myself, wherein she told him, He little knew, what might quickly befall some of his Posterity.[76]

Little Joseph's death occurred just after the Salem Witch Trials ended. Although Mather had argued against some elements of the proceedings, he had defended the trials in general. Certainly, he did not doubt the reality of witchcraft. Did the curse of a witch offer a better explanation of how such a beautifully designed part of the human anatomy (Ray had written of how sphincters demonstrated God's wisdom) could go so terribly wrong?

The fact that even Mather appealed to a different explanation than God's unmediated will provides an eloquent hint that shifting explanations toward secondary causes and natural laws (once witches were set aside as an option) might allow Americans to absolve God (or even dispense with God altogether) when trying to make sense of child illness and loss. Nearly a century later, in 1767, the professor and demonstrator royal in anatomy and surgery at Rouen, France, Claude-Nicolas Le Cat, described the "monstrosities" with which Joseph Mather was born as due to some kind of dropsy that occurred in the womb, which then interfered with the course of development: "All these mysteries, which one would have thought impenetrable," Le Cat wrote, "are easily accounted for." (Dropsy is swelling of tissues, or edema. "Monstrosities" was the term used for congenital abnormalities, the study of which was called teratology.)[77]

This possibility of secondary causes taking the place of both witches and God is suggested by Boyle's confession that one must not expect God to intervene in natural law to save individuals from their missteps and ignorance. As historian of science John Hedley Brooke notes, Boyle did not think individuals should expect God to suspend the law of gravity when, for example, someone fell over a cliff.[78] As we have seen, the members of the Royal Society were well aware that explanations that rooted suffering and affliction in secondary causes might obscure the sovereignty of God as the first cause. This was one of the reasons that both Ray and Mather emphasized the purposeful parts of animals and plants, since they were certain that only God's close, personal design could account for such phenomena. Attention to things that could only be explained via the direct intervention of God was supposed to remind Christians of God's close, personal control over human lives. They were certain that, properly pursued, natural history and anatomy would be the perfect support for "right thoughts in sad hours."

As American naturalists, ministers, and physicians linked the study of nature to Christian explanations of the world, they recorded their visions of "right thoughts in sad hours" and "the duty of good men" in print. Meanwhile, parents without access to printing presses experienced child loss in ways and at scales that New Englanders of European descent did not. Slavery permitted men, rather than God, to snatch children away from parents. Epidemics, warfare, and famine devastated Indigenous communities. For generations, those who converted to Christianity had to wonder why God continued to punish even "praying Indians" with death and disease while the descendants of Europeans prospered and their population grew. Confronted by the high mortality rates of his people, Samson Occom, a member of the Mohegan nation, a Presbyterian minister, and the first of his people to publish in English, confessed that "Some Times I am ready to Conclude that [Indians] are under great Judgment and Curse from God."[79]

Mather applied God's two books (scripture and nature) to this mystery as well. In 1702 Mather explicitly described the horrific epidemic of 1616–1619 that swept through the region's Indigenous communities as God's means of making way for the English. Mather wrote of "the wonderful providence of God" in bringing the pilgrims to "a country wonderfully prepared for their entertainment, by a sweeping *mortality* that had lately been among the natives." He recounted how "the Indians in these parts had newly, even about a year or two before, been visited with such a prodigious pestilence, as carried away not a *tenth*, but *nine parts of ten*, (yea, 'tis said, *nineteen* of *twenty*) among them: so that the woods were almost cleared of those pernicious creatures, to make room for a *better growth*."[80] But, as historian David Jones notes, Mather did not mean that God acted directly to "make Indians vanish in a puff of smoke. Instead, providence was manifested through natural processes, or secondary causes."[81] The description of natural processes, or secondary causes, via the new sciences could be used, then, to draw quite different lessons about whether anything could or should be done about high mortality rates, depending on whose children were dying. Drawing upon Bacon's vision of how new knowledge might serve for "the relief of man's estate," Mather imagined saving settlers' children from smallpox, measles, and other ailments. On the other hand, he imagined explanations of Indigenous mortality that roped the loss of some children into visions of settler expansion, improved insights into the natural processes through God-governed creation, and the progress of Christian civilization. Some parents' child loss, in other words, became the inevitable result of divinely governed, natural processes. Physicians should still try to save Indigenous children with smallpox inoculation and missionaries should still try to save Indigenous souls with Christianity, but one could not, of course, change nature's laws.

The legacies of these narratives would be long-lasting, remaining even after some Americans increasingly appealed to the Book of Nature, rather than scripture, to explain the world. In 1928, a full two centuries after Mather told his tales of pestilence making "room for a better growth," Oregon historian and governor Leslie M. Scott described how diseases like smallpox had prepared the region for white settlers by decimating "rival races." "Throughout the entire West," Scott wrote, "the Indians were victims, but perhaps nowhere else so badly as in the Pacific Northwest; and nowhere else were the results so good for the whites." By the time Scott wrote, one might appeal to the benevolence of God, nature, or some complex mix of both to explain such things. "Nature or the world's advancement or an all-seeing providence," Scott wrote, "was preparing for the coming of the settlers."[82] This ironic tendency within American science, religion, and medicine to imagine saving some children (on the grounds one might learn secondary causes and improve medical care) while ignoring the high mortality rates of other children (by seeing those deaths as natural) will be a persistent theme.

Chapter 2

THE LAWS OF NATURE AND OF NATURE'S GOD

IN 1736 BENJAMIN FRANKLIN'S "fine boy of four years old" died of smallpox, "taken in the common way." By that time, Boylston's numbers had long since changed Franklin's mind about the efficacy of inoculation. The lower mortality rates of inoculated individuals had convinced Franklin that Mather and Boylston's campaign had been both wise and good. Indeed, he had fully intended to have the boy inoculated, but physicians rightly assumed a child best be in good health during this intentional exposure to so deadly a disease, and they had delayed on account of the boy's chronic diarrhea. In writing of Franky's death, Franklin said nothing about resigning himself to God's will. Instead, he blamed himself for not inoculating his little boy in time. "I long regretted bitterly and still regret," he wrote decades later in his autobiography, "that I had not given it to him by inoculation. This I mention for the sake of the parents who omit that operation, on the supposition that they should never forgive themselves if a child died under it; my example showing that the regret may be the same either way, and that, therefore, the safer should be chosen."[1]

Thirty-six years after Franky's death, Franklin confessed to his sister Jane Mecom that he still could not think of his dead boy "without a sigh." But it was regret, rather than an effort to bend his heart to resignation to God's will or long musings on why God allowed such things in the first place, that formed the dominant chord in Franklin's few comments on his boy's death.[2] Both Mather and Franklin believed in God. They both believed in the potential power of the new sciences. And they both believed in the possibility of progress. But in many ways, Franklin and Mather stood on either side of chasms carved out by Americans' efforts to reconcile God's governance with natural law, bridged rather precariously by their agreement that ameliorating suffering on earth was not only possible but divinely approved. Those chasms were of variable depths and kinds.

They arose from debates over whether God's governance is close or distant, the relative authority of the Book of Scripture and Book of Nature, whether affliction is sent directly by God or arose via some mix of delegated natural laws and the actions of fallible humans, and what must be believed (and why) about a child's fate when medicine, inoculation, or prayer failed.

That Curious Engine of Divine Workmanship

Two years before Franky died, in 1734, Franklin had sat down at his desk and reflected on a series of questions raised by the terrible commonness of child loss. Although the problems were old, the mechanical philosophy's version of the design argument allowed the challenges posed by child loss to be posed in new forms: If God was all-wise, all-powerful, and all-good, then why did this wise God allow the littlest, most precious "machines" (the bodies of infants and children) to sometimes break in such terrible ways? Why didn't God use his limitless power to prevent the horrible deaths of so many children from anatomical malformations, accidents, or infectious disease? Wrestling with these questions, Franklin composed a brief essay entitled "The Death of Infants." He began by noting that "one half of Mankind, which are born into this World, die, before they arrive to the age of Sixteen," a truth that could be confirmed by "even a cursory view of any common Burial-place."[3] He then wrote:

> Let us now contemplate the Body of an Infant, that curious Engine of Divine Workmanship. What a rich and artful Structure of Flesh upon the solid and well compacted Foundation of Bones! What curious Joints and Hinges, on which the Limbs are moved to and fro! What an inconceivable Variety of Nerves, Veins, Arteries, Fibres and little invisible parts are found in every Member! What various Fluids, Blood and Juices run thro' and agitate the innumerable slender Tubes, the hollow Strings and Strainers of the Body! What millions of folding Doors are fixed within, to stop those red or transparent Rivulets in their course, either to prevent their Return backwards, or else as a Means to swell the Muscles and move the Limbs! What endless contrivances to secure Life, to nourish Nature, and to propagate the same to future Animals!

Franklin followed this very Matheresque description of the extraordinarily purposeful parts of the infant frame with the following demand: Why should "so wise, so good and merciful a Creator ... produce *Myriads* of such exquisite Machines ... but to be deposited in the dark Chambers of the Grave?" "Should,"

Franklin demanded, "an able and expert Artificer employ all his Time and his Skill in contriving and framing an exquisite Piece of *Clock-work*, which, when he had brought it to the utmost Perfection Wit and Art were capable of, and just set it a-going, he should suddenly dash it to pieces; would not every wise Man naturally infer, that his intense Application had disturb'd his Brain and impair'd his Reason?"

Franklin did not answer his question by declaring God insane. Instead, he wrote that the death of infants proved, by the dictates of reason alone, the existence of heaven. For when we reflect, he wrote, "on the vast Numbers of Infants, that just struggle into Life, then weep and die," and that no just, wise, and infinite Being could "create to no end," then it must be concluded "that those animated Machines, those *Men* in miniature, who know no Difference between Good and Evil . . . were made for good and wise Designs and Purposes, which Purposes, and Designs transcend all the Limits of our Ideas and all our present Capacities to conceive." Franklin insisted that to imagine the death of so many perfectly made "animated machines" for naught was absurd. Heaven *must* exist, not on scriptural grounds, but because otherwise no justification existed for the fact so many of these tiny, beautifully designed machines broke. And that was an unbearable thought. Indeed, the problem of the death of perfectly designed infants may have been what drew Franklin back from the precipice of unbelief, for this is apparently the only argument he ever gave for belief in a future state.

Mather found proof of a close, personal God and the promise of heaven in both the Christian revelation and the design of human anatomy. Franklin was less certain. In 1790, at the age of eighty-five, Franklin replied as follows to minister and president of Yale College Ezra Stiles's request to know something about his religion: "Here is my Creed. I believe in one God, Creator of the Universe. That He governs it by His Providence. That he ought to be worshipped. That the most acceptable Service we render to him, is doing Good to his other Children. That the Soul of Man is immortal, and will be treated with Justice in another Life respecting its Conduct in this. . . . As for Jesus of Nazareth . . . I think the system of Morals and Religion as he left them to us, the best the World ever saw . . . but I have . . . some Doubts to his Divinity; though' it is a Question I do not dogmatism upon, having never studied it, and think it is needless to busy myself with it now, where I expect soon an Opportunity of knowing the Truth with less Trouble."[4]

Franklin believed in God but clearly repudiated some of the fundamental tenets of Mather's brand of Christianity. He was not alone. Increasing exposure to non-Christian understandings of the world, ethical rebellions against orthodox Christianity (especially the doctrine of eternal damnation), political fights against the influence of ministers, and the quandary of the "machine-gone-wrong" all posed profound challenges to the Royal Society's confident faith that the new sciences would be an ally in the fight to establish consensus about God. Some asked, for example, why the beautiful design of the woodpecker's tongue demonstrated God's close, personal control of human society. God might just as well, some insisted, have designed the woodpecker long ago and departed. The possibility of this more deist (rather than theist) interpretation of the new sciences had always lurked. Indeed, Franklin described how his own reading of natural theology books at the age of fifteen had led him *away* from belief in a close, personal God: "The arguments of the Deists, which were quoted to be refuted, appeared to me much stronger than the refutations; in short, I soon became a thorough Deist."[5] It was not, Franklin admitted, the effect that Newton, Boyle, Ray, Mather, or any other seventeenth-century proponent of the new sciences had intended.

The appeal of distancing God's governance over creation and human lives (even as they spoke of God's benevolent design of natural laws) arose from various quarters. With the extraordinary feat of Newton in mind, individuals like Franklin began measuring God's goodness, justice, and rationality by more human standards. The argument (often repeated by Mather) that humans could not know why God permitted great suffering was losing its appeal among individuals increasingly confident in the possibilities of human reason. The idea that God wanted humans to be happy on earth as well as in heaven was also a clear departure from sermons about the inevitability of great affliction. Some became deists in an effort to solve the puzzle of why God permitted disease, famine, earthquakes, war, and child loss. God had not actually sent these things at all, deists claimed. Their distant, law-giving creator was entirely uninvolved in the day-to-day operation of the world, including any and all suffering.[6]

It is important to note that very few thought dispensing with a creator altogether a rational option. Historian James Turner has described how unbelief "seemed almost palpably absurd" well into the nineteenth century. The Protestant Reformation of the sixteenth century had changed much, of course, and one individual's beliefs might look heretical to another. But the proposition that there is no God or that one cannot know one way or the other was virtually unthinkable. Atheists might haunt dreams, but in reality, they were rare. Indeed, the design argument ensured that unbelief was not only heretical but *irrational*. As a result, notes Turner, "America does not seem to have harbored a single individual before the nineteenth century who disbelieved in God." After all, "natural laws themselves presupposed a divine Lawgiver."[7] The Scottish philosopher David Hume had asked his damning questions: Who designed the designer? Why did the designer have to be good when all experience seemed to indicate otherwise? And the French satirist Voltaire was ridiculing natural theology's implicit conclusion that "this is the best of all possible worlds." But even Voltaire insisted that his compatriot, the physician Julien Offray de La Mettrie, who defended atheism in a 1743 book entitled *Man a Machine*, must be *fou* (crazy).[8]

Although Franklin flirted with deism as a young man, he later wrote in his autobiography that he eventually abandoned deism on the grounds that "though it might be true," the stance "was not very useful."[9] And above all, Franklin wanted ideas to be useful. In the words of historian Roy Porter, the eighteenth-century "Enlightenment . . . translated the ultimate question 'How can I be saved?' into the pragmatic 'How can I be happy?'"[10] Franklin also asked, "How can *we* be happy?" He believed that only in doing good for one's fellows could one be truly content. In a letter to Hume, Franklin tried to figure out what to call this "certain Interest too little thought of by selfish Man, and scarce ever mention'd, so that we hardly have a Name for it." He tried out the phrases "the *Interest of Humanity*, or common Good of Mankind."[11]

Franklin tried to answer the question "How can *we* be happy?" with a drive that still astonishes. Within medicine, Franklin clearly believed that the ability to prevent and treat disease could be improved. When his brother John, afflicted with bladder stone and urinary retention, longed for a flexible catheter rather than the rigid metal ones of the day, Franklin imagined how one might be made and rushed to the silversmith with his detailed design ("sitting by till it was finished"). His letter to his brother was full of potential ideas for how to improve the device further. Convinced that lead caused the "drybellyache" (first by the pain in his own hands when handling hot leaded type, and then by studying the trades of patients at a hospital), Franklin lamented the fact that the correlation had apparently been known for sixty years: Why didn't people *do something*? He helped found the first hospital in the colonies (in the early 1750s) and the first scientific society outside of Europe (the American Philosophical Society).[12] Perhaps most famously, he studied the laws of electricity and then imagined how the fact that iron conducted electricity might be harnessed to prevent suffering. Would not pointed rods, he wondered, "probably draw the electrical fire silently out of a cloud before it came nigh enough to strike, and thereby secure us from that most sudden and terrible mischief!"[13] (Lightning rods were the first, and for a long time the only, application of the new sciences that saved lives.)

Franklin's fellow natural philosophers clearly valued his energy and vision. Joseph Banks, president of the Royal Society of London from 1778 to 1820, was upset when Franklin's attention increasingly diverted from science to politics. Franklin replied, "Be assured that I long earnestly for a Return of those peaceful

Times when I could sit down in sweet Society with my English philosophic Friends, communicating to each other new Discoveries, and proposing Improvements of old ones, all tending to extend the Power of Man over Matter, avert or diminish the Evils he is subject to, or augment the Number of his Enjoyments."[14] In fact, the political campaigns and the visions of extending the power of man over matter were tightly intertwined: both reflected a belief in man's duty to improve the human condition on earth, whether one hoped for a better life in heaven or not. To Franklin, Mather's belief that true worship entailed doing good for one's fellow human beings became the whole point of religion: "The most acceptable service to God," Franklin wrote in his *Autobiography*, "was the doing good to man."[15]

Although they believed very different things about God, Franklin and Mather agreed that humans could ameliorate human suffering on earth. Indeed, Franklin later attributed his own commitment to "doing good" to Mather's influence, especially his reading of Mather's *Bonifacius: An Essay Upon the Good*. He also attributed his love of science to Mather's sermons on Newton and reading *The Christian Philosopher*.[16] This common ground regarding the methods and purpose of the new sciences, despite profound theological differences, foreshadowed the ways in which American science would flourish amid religious diversity for generations. Some proponents of the new sciences attributed human suffering and ignorance to the fall of man and saw science as the best means of recovering that lost knowledge and, in doing so, preparing the earth for Christ's return. Others, as we will see, abandoned literal interpretations of the fall of man and the second coming while upholding faith in the divinely approved power of science to ameliorate suffering. But both agreed that, first, progress via human effort was possible, and second, that the new sciences provided an important means of achieving that progress.

Thus, although Franklin focused on the production of heaven on earth for its own sake, rather in preparation for Christ's promised return, his passionate faith in progress tapped into the values and beliefs of many American Christians. Turner has described, for example, how popular, postmillennialist interpretations of the Book of Revelation (examined in the previous chapter) bolstered belief in the possibility of ameliorating suffering. The millennium (i.e., the return of Christ and the final establishment of the Kingdom of God) "would emerge, not all of a sudden, but as a gradual perfecting of earthly life through human effort, inspired by divine grace."[17] Indeed, this vision of a progressive direction to history (which imagined humans working in concert with God's purposes) inspired abolitionists to try to do their part to inaugurate the Kingdom of God on earth by fighting to end slavery. Meanwhile, the extraordinary discoveries of men like Newton via the new sciences were supposed to provide a map for how to achieve progress in the material realm, including medicine. This is all important context for understanding why, when Americans founded their first scientific and medical societies, they did not view the assumptions, methods, and goals of these societies as in conflict with Christianity.

Franklin instilled Bacon's and Mather's calls for Christians to both study nature and do good within the goals of the American Philosophical Society. The society would, Franklin wrote, pursue "all philosophical Experiments that let Light into the Nature of Things, tend to increase the Power of Man over Matter, and multiply the Conveniences or Pleasures of Life."[18] Meanwhile, pervasive belief in divinely approved progress via human effort and the strong tradition of natural theology provided a context in which individuals with quite different concepts of God could participate in the society. John Hedley Brooke notes that natural theology did not demand that one define the closeness or distance of the designer. As a result, believers in more deist (God is distant) visions of God could study nature alongside those with more theist (God is close) visions while avoiding acrimonious theological debate.[19]

It is important to remember that an alternative to belief in a progressive shape to human history on earth did exist within Christianity, one that will eventually play a key role in our story. In contrast to *post*millennialists who argued that the Book of Revelation predicted increasing progress on earth, *pre*millennialists believed that Christ's thousand-year reign would follow a period of increased suffering on earth (thus Christ returned *prior* to the arrival of better things). Turner notes that the view that suffering would increase over time rather than decrease proved too pessimistic a vision for most American Christians, at least for the time being. As historian James Moorhead notes, by the 1790s American Christianity solidified into "an understanding of history as gradual improvement according to rational laws that human beings could learn and use."[20] This postmillennialist interpretation of the shape of human history dominated stances on the relationship between science, medicine, and religion until the First World War and provided the theological context in which many Americans' support for science flourished.

Perhaps because he knew his American audience so well, Franklin explicitly cited God as the source of new discoveries during his campaign to convince parents to inoculate their children against smallpox. When, more than twenty years after Franky's death, Franklin placed the colonies' first epidemiological table (based on mortality numbers from the smallpox epidemic in Boston of 1753–1754) before American parents, he accompanied the numbers (translated to percentages, the mortality rates were 9.3 percent for those not inoculated and 1.4 percent for inoculated patients) with the following words: "Surely parents will no longer refuse to accept and thankfully use a discovery GOD in his mercy has been pleased to bless mankind with."[21] In other words, he called upon numbers *and* faith in a particular relationship between new discoveries and God's governance to convince American parents that creating a small wound in a child's arm and placing smallpox pus within was a rational, right, and pious thing to do.

Beyond the Bounds of Reason

In 1781 the three-year-old son of a young Quaker mother who lived across the street from Franklin died of smallpox. After William died, Lowry Wister composed desperate appeals to God in her diary. She wished to resign herself to God's will, but she had fearful doubts whether she had been right to follow the doctor's instructions and give her little boy two grains of tartar emetic (a purgative that caused intense vomiting). "Oh most merciful Father support me," she prayed, "the recollection of my sweet infants sufferings rend my very soul, and it is thou alone Oh! my God that canst speak peace unto me, help me Oh! most gracious Lord for I am very weak I have no strength of my own." Wister then wrote that her reflections on her "sweet child's illness" were painfully increased by her doubts regarding the medicine given to him. This line was followed by another prayer: "Oh! gracious God look down upon me, have compassion on my sufferings thou knowest they are great." Wister's laments were interspersed with hope of heaven—"But his spirit is I firmly trust, rejoicing in the presence of Jesus his Saviour, and in that firm blessed hope rests my chief consolation"—which then gave way again to more laments—"help me Oh! most gracious God for I am poor, have compassion upon me for I am very weak and low."[22] Wister's diaries illustrate the profound tension between belief in God's all-sovereign providence and the fact that physicians (whether they had access to the new sciences or not) were, after all, mere men, subject to all the follies, whims, temptations, ignorance, and sins of fallen creatures.

Mather's and Franklin's faith that knowledge could be improved inspired not only their visions of progress but also their actions when their children died. With access to societies and clubs bent on *doing*, Franklin had channeled his grief into a campaign to improve Americans' ability to prevent child loss in the future. Indeed, his regret for not inoculating Franky in time had driven his search for someone who could write a pamphlet (the epidemiological table described in the last section appeared in a pamphlet on inoculation by Dr. Heberden) with both the evidence in support of inoculation and directions for how parents could inoculate their children safely. However, not all Americans could, in the wake of child loss, converse with a physician on an equal footing, spend their time composing prefaces and compiling epidemiological tables, or obtain access to printing presses, whether they believed children could be saved or not. As a woman in colonial New England, Wister could not channel her grief into building institutions aimed at increasing humanity's control over nature. As historians of science have demonstrated, European traditions of science grew directly out of a monastic and clerical culture that excluded women. A consensus that God commanded the husband to rule over the wife, designed women to bear children, and made women "by nature" emotional rather than rational meant that proponents of the new sciences did not think women could even *do* science. Furthermore, the particular system the Royal Society developed to produce "matters of fact" (trusting each other, for example, to produce honest accounts of experiments since not all experiments could be replicated) meant that those stereotyped as beguiling, emotional, and untrustworthy could not participate in the production of natural knowledge. An exceptional, wealthy woman could serve as both patron and inspiration, but she could not show up at either a Royal Society or American Philosophical Society meeting and be treated as an intellectual equal.[23]

Indeed, the new sciences so admired by Wister's neighbor Franklin generally reinforced the categories by which women (and many men) were excluded from institutions, academies, societies, and soon universities. For once the mechanization and materialization of nature was applied to supposed physical, mental, and emotional differences between men and women, the new sciences tightened the boundaries around what mothers could think, feel, and do in the wake of child loss.[24] Thus, even as the mechanistic, material explanations adopted by the new sciences made it possible to imagine new ideas about God, nature, suffering, and salvation, those same explanations reified boundaries around who precisely had the rational capacity to study and understand both God and nature.

Amid the very gendered nature of expectations in colonial, and then revolutionary, America, even those who abandoned biblical explanations of the world altogether simply rediscovered traditional boundaries between a "male sphere" and a "female sphere" in nature. Thomas Jefferson believed, for example, that women were *by nature* best suited to the private, domestic realm, and that therefore their "pursuit of happiness" must occur in the home. When he imagined "equality" for (white) women, he proposed "destroying artificial barriers to their natural domesticity—to their service to men and their care and protection by men." When he imagined equality for (white) men, by contrast, he meant removing barriers to economic and political freedom and participation in public life.[25] In other words, the same new sciences through which Mather, Franklin, and Jefferson imagined the world might be improved were used to justify the exclusion of many Americans from participating in these new visions of progress.

Take Franklin's sister Jane Mecom: Mecom and her brother's life could hardly have been more different. As historian Jill Lepore notes, Franklin "ran away from home when he was seventeen. [Mecom] never left. . . . He became a printer, a philosopher, and a statesman. She became a wife, a mother, and a widow."[26] As the New England colonies were transformed into the United States of America, Mecom bore twelve children. Only one, a daughter, was still alive when Mecom died in 1794. But unlike her brother, Mecom could not focus on building new societies or hospitals in the wake of great loss. That was not the kind of thing a woman was supposed to do.

In her small *Book of Ages* (a list of the births and deaths of her children), Mecom added a line from the Book of Job after the date of her daughter Polly's death: "The Lord giveth and the Lord taketh away." She then wrote, "Oh may I never be so Rebellious as to Refuse Acquiescing & saying from my heart Blessed be the Name of the Lord." She had now buried eleven of her twelve children and had urged her heart to resignation each time. We have a few hints that Mecom, like her brother, was tempted by rebellion against orthodox explanations of such great affliction. Faced with the apparent randomness of all those deaths, she questioned ministers' explanations of child loss: "I think there was hardly ever so unfortunate a family. I am not willing to think it is all owing to misconduct. I have had some children that seemed to be doing well till they were taken off by death."[27] When Polly died, Mecom wrote the following to her brother: "Sorrows roll upon me like the waves of the sea. I am hardly allowed time to fetch my breath. I am broken with breach upon breach, and I have now, in the first flow of my grief, been almost ready to say, 'What have I more?' But God forbid, that I should indulge that thought, though I have lost another child. God is sovereign, and I submit."[28] Lepore explains that Mecom "placed her faith in Providence. [Franklin] placed his faith in man." Long-standing beliefs regarding the separate, God-given (and thus natural) roles of men and women reinforced the very different options Mecom and Franklin had for imagining the relationship between nature, God, and human agency.

Some women could, in this world, serve as witnesses to nature's demonstration of God's design by drawing and painting plants and animals. Women's supposedly natural "tender feeling and quiet intuition" meant they were the best caretakers of children's moral and religious development (note these were the opposite of the rational, "masculine virtues" deemed necessary for doing science). Thus, natural theology and natural history were relevant to their sphere. But they could not engage in the interpretive work required to produce rational accounts of nature. Women read, for example, in a popular conduct book of the day, *The Whole Duty of a Woman*, that "it is not for thee, O woman, to undergo the perils of the deep, to dig in the hollow mines of the earth, to trace the dark springs of science, or to number the thick stars of the heavens.... Thy kingdom is thine own house and thy government the care of thy family."[29]

Critics of this exclusion of women from public and intellectual life did exist. Mary Wollstonecraft, for example, published *Vindication of the Rights of Women*, in which she argued that women deserved both equal rights and education, in 1792. Over the next century, however, belief in the virtues and necessity of a *natural* division of separate spheres hardened rather than weakened, especially among the middle class. Meanwhile, given the fact women like Mecom and Wister could not join the scientific institutions aimed at creating a "heaven on earth," little incentive existed for them to imagine that God had delegated so much to fallible men.

And what of mothers and fathers whom American society deemed outside the boundaries of citizenship and even humanity? What, for example, could a young woman brought to American shores to be sold into slavery believe in the face of child illness and loss? We have fewer records with which to answer this question, for obvious reasons. But we do have the poems of a mother who appealed to both of God's books, first to make sense of the great evil of slavery and then to puncture the boundaries others were establishing around her own children. Phillis Wheatley (we do not know the name her parents gave her) was kidnapped from her home in Africa at the age of seven, survived the journey across the Atlantic, and was then purchased by the Wheatley family. In an uncommon (and eventually illegal in some states) move, the Wheatleys taught Phillis to read and write. When they discovered her writing poetry, they even took her to England, where she found a patroness and thus access to a publisher. The evangelical abolitionist community embraced Wheatley as evidence of the common humanity of Africans. (As we will see, given the use of both scripture and the new sciences to argue otherwise, yes, the point had to be made.) Meanwhile, news that an African could compose poetry inspired such skepticism that a board of eighteen "respectable" men called Wheatley before them to test whether she had actually written the poems. They concluded, with wonder, that she had.[30]

Many of Wheatley's poems are funeral poems composed at the request of mourners. An early nineteenth-century memoirist described these poems as songs with "soft tones of sympathy in the affliction occasioned by domestic bereavement."[31] The poems repeat, again and again, the Christian's duty to resign oneself to God's will, even as Wheatley also captures the temptation to decry what God has decided to do. In her poem "On the Death of J.C.—An Infant," she describes the scene at the bedside of a dead baby:

> The tear of sorrow flows from every eye,
> Groans answer groans, and sighs to sigh reply.
> What sudden pangs shot through each aching heart,
> When Death, thy messenger, despatched his dart![32]

Wheatley imagines the parent, falling "prostate, withered, languid and forlorn," asking whence their child flies. Then, a cherub reminds the grief-stricken mother or father of the promise of heaven and the limitations of human understanding:

> Shall not the intelligence your grief restrain,
> And turn the mournful to the cheerful strain?
> Cease your complaints, suspend each rising sigh,
> Cease to accuse the Ruler of the sky.
> Parents, no more indulge the falling tear:
> ... Enough—forever cease your murmuring breath;
> Not as a foe, but friend, converse with Death,
> Since to the port of happiness unknown
> He brought that treasure which you call your own.
> The gift of heaven entrusted to your hand
> Cheerful resign at the divine command:
> Not at your bar must Sovereign Wisdom stand.[33]

Wheatley knew how hard it was to resign oneself to a world in which God sent such great affliction. "To ease the anguish of the parent's heart," she asks in another poem, "What shall my sympathizing verse impart? / Where is the balm to heal so deep a wound? / Where shall a sovereign remedy be found?" But her answering refrain is both varied and constant: "Be to heaven resigned."[34] Amid grief, the devout parent must take refuge in the promise of heaven, where there would be an end to all affliction and an eternal reunion with loved ones. They must bend their will to virtue, to ensure a path to heaven, and thank God for taking children before a life of temptation and vice might threaten their salvation. In "On the Death of a Young Lady of Five Years of Age," Wheatley writes:

> Why then, fond parents, why those fruitless groans?
> Restrain your tears, and ease your plaintive moans.
> Freed from a world of sin, and snares, and pain,
> Why would you wish your daughter back again?[35]

Though most of her appeals rely on the scriptural promises to remind her grieving readers of heaven and God's providence, Wheatley also turns to God's other book, the Book of Nature. There, she finds evidence of the promise of heaven in the "wondrous works" of the Almighty reflected in the cosmos, trees, flowers, and the human frame, from which "What Power, what Wisdom, and what Goodness shine!"[36] She, too, had heard of a Godless alternative. In "Thoughts on the Works of Providence," she writes:

> The Atheist sure no more can boast aloud,
> Of chance, or nature, and exclude the God;
> As if the clay, without the potter's aid,
> Should rise in various forms and shapes self-made,
> Or worlds above, with orb o'er orb profound,
> Self-moved, could run the everlasting round.
> It cannot be—unerring Wisdom guides
> With eye propitious, and o'er all presides.[37]

While some, including Franklin, found solace in distancing God from the world and reserving agency solely for human beings, Wheatley emphasized the solace of God's close, personal control over creation and human lives. Clearly, she had heard of atheists who boasted that chance alone could explain the cosmos, trees, flowers, and the human frame. She also knew naturalists' answer: no clay could take the shape of a bowl or pot without the potter's aid. Wheatley had faith, based on both nature and scripture, that God ruled over all. She believed in scripture's message that all suffering is redemptive and that God brought good from all evil, even slavery: "'Twas mercy brought me from my *Pagan* land," she wrote, for it was thus that she had learned of God and of Christ, the Savior.[38] Historian Henry Louis Gates Jr. notes that twentieth-century civil rights activists criticized Wheatley for imagining that a benevolent God used slavery to spread Christianity. But Wheatley was immersed in a worldview in which God *must* have a reason for permitting such tremendous suffering, and in which to imagine otherwise—that great affliction had no purpose—was unimaginable.[39]

The trouble was, of course, that defenders of slavery used this worldview to their advantage. US senator James Barbour insisted in 1820 that "however dark and inscrutable may be the ways of heaven, who is he that arrogantly presumes to arraign them?" Slavery, Barbour argued, was "a link in that great concatenation which is permitted by omnipotent power and goodness that must issue in universal good." Others cited the presence of slavery in the Bible or pointed out that nowhere had Jesus explicitly condemned slavery.[40]

Christian abolitionists countered such arguments by insisting that surely explicit biblical condemnation of slavery was not needed given Christ's messages of "love thy neighbor as thyself" and the golden rule ("Therefore all things whatsoever ye would that men should do to you: do ye even so to them").[41] Clearly, for abolitionists, belief in a close, personal God did not mean slavery must be accepted. "Wheatley, herself a victim of this evil," notes scholar Jeffrey Bilbro, "affirms that God has miraculously shown his mercy to her through other people's sin and yet she does not excuse this sin: a difficult and complex declaration."[42] Just as St. Basil had long since urged pious Christians to view disease as something that must be addressed via both prayer and action, Wheatley composed prayers urging resignation when children died, while insisting that slavery must be ended.

Wheatley knew what was required for slavery to end: First, she emphasized that children in slavery were *taken by men*. Ministers often exhorted parents to remember that children belonged to the Lord and must be given back at his will. Of a child "around which natural affection wreaths and entwines itself so closely," warned one minister, "I call it '*mine*'— but it is *not* mine, but *God's*. . . . Our offspring, then, are the property of God." The Christian parent was supposed to imagine the dead child whispering: "*Be still, and know that I am with God!*"[43] Yet in her poems Wheatley emphasized that slavery demanded American mothers and fathers resign themselves to God for something humans clearly did. Slavery commanded parents, under threat of human violence, to give their children to other human beings. She described human beings, not God, snatching and seizing babies from their parent's breasts. In her poem "To the Right Honourable William, Earl of Dartmouth," a reply to his question whence sprung her love of freedom and "wishes for the common good" Wheatley asks him to imagine:

> What pangs excruciating must molest,
> What sorrows labour in my parent's breasts
> Steel'd was that soul and by no misery mov'd
> That from a father seiz'd his babe belov'd:
> Such, such was my case. And can I then but pray
> Others may never feel tyrannic sway?[44]

Second, Wheatley challenged her readers (most of whom would have lost children of their own) to put themselves in the place of enslaved parents. She demanded that readers both imagine her parents' suffering grief as equal to their own and also apply the golden rule to *all* mothers and fathers.

Wheatley lived, however, in a world in which those in power called upon the Book of Scripture, Book of Nature, or (most often) both to repudiate these demands as unscientific and impious. By the time Wheatley was writing, the Swedish physician Carl Linnaeus was imagining that different human "varieties" could be both accurately described and placed on a hierarchy according to different material, moral, and mental qualities. In classifying Europeans and Africans as different varieties, based on presumed physical characteristics, Linnaeus described Europeans as "sanguine" or optimistic and Africans as "phlegmatic" or unemotional. In 1787 Christoph Meiners went even further and described Africans as more similar to animals than to (European) human beings. Indeed, he alleged they were less sensitive to pain and their wounds healed more quickly. Their females, he wrote, gave birth without pain and "as easily as wild beasts," and they could bear pain "as if they had no human, barely animal, feeling." The existence of slavery meant the stakes in adopting such claims were high.[45]

It is important to note that initially it wasn't evident whether the mechanical philosophy and new sciences would bolster justifications of slavery or undermine them. When, for example, Virginia-born physician John Mitchell developed a mechanistic explanation of the origin of different skin colors (based on the refraction of light upon different densities of skin), he declared that the difference showed God's benevolence in rendering African skin "more insensible" to heat and humidity in a hot, humid climate. The abolitionists Olaudah Equiano and David Rittenhouse cited Mitchell's work as evidence that science showed "the sin of slavery" had doomed Africans "merely because *their* bodies may be disposed to reflect or absorb the rays of light" in a different way. But defenders of slavery cited Mitchell's work in order to claim that Africans and their descendants possessed a natural, physical hardiness that made them impervious to violence and immune to external injuries. In other words, they used Mitchell's mechanistic explanation of skin color to justify both slavery and its violence as natural and (since God created nature) divinely approved.[46] The sciences of anatomy and classification were thus used to erect boundaries around whose children even suffered in the first place. These moves would have profound implications for who the powerful imagined might benefit from visions of earthly salvation.

The Laws Which Bind

Like many Americans, Thomas Jefferson wrestled with the problem of suffering as his own children (some acknowledged, some not) grew up and died around him. In 1784 Jefferson was in Europe negotiating treaties on behalf of a new entity called the United States of America. While he was abroad, his two-year-old daughter Lucy died from the "hooping cough."

Also known as "whooping cough," "chin-cough," or "tussis convulsiva" (in modern vaccination schedules, it's called "pertussis" and is the *P* in the DTP vaccine), the "hoop" referred to the sound that accompanied the child's cough as they struggled to bring air into their lungs. In 1822 the physician Benjamin Waterhouse warned that the American public was "not sufficiently apprized of the *violence*, and of the *fatality* of this '*terrible distemper.*'" Waterhouse knew of an instance where "the vehemence of a cough of this kind" broke a child's vertebrae through the middle. He described the wearing effect on the entire alimentary canal, sometimes causing a prolapsed rectum from the force of the heaves. Sometimes, breath simply could not be drawn, "in consequence, of which, the patients are, as it were, suffocated." (In some parts of America the disease was called the "choak.")[47]

Waterhouse chronicled the increasing mortality from the whooping cough evident in the London mortality bills and cited a claim by Dr. Watt that "next to small pox, *formerly*, and the measles *now*, CHIN-COUGH IS THE MOST FATAL DISEASE TO WHICH CHILDREN ARE LIABLE." He also provided an account of accepted treatments: blistering, emetics, and bloodletting (based on the prevalent theories that "two inflammatory affections cannot exist" and that it was useful to deprive the body's power to "do mischief" when it was disordered). He admitted that he preferred relying on the body's self-healing power even as he prescribed the usual purges and warned parents and young physicians against most so-called heroic treatments, since in their effort to do *something*, they often did too much.[48]

Jefferson's daughter Lucy had been attended by Dr. James Currie, but on October 13, 1784, Elizabeth Wayles Eppes conveyed the terrible news that the doctor's efforts had been in vain (for her own daughter too): "Its impossible to paint the anguish of my heart on this melancholy occasion. A most unfortunate Hooping cough has deprived you, and us of two sweet Lucys, within a week. Ours was the first that fell a sacrifice. She was thrown into violent convulsions linger'd out a week and then expired. Your dear angel was confined a week to her bed, her sufferings were great though nothing like a fit. She retain'd her senses perfectly, calld me a few moments before she died, and asked distinctly for water."[49] Dr. Currie also wrote, "I was calld too late to do any thing but procrastinate the fate of the poor Innocent."[50] (If he had been called earlier, Currie's treatment would have been to place Lucy in a cold water bath, for he held that high fever prevented the perspiration required to "abridge" the disorder.)[51]

Of Jefferson and his wife Martha's six children, only one, their first daughter, lived past the age of twenty-six. (Martha had already lost a son, aged four, from a previous marriage.) A daughter, Jane, lived for less than eighteen months, and a son for just seventeen days. A daughter named Mary lived six years. And two daughters, both named Lucy and born just two years apart, lived less than two years. We know less about the children Jefferson fathered with Sally Hemmings, although historians suspect (based on contemporary mortality rates) that some died in infancy. Indeed, the absence of these children from Jefferson's correspondence is a profound reflection of the fact that, as we will see, he drew tight boundaries around whose children and whose suffering mattered most.

As he grieved, Jefferson's explanation of the origin, meaning, and best response to suffering was in many ways very different from that of Mather. Mather had urged his parishioners to turn away from earthly comforts and toward a personal God when their loved ones died. Jefferson, by contrast, refused to find solace for his grief in the idea that a close, personal God had chosen to steal first Lucy and then her mother, Martha (who died four months later), away from him. Martha's death had almost been too much. Their surviving daughter, Patsy, wrote that her father was led out of the room in "almost a state of insensibility. . . . He kept his room for three weeks," after which, during long rides on horseback together, Patsy became "a solitary witness to many a violent burst of grief."[52] Eventually, Jefferson found solace solely in a renewed commitment to natural philosophy and natural history, and the strict uniformity of natural laws. His God designed natural laws and departed. We find, Jefferson wrote, in nature's grand uniformity pleasures "ever in our power, always leading us to something new, never cloying, we ride, serene and sublime, above the concerns of this mortal world, contemplating truth and nature, matter and motion, the laws which bind up their existence, and that eternal being who made and bound them up by these laws."[53]

Jefferson was not alone. Skeptics, most famously Hume, were composing damning criticisms of those who used the design argument to justify belief in God's close, personal control by insisting that the suffering of both humans and animals could not be reconciled with such a God (at least not if God was presumed to be benevolent). In his *Dialogues Concerning Natural Religion*, Hume demanded that men observe the suffering in the world with clear eyes: "Look round this universe.... The whole presents nothing but the idea of a blind Nature, impregnated by a great vivifying principle, and pouring forth from her lap, without discernment or parental care, her maimed and abortive children!" If one wished to make conclusions about God based on observing nature, Hume implied, surely benevolence was the last attribute one could give to the deity who created such a world.[54] This was not necessarily an argument that a creator-god did not exist. Rather, Hume was insisting that one could not demonstrate the existence of the close, benevolent God of most Christians from the study of nature. Jefferson's view, expressed in a letter to John Adams in 1823, was "(without appeal to revelation) that when we take a view of the universe, in its parts, general or particular, it is impossible for the human mind not to perceive and feel a conviction of design, consummate skill, and indefinite power in every atom of its composition."[55] Jefferson's designer was not, however, the close, personal God of Christianity. Belief in that kind of God, Jefferson was certain, did not make adequate sense of the world.

As we have seen, natural theology could accommodate moves to delegate God's governance to natural laws to a point. Thomas Balguy's *Divine Benevolence Asserted; and Vindicated from the Objections of Ancient and Modern Sceptics* argued, for example, that God's goodness must be located (and recovered amid great affliction) in the fact God was not "capricious." That is why, Balguy asserted, the course of nature is uniform. Whatever events befall us, good or bad, arise from certain general principles in the constitution and government of the universe. Like his seventeenth-century predecessors, Balguy saw a great benefit in describing God as governing though uniform, natural laws, for one could then imagine that human beings might learn those laws in the interest of ameliorating suffering. "We may observe," he wrote, "that, as knowledge increases in the world, the number of *remedies* increases, both against outward pains and sickness; and that the pains men suffer, make them more cautious, and more attentive to discover *new* remedies." Such discoveries were only possible, in Balguy's view, if God did not act via "capricious," miraculous intervention.[56]

In some ways, of course, Balguy's emphasis on nature's uniformity as the means of improving knowledge mapped perfectly onto long-standing traditions (examined in chapter 1) of physicians studying the Book of Nature to ameliorate human suffering. The distinction between the first cause (God) and secondary causes (the natural laws and material mechanisms according to which nature generally operated) had allowed ministers to support physicians while reminding patients to ultimately look to God for their cure. We have seen how, upon the death of children, Mather referred to this distinction when he wrote, "It may be one of thy Delights is put into the Ground: O look up, look above

Chance, look above all second Causes." Mather was reminding bereaved parents that although sickness might be explained via secondary causes, all affliction ultimately came from God, "the first Cause of all."[57]

We have also seen how proponents of the new sciences argued that imagining nature as driven by uniform natural laws provided an important means of regaining knowledge about nature lost at the fall of man. Boyle insisted, for example, that "a deeper insight into nature may enable men to apply the physiological discoveries made by it ... to the advancement and improvement of physick [medicine]."[58] Sydenham, the famous seventeenth-century physician, also insisted that a belief in the uniformity of natural laws, or secondary causes, was the only basis through which medicine could be improved. Medicine stagnated, Sydenham argued, because men saw disease as the "confused and irregular operations of disordered and debilitated nature," which seemed to make studying the causes of disease pointless. But if, like the mechanisms of a waterwheel, nature proceeded via uniform and consistent systems of cause and effect, then an observant physician could learn the regular course of disease. More importantly, they might learn the steps nature (carefully defined as a "certain assemblage of natural causes, destitute of understanding, disposed by the Supreme Being") takes to overcome the disease, and discern when the physician might aid nature's healing efforts.[59]

The early founders of the new sciences were well aware of the potential dangers lurking within such moves. They might focus so much on secondary causes and earthly progress that they forget to acknowledge the presence and sovereignty of the first cause (God). But they were also clearly tempted by what a division of causes might offer to the attempt to explain human experience on Earth. This temptation existed, for example, in the fact that even Boyle distanced God's hand from specific, individual afflictions at times.[60] Clearly Boyle, Ray, and Mather saw potential benefits in explaining some incidents of suffering primarily via secondary causes. But they were also certain that both scripture and the design argument would prevent secondary causes from pushing God to a distance by explaining too much.

As the assumptions of the new sciences spread across the Atlantic, some kept God's hand close to incidents of both joy and suffering. When Jefferson's daughter Polly died, Abigail Adams expressed sympathy by noting she, too, knew how "agonizing" were the "pangs of separation." But she hoped that Jefferson might "derive comfort and consolation in this Day of your sorrow and affliction, from that only source calculated to heal the wounded heart—a firm belief in the Being, perfections and attributes of God." "I have tasted the bitter cup," she wrote, "and bow with reverence, and humility before the great Dispenser of it, without whose permission, and over ruling providence; not a sparrow falls to the ground."[61]

Adams was citing a passage from Matthew 10:29–31, in which Christ demanded of his apostles: "Are not two sparrows sold for a farthing? and one of them shall not fall on the ground without your Father. But the very hairs of your head are all numbered. Fear ye not therefore, ye are of more value than many sparrows."[62] Given Christians' belief that God is "of infinite power, wisdom, and goodness," the message of the "falling sparrow" passage was supposed to be clear: Human beings were under God's watchful care. Amid tremendous diversity, most Christians agreed: All suffering came directly from God, and therefore a divine purpose existed in every affliction, whether fallible human beings could discern that purpose or not. In this world, the appropriate response to affliction was widely agreed on, even if it needed constant repeating: Trust in God's will and the possibility of reunion with loved ones in heaven. This call for trust in God did not necessarily equate to inaction in the face of suffering (like adherents of most religious traditions, Christians practiced charity and medical care), but the idea that all suffering was part of God's plan provided meaning and purpose to the incidents of affliction that inspired such action.

Jefferson interpreted the fall of sparrows (and the loss of individual children) quite differently. He fully believed that a benevolent first cause governed the universe via natural laws but suspected that the designer's attention was very distant from each individual affliction. Despite these different interpretations of the relationship between natural law and God's governance, Jefferson did not abandon faith that suffering was ultimately redemptive in some way. And that meant he could find common ground with devout physicians like Benjamin Rush, who believed in both a personal God and prayer. Jefferson emphasized God's design of natural laws, while Rush emphasized God's close, personal attention to human lives, but both could use the word *providence* while avoiding detailed discussions about what they meant. Jefferson could write to a more orthodox Christian like Rush about yellow fever, for example, as follows: "When great evils happen, I am in the habit of looking out for what good may arise from them as consolations to us, and Providence has in fact so established the order of things, as that most evils are the means of producing some good."[63] Rush would certainly have agreed (though he probably would have written that, given God's benevolence, wisdom and power, *all* evils must produce much good). Rush, however, believed God had used him *specifically* to discover a cure for yellow fever during the 1793 epidemic in Philadelphia.[64] This was the kind of close, personal providence that Jefferson refused to countenance.

While Jefferson did claim, like Rush, to derive comfort from the view that

God's governance was benevolent and wise, his God was not the God of orthodox Christianity. He famously created a version of the Bible that removed all reference to divine intervention in natural and human affairs. He believed a benevolence to the system existed, but only if the universe was examined from an appropriate distance. When Jefferson wrote "it is impossible for the human mind not to perceive and feel a conviction of design, consummate skill, and indefinite power in every atom of its composition," that claim did not mean one must see design benevolence and purpose in the death of one's beloved wife or child. Jefferson emphasized God's goodness in the design of the cosmos, rather than the detailed fates of swallows or human lives. In other words, he turned Cotton Mather's commitment to the study of natural law into the best and only witness of benevolent governance while absolving God of the terrible details those laws sometimes produced.

Jefferson's and Franklin's credentials as Christians were, of course, questionable at best even in their own day. Writing to Rush, Jefferson defined being a Christian as separate from belief in the divinity of Christ: "I am a Christian, in the only sense in which he [Jesus] wished any one to be; sincerely attached to his doctrines, in preference to all others; ascribing to himself every human excellence, & believing he never claimed any other."[65] Jefferson argued that trust in natural laws and Christ's goodness were redemptive, emancipatory, and ameliorative enough. He denied the miracles attributed to Christ and repudiated the doctrines of original sin and eternal damnation but believed Christ's ethical teachings best for founding a republic and encouraging science. Despite these radical adjustments of Christianity, both Franklin's and Jefferson's belief in progress mapped well onto many Christians' hopes of improvements on earth thanks to postmillennialist interpretations of scripture. Indeed, as Turner notes, postmillennialist Christians turned Jefferson's rationalist hope for progress "into the divine juggernaut of history."[66]

Whose children would be included in this "divine juggernaut of history," and whose might be crushed in the name of progress? How Jefferson answered this question shows yet again how those who argued that discerning natural laws would ameliorate suffering cited those same natural laws to justify ignoring the suffering of some. For although he expressed skepticism of grand classification systems like that of Linnaeus, Jefferson, too, explicitly linked apparent physical, material differences to presumed mental and moral characteristics. Having confidently classified human beings based on his assessment of their mental and moral ability to progress (according to his own criteria, of course), Jefferson decided that Indigenous Americans possessed an innate capacity to advance via the adoption of agriculture and industry. Their high mortality, he argued, was due to their tribal social structure and political organization rather than an innate, material inferiority. This explanation meant, of course, that Indigenous communities must abandon their traditional beliefs and behaviors in order to avoid annihilation in the wake of European expansion. As he explained to the chiefs of the Shawnee Nation in 1807: "When the white people first came to this land, they were few, and you were many: now we are many, and you are few. And why? Because, by cultivating the earth, we produce plenty to raise our children, while yours, during a part of every year, suffer for want of food, are forced to eat unwholesome things, are exposed to the weather in your hunting camps, get diseases and die." Jefferson's vision of progress and his explanation of Indigenous suffering placed a choice before Indigenous communities: adopt the ways of the white man and flourish or continue in the old ways and "disappear from the earth."[67]

Ironically, these explanations meant that white Americans could see themselves and their policies as benevolent even when those policies caused sickness and death. After Jefferson's acculturation policies failed due to both conflict with settlers and Indigenous resistance, the federal government decided that allowing Indigenous Americans to live and raise their children east of the Mississippi was no longer an option. In 1820 President Andrew Jackson informed the chiefs of the Choctaw that "their Great Father" wished them to move west, where they would be happy. If they did not move, he warned, "they may be lost forever." Ten years later, congress passed the Indian Removal Act. That act led to the Trail of Tears and the deaths of thousands of Cherokee, Muskogee, Chickasaw, Choctaw, and Seminole adults and children. Meanwhile, Jefferson's explanation of Indigenous mortality allowed many white Americans to absolve both the federal government and themselves of blame for such suffering.[68]

The implication in these narratives—that in designing the world just so, God or nature preferred some children over others—did not go unnoticed. In 1789 a group of Oneidas and Tuscaroras (in present-day New York) asked a Presbyterian minister, Samuel Kirkland, whether "the displeasure of Heaven in a singular manner rested upon the Indian Nations" and "whether this displeasure of Heaven or curse of God was not inflicted on them for some great sins against God committed by their forefathers." Missionaries like John Sergeant Jr. replied to such queries by insisting that God left tribes "to destruction" because "of their national sins of drunkenness, idleness and the like, and gave their country to a more industrious and virtuous people." But the Stockbridge Mohicans demanded to know "How can god be just in making such a difference between Indians and white people?" An Oneida headman asked a minister why the white people seemed "to be the Lord's favorites" despite "the conduct of a great many of them, and of their great chiefs too."[69] Some even turned such questions into a confident repudiation of claims of white superiority. During a visit to London in the early 1840s, Neu-mon-ya, the war chief of an Ioway delegation, replied as follows to an English minister who demanded that the Ioway acknowledge smallpox as a divine punishment: "If the Great Spirit sent the small pox into our country to destroy us, we believe it was to punish us for listening to the false promises of white men. It is white man's disease, and no doubt it was sent amongst white people to punish *them* for their sins."[70]

Thomas Jefferson did not talk about smallpox as a divine punishment. But he did place the blame for high mortality from smallpox squarely in the hands of Indigenous leaders who did not adopt white ways. During a meeting with Chief Little Turtle of the Miami in the winter of 1801–1802, Jefferson spoke of how "the *Great Spirit* had made a donation to the enlightened White Men; first to one [Dr. Jenner] in England, and from him to *one* [Waterhouse] in Boston, of the means to prevent them from ever having the small-pox (which had occasioned great fatality among the race)." (Chief Little Turtle agreed to be vaccinated by Rev. Dr. Gantt, and Jefferson gave him both vaccine matter and Waterhouse's written instructions.)[71]

Jefferson used the British physician Edward Jenner's development of smallpox vaccination (which, beginning in 1796, used cowpox to confer immunity rather than smallpox) as evidence of the superiority of European science and medicine over Indigenous knowledge systems. As historian Andrew Isenberg notes, although policies like the Indian Vaccination Act of 1832 arose from an effort to prevent Indigenous suffering, they also provided rationales for white expansion. Proponents of the act assumed that spreading the benefits of medical science via vaccination would convince Indigenous Americans of the superiority of European scientific knowledge and the benefits of "civilization" (fear of smallpox spreading to white settlements was also of grave concern).[72] Indeed, some linked vaccination campaigns explicitly to Christian conversion efforts. An observer of Jefferson's meeting with Chief Little Turtle commented that, in preserving their lives via vaccination, introducing the spinning wheel, and teaching them to cultivate the soil, Jefferson "has thus prepared them to receive the beneficent principles of the Christian religion."[73] (Most white Americans believed science, Christianity, and civilization went together.)

In actual fact, of course, neither inoculation nor vaccination had been discovered via mechanistic insights into how either practice actually worked, and inoculation had been practiced for centuries by non-Christians. But composing histories that acknowledged African contributions to the discovery of smallpox inoculation did not map onto Jefferson's assumptions about the world. African reason, he wrote, is "much inferior, as I think one could scarcely be found capable of tracing and comprehending the investigations of Euclid." Indeed, confident of his own rationality, Jefferson claimed that science justified relegating mothers, fathers, and children of African descent outside the bounds of concern. He argued, for example, that "natural," material differences between Africans and Europeans meant they felt suffering differently. We know this in part because of his response to Wheatley's poems. After the secretary of the first French mission to the United States, François, marquis de Barbé-Marbois, asked Jefferson about America's "slave" poet, Jefferson replied with a description of what he called the "real distinctions which nature has made" between different groups of human beings. Of Africans and the descendants of Africans, he wrote, "Their griefs are transient. Those numberless afflictions, which render it doubtful whether heaven has given life to us in mercy or in wrath, are less felt, and sooner forgotten with them. . . . Misery is often the parent of the most affecting touches in poetry. Among the blacks is misery enough, God knows, but no poetry. . . . Religion indeed has produced a Phyllis Whately [sic]; but it could not produce a poet. The compositions published under her name are beneath the dignity of criticism."[74]

Jefferson insisted that, given these supposed "real distinctions" between races, the abolition of slavery (which he believed inevitable) must be followed by the forced colonization of all freedmen to either Africa or the Caribbean. Otherwise, he argued, miscegenation, degeneration, and race warfare would result, and the American experiment would fail. He proposed emancipation schemes in which children would be born free, raised by their parents, educated "at the public expense" (rather than at the expense of their parents' owners), and then colonized *out* of the country. In an 1824 version of such a scheme he gave a price to these "new-born infants" while noting enslavers would be owed "due compensation" for their property loss: "say twelve dollars and fifty cents" each.[75]

Having classified Sally Hemmings, some of his own children, and every other individual of African descent, as both distinct and inferior beings *by nature*, Jefferson excluded more than a million human beings from his vision of the pursuit of life, liberty, and happiness. He cited "natural," material distinctions to place certain children beyond the bounds of earthly progress entirely, even as he mourned his own, white children. He did so by linking physical differences like skin color to presumed mental and moral differences, including the capacity to experience the same degree of physical, mental, and emotional pain. In retrospect, the self-interested quality of these conclusions for a man who "owned" other human beings is obvious and thus, ironically, precisely the kind of moves that the methods of the new sciences were supposed to prevent.

In August 1791 the Black astronomer, mathematician, and clockmaker Benjamin Banneker called Jefferson to task for his confident talk of natural distinctions between races. Banneker argued that God "hath not only made us all of one flesh, but that he hath also, without partiality, afforded us all the same sensations and endowed us with all the same faculties."[76] In writing as a student of "the secrets of nature" that all races had the same sensations and faculties, Banneker met Jefferson on grounds white Europeans had claimed for themselves: the evidence of mechanistic anatomy, physiology, and natural history. And yet, as historian Ian Frederick Finseth points out, giving natural science increasing authority in the debate over human difference could backfire when science and medicine were pursued through the lens of hierarchy and prejudice. As we will see, this danger existed whether that science was tightly tied to demonstrations of God's governance or not.[77]

THE HAND THAT MADE THEE IS DIVINE

Jefferson's good friend the physician Benjamin Waterhouse knew the suffering caused by child diseases as well as anyone. We met Waterhouse in the previous section describing the horrific suffering caused by whooping cough. During the worst moments, the "patient appears to gasp, and strain, and hiccup," Waterhouse wrote, "and strangle rather than cough. In this painful action, the veins of the face, as well as the eyes appear swelled; the countenance becomes very dark; the eyes are turned upwards, and there seems danger of a fatal strangulation." As Waterhouse knew all too well, sometimes the disease did indeed strangle children to death.[78]

Waterhouse believed in a very different kind of Christianity than Mather's. Having been raised a Quaker, he drifted toward Unitarianism (a more liberal version of Christianity examined in the next chapter). Jefferson praised Waterhouse for agreeing that Christ's most important message was that "to love God with all thy heart, & thy neighbor as thyself, is the sum of religion."[79] None of this meant that either Jefferson or Waterhouse gave up the design argument or belief in a creator. Even in a treatise on whooping cough, Waterhouse piously described the vocal tube as follows: "Compare it with the best, and most perfect result of human contrivance, in forming a musical instrument, and confess that the hand that made thee is divine!" Like Mather, Waterhouse believed that the purposeful design evident via both natural history and human anatomy proved God's benevolence. Waterhouse also argued that God had provided humans with the capacity to discern the laws of nature and improve medicine.[80]

Waterhouse thus echoed an assumption that went back to Bacon and the

Royal Society of London: the study of secondary causes was the benevolently provided means through which human beings could ameliorate the human condition. Indeed, Waterhouse was confident that, perhaps with the exception of "lues venerea" (syphilis) and cancer, every disease had a natural course and a natural cure that could be discovered by human beings. He spoke of surgeons, for example, giving "nature an opportunity of exerting those powers with which she is invested by the Creator" and of physicians aiding the body's "vivifying spirit" (or *vis medicatrix naturae*) in order to return individuals to an "ordered" state. Meanwhile, Waterhouse adopted new means of preventing children's systems from becoming disordered in the first place. In 1800, with Jefferson's aid, he brought Jenner's practice of smallpox vaccination (using cowpox rather than smallpox to inoculate) to the United States. He had so much faith in the procedure that he vaccinated his own children first, then "challenged" their bodies with actual smallpox.[81]

Did physicians think such efforts might be so successful that, accumulated at thousands of bedsides, they could actually lower child mortality rates? And if so, what would be the best means of doing so and whose children would be the target of such efforts? To imagine that one might move child mortality rates required thinking of affliction on a population-level scale, rather than focusing on individual suffering. Royal Society member John Graunt is credited with one of the first published works (in 1662) to examine bills of mortality as a means of imagining that "premature death" could be both tracked and prevented. Yet nearly a century later, when remarking in 1751 on American population growth rates and assuming eight children to each family ("of which if one half grow up"), Franklin did not factor in any potential future reduction in the child mortality rate. He does not seem to have imagined, despite the precedent of smallpox inoculation, that even a small part of one half of the doomed "little machines" could be prevented from dying, whether one learned the natural laws governing these numbers or not.[82] By the late eighteenth century, however, some were studying population-level mortality rates on the grounds that they too might be brought within the explanatory framework of natural laws and perhaps even controlled. Indeed, France's Nicolas de Condorcet and Britain's William Godwin even proposed that improved knowledge (and better social and political organization) would help humanity create environments in which death itself might be prevented and humans live forever.[83]

Such grand hopes raised questions about the limits of human agency that had profound implications for how to think about child mortality and what to do about it. The British clergyman Thomas Malthus firmly believed, for example, that one must try to ameliorate human suffering as much as possible, but the fact that some seemed to imagine no limits to human progress made him anxious. In his 1798 book, *An Essay on the Principle of Population; or, A View of Its Past and Present Effects on Human Happiness, with an Inquiry into Our Prospects Respecting the Future Removal or Mitigation of the Evils Which It Occasions*, Malthus criticized the "heaven on earth" visions of Condorcet and Godwin by arguing that remaking the world was not so easy. Malthus was particularly concerned by a set of poor law reforms being proposed in Britain in the wake of the French Revolution of 1789. He argued that "fixed laws of our nature"—namely, the demand for food and the "passion between the sexes"—would inevitably interfere with grand attempts to reduce poverty. Population inevitably increased, Malthus argued, at a faster rate than food. This meant that there must *always* be "a strong check on population, from the difficulty of acquiring food." War, famine, poverty, epidemics, infanticide: all could be explained as the inevitable result of the constant and merciless action of "fixed laws."[84]

We have seen how even a deist like Jefferson believed that most evils "are the means of producing some good." For a theist like Malthus, what good could the suffering caused by this system of inevitable checks on population serve? Malthus replied by insisting that these fixed laws of nature were in fact the benevolent "instruments employed by the Deity in admonishing us to avoid any mode of conduct which is not suited to our being, and will consequently injure our happiness." Without, for example, the threat of famine (from population and food increasing at such different rates) humanity would have "no motive . . . sufficiently strong to overcome the acknowledged indolence of man, and make him proceed in the cultivation of the soil." Malthus insisted that, viewed from the right perspective, these seemingly merciless laws produced a great overbalance of good. Indeed, he thought he was *recovering* God's benevolence, wisdom, and power, despite the existence of so much suffering, by explaining the laws through which the creator chose to work.[85]

Although he would often be accused of favoring high child mortality as a population check, in fact Malthus believed that humans must try to reduce premature death. Indeed, hope in progress by human effort inspired Malthus's firm stance on the relation between natural law and the possibility of miracles. Exertion, he argued, must be directed by reason. This was why the supreme being acted always according to general laws. The constancy of the laws of nature and the certainty with which humans could expect the same effects from the same causes provided a framework within which human beings could apply their God-given reason to the amelioration of suffering. The constancy of natural laws, he wrote, provided "the foundation of the industry and foresight of the husbandman, the indefatigable ingenuity of the artificer, the skillful researches of the physician and anatomist, and the watchful observation and patient investigation of the natural philosopher. To this constancy we owe all the greatest and noblest efforts of intellect. To this constancy we owe the immortal mind of a Newton."[86] What Malthus could not support were reform policies that in his view would just create more suffering in the long run because they ignored natural laws. Policies, he urged, must be guided by a rational, clear-eyed assessment of how God had designed the world.

Malthus's explanations of suffering permeated natural history through its inclusion in the most influential update of the design argument, a book composed by British clergyman William Paley and published in 1802: *Natural Theology; or, Evidences of the Existence and Attributes of the Deity, Collected from the Appearances of Nature*. In a famous opening paragraph, Paley asked his readers to compare the origin of a stone found on the ground with that of a watch. The purposeful parts of the watch—designed to tell time—led to the obvious conclusion that the watch had been created by a watchmaker. Paley quickly moved from the conclusions drawn from the purposeful parts of the watch to what must be concluded from the purposeful parts of the anatomy of the human eye, which he in turn compared with the purposeful parts of a telescope. Purpose required design, and what a grand designer the wondrous, purposeful parts of animals and plants revealed. To Paley, the lesson about God's governance was clear: Nature, combined with the evidence of the New Testament, demonstrated God's close attention to and governance over creation, from the fall of the sparrow to the loss of a much beloved child. Searching for purpose in every affliction, including poverty, Paley cited Malthus's work as evidence that even natural laws that seemed to cause great suffering demonstrated God's benevolent design of the world. Malthus's "laws of population," Paley argued, revealed God's purposes in designing a creation in which population increased at a higher rate than food.[87]

The fact that Paley's *Natural Theology* was required reading for generations of American college students meant Malthus's explanation of suffering as due to fixed laws became widely known in the United States, including in its medical schools. Indeed, Paley's pious update of natural theology formed the context in which both natural history and anatomy were done for generations. One scholar notes that "more American students before 1850 read Paley than any other moral philosopher."[88] By the 1830s Paley's and Malthus's talk of a beneficent, divinely designed "struggle for room and food" that arose from the "fixed laws of our nature" had permeated British and American natural history, medicine, and political economy.

Whether Malthus was right and what precisely Malthusian laws meant for responding to different kinds of child mortality was hotly debated. In his 1822 treatise on whooping cough, for example, Waterhouse lamented that Malthus's theory of a "redundancy of population" might be used to justify inaction in the face of child illness. "Half the human race die under ten years of age," Waterhouse wrote. "Is this the inevitable consequence of our existence?" Surely it could not be the plan of providence "to require the destruction of one third of mankind before they attain the age of two years?" Surely no "restrictive law" existed that meant infant life must be destroyed to check excessive population! It was an unbearable thought, completely irreconcilable with God's goodness. Instead, Waterhouse insisted that both infantile disease and early death must be the result solely of "unnatural management." (He also wrote, in contrast to Malthus and with great confidence, that "like destruction is not observable among the young of the brute creation," a point that we will soon see countered by Darwin and others.) Certain, then, that things might be managed better, Waterhouse urged his readers to work to prevent such great destruction among the young.[89]

Two decades later, in the 1840s, a Scottish physician named Andrew Combe expressed concern that Malthus's laws were being cited to justify complacency in the face of child mortality. Combe placed two options before the reader: Either the high mortality among the young "constitutes a necessary part of the arrangements of Divine Providence which man can do nothing to modify," or it proceeded "chiefly from secondary causes purposely left, to a considerable extent, under our own control." Notice that Combe agreed with Malthus that God governed via secondary causes; but he insisted that learning the natural laws that governed the infant "constitution" could reveal the results of Malthus's supposed natural laws (including high mortality rates) to be artificial results of ignorance and poor management. Both Malthus and Combe tried to get divine authority behind their respective positions. Combe argued, for example, that it was "impious to suppose that it enters into the design of an all-wise and all-benevolent Creator, for so many children to be necessarily born to early death." Insofar as "we shall discover and fulfill the laws which the Creator has established for our guidance and preservation," he insisted, "so would the waste of infant life decline."[90]

Combe's 1840 *Treatise on the Physiological and Moral Management of Infancy* tried to convince parents that they had a "direct duty to study the nature of the infant economy, and discover the causes of the diseases by which life is endangered." Only when parents saw illness as the result of purely natural cause and effect, he insisted, would they be able to keep their infants alive, for they would base treatment on "*the nature and laws of the infant constitution.*" Indeed, Combe argued that once men discovered the natural laws governing the infant constitution, "the rules of conduct deducible from them come before us stamped with more than human authority." Accurate knowledge about nature, Combe wrote, provided "indirect, intimations of the Divine will." Combe thus explicitly argued that God had arranged the structure and laws of the infant body "for our welfare and advantage" and that humanity must study these laws to avoid bringing disease and suffering upon its children.[91]

Combe's firm belief that God governed the world via natural laws rather than miracles need not place him in conflict with orthodox Protestantism. Many Protestants had long since relegated miracles to the time of Christ's apostles and earlier (a move that allowed them to distinguish themselves from Catholics' belief in miracle-dispensing saints and healing shrines). Both Combe and Malthus justified their arguments against the idea that God governs via miraculous intervention by emphasizing God's benevolent provision of science as the only possible means of progressive human agency. Malthus argued that only if natural laws were uniform (i.e., unbroken and consistent) could human beings trust that a particular action could have a desired effect. Combe agreed. Indeed, to behave otherwise, Combe wrote, was to shut one's eyes to the means by which God governs and then to presumptuously pray that he will alter the order of nature in one's favor, thereby removing "the consequence of our deliberate disobedience."[92]

It soon became clear, however, that Malthus's description of the world could be used to demand resignation to certain kinds of suffering. Malthus never argued that physicians should not try to save children from illness and death. Both Paley and Malthus insisted that the fact that "population always treads on the heels of improvement" must not dishearten endeavors for "public service."[93] But Malthus did argue against policies that he believed encouraged early marriages and thus produced more children than parents could feed. (Of course, as Malthus's critics pointed out, avoiding policies that assured young, poor parents that their children wouldn't starve meant that, inevitably, in times of famine or hardship, children starved.) Combe, by contrast, argued that human beings could actually move child mortality numbers (which he put at between a third and a half of all children in the first five years of life) down permanently rather than be trapped in an inevitable Malthusian cycle of increased population followed by increased mortality. He did so by, first, insisting (like Waterhouse) that high mortality rates among the young did not occur in animals most similar to man (such animals, Combe argued, are guided by an unerring instinct that keeps their progeny safe), and second, by comparing mortality rates under different conditions. In the 1780s Dr. Joseph Clarke of the Lying-in-Hospital of Dublin had decreased the mortality of infants from convulsive disease from one in six to one in twenty, for example, simply by improving the ventilation in the wards. For Combe, that simple fact provided clear evidence that even more progress could be made if parents and physicians only knew more: "What a lesson of vigilance and inquiry ought not such occurrences to convey," he concluded, "when, even now, with all our boasted improvements, *every tenth infant still perishes within a month of its birth?*"[94]

For some Americans, Combe's book provided a powerful justification of the belief that child mortality rates might be permanently lowered via human action. For years Elizabeth Cady Stanton watched children around her: "On all sides, ill half the time, pale and peevish, dying early, having no joy in life. I heard parents complaining of weary days and sleepless nights, while each child, in turn, ran the gauntlet of red gum, jaundice, whooping cough, chicken-pox, mumps, measles, scarlet fever, and fits." She wrote with dismay that they all seemed to think such things "were a part of the eternal plan–that Providence had a kind of Pandora's box, from which he scattered these venerable diseases most liberally among those whom he especially loved." When her son Daniel was born in March 1842, she wrote, "Having gone through the ordeal of bearing a child, I was determined to keep him." She read everything she could on the subject of "infant management" but could not find much. Americans, Stanton lamented, asked expert horticulturists how to garden under the assumption there were laws governing plant health that could be known and improved. Yet they did not do the same with their babies. Stanton found, she wrote, just "one powerful ray of light" that illuminated the darkness: Andrew Combe's *Treatise on the Physiological and Moral Management of Infancy*.[95]

It soon became clear, however, that Waterhouse and Combe were right to fear that Malthus's efforts to bring child mortality within the realm of nature's laws might be used to justify inaction in the face of certain kinds of suffering. Some Americans were quite willing to imagine that their own children might be saved by a better knowledge of the natural laws governing the "infant constitution" while citing those same laws to justify inaction for the children of others. When the books and ideas examined above crossed to America's shores, they entered a highly racialized society. As historian John S. Haller notes, "almost the whole of scientific thought in both America and Europe in the decades before Darwin accepted race inferiority, irrespective of whether races sprang from a single original pair or were created separately."[96] (As we will see, this was true even of many abolitionists.) Within this world, Malthus's statements about an inevitable, natural, divinely designed "struggle for existence" became justifications for doing nothing in the face of both Indigenous and Black communities' suffering.

The theologian Horace Bushnell, for example, argued against slavery as a moral evil but then cited Malthus's work to claim that manumission would, in giving free reign to the inevitable "struggle for existence," lead to the extinction of nonwhite races. Bushnell did not imply that anything could or should be done to prevent either the struggle or extinction. Instead, he saw the hand of "the Almighty Himself" at work in both. "Since we must all die," Bushnell wrote, "why should it grieve us, that a stock thousands of years behind, in the scale of culture, should die with fewer and still fewer children to succeed, till finally the whole succession remains in the more cultivated race?" Others cited Malthus to argue that allowing slavery in new states would displace white homesteaders. "In a political, a moral, an intellectual, and a religious point of view," wrote one Malthus-reading opponent of allowing slavery in Missouri, "is not a white population better than a black one?" Meanwhile, defenders of slavery used Malthus's work to argue that "slave economies" produced *less* suffering than "free-labor" systems by controlling reproduction. (An ardent abolitionist, Malthus was dismayed by these uses of his essay.[97])

A physician named Jesse Chickering tried to counter these uses of Malthus's work by pointing out that prejudice, not nature, caused differential mortality rates. Indeed, the depth of that prejudice inspired Chickering to agree that Black men, women, and children should be "returned" to Africa in order to "save" the race. "The free coloured population of the United States," he wrote in 1853, "seem doomed to perish in a state of freedom, falsely so called, among a people whose feelings and whose prejudices loathe fellowship with them, and where the whole structure of society is set against their equal participation in the blessings of the land."[98]

The idea that differential mortality rates originated in prejudice (rather than God's design or natural laws) proved less influential in nineteenth-century American science and medicine. Whether they highlighted God's design, nature's laws, or both, plenty of Americans explained high mortality within both Black and Indigenous communities as the inevitable cost of "progress." In doing so they used supposed natural laws to justify ignoring the suffering of some. In other words, belief in progress via a law-bound struggle for existence was used to attribute the high child mortality of some Americans to the benevolent laws of nature rather than the actions and beliefs of human beings (as we have seen, Jefferson was willing to blame the actions and beliefs of Indigenous Americans who did not adopt European ways). Thus, even as belief in the possibility of lowering child mortality via the study of natural laws expanded, some cited those same laws as reasons *not* to actively intervene in the mortality rates of some communities.

Chapter 3

A World Made for Melioration

> Step the meek birds where erst they ranged,
> The wintry garden lies unchanged,
> The brook into the stream runs on,
> But the deep-eyed Boy is gone.
>
> —Ralph Waldo Emerson, from "Threnody"

On a Sunday in January 1842 Ralph Waldo Emerson took his five-year-old son Waldo to see Concord's new church and organ. Waldo fell ill with scarlet fever the following evening and was dead by Thursday evening. His father sat at his desk that same night to notify family members and friends. Brief messages went out from the Emerson house: "Our darling is dead," began one. "My little Waldo died this evening," began another. To Thomas Carlyle, Emerson wrote, "From a perfect health & as happy a life & as happy influences as ever child enjoyed, he was hurried out of my arms in three short days."[1]

The disease that took Waldo was one of the most feared childhood killers of the nineteenth century. It was called "the dread scourge of children." Parents watched helplessly as the sudden onset of a painful sore throat was followed by vomiting, chills, and high fever, and then the bright red rash that gave the illness its name. Scarlatina was acute, contagious, and thought to sometimes cause permanent hearing loss. Physicians described the worst cases descending into convulsions, delirium, and death. And it all happened with such cruel swiftness. The physician Caspar Morris, reflecting on twenty-five years' observation of the disease, wrote, "No disease had so unexpected or violent an onset. The parent sits down in the evening in the happy centre of a group of smiling objects of affection, his heart swelling with delightful anticipation as his eye glances around the circle; and ere the next return of the same weekly period, half of them slumber in the embrace of death."[2]

Attempts to apply lessons learned from smallpox and prevent the worst ravages of scarlet fever via vaccination failed. Although many physicians believed that the disease spread by contagion (as opposed to atmospheric changes), no one knew what caused the terrible symptoms. Germ theory, which would provide the framework for identifying the real culprit (a bacteria called *Streptococcus pyogenes*) was more than a generation away. Morris reported that "teluric emanations, sidereal influences, animalculae, and fungi" had all been invoked. Amid uncertainty regarding cause, treatment varied. Morris recommended capsicum, Dover's powder, carbonate of soda, sweet spirit of nitre, quinine, all in careful moderation and at the right stage of illness. Give the child cool drinks, cool air, a light diet, he wrote, and avoid "depletion treatments" (bloodletting and emetics). Alas, Morris lamented, many physicians had hurried children needlessly to the tomb through the well-meant but ill-directed use of violent purgatives and calomel, a common mercury-based medicine.[3]

Thanks to the mechanical philosophy's emphasis on dissection (especially the study of morbid anatomy, or the correlation of symptoms with postmortem lesions and patterns), early nineteenth-century physicians were much better than their predecessors at diagnosing some ailments. But with the possible exception of convincing some physicians that mercury might not be the safest of medicines, the methods of the new sciences had thus far contributed little to the "relief of man's estate" in the medical realm. Those convinced that progress was possible and child mortality rates might be lowered had to look elsewhere for evidence that faith in progress was warranted. Unable to call on progress in medicine and wary of relying so much on orthodox Christianity, some Americans called upon radical new interpretations of how to read both the Bible and the geological past. Together, these led to tightly intertwined new narratives of the past to justify belief in a better future. Once again, whether these new justifications of belief in progress would include all Americans' children within their visions of a heavenly future on earth was not necessarily clear.

Transcending the Earthly Sphere

Waldo's mother, Lydia, knew what she was supposed to feel, think, and write in the wake of Waldo's death. She wrote of how, for a few hours after Waldo died, she succeeded in viewing his death through the lofty, abstract, religious plane of the "greater good" promised by divine providence, namely eternal life in heaven. But during the night, when all was quiet and she slept with her three-year-old daughter, Ellen, by her side, "grief desolating grief came over [her] like a flood—and [she] feared that the charm of earthly life was forever destroyed."

She wrote, "I saw not how I could ever feel happy again. I thought of the words 'Time brings such wondrous easing' and believed Time could bring no easing to us."[4] In the letters her husband wrote to friends and family, Emerson made no mention of an afterlife or of heaven. Emerson could take consolation, he wrote, only in the memory that little Waldo had "never been degraded by us or by any, no soil has stained him, he has been treated with respect & religion almost." Just once, Emerson's words harkened back to the kind of biblical language used by ministers like Mather. Two weeks after they lost Waldo, Emerson wrote to Lydia, "And now how art thou, Sad wifey? Have the clouds yet broken, & let in the sunlight? Alas! Alas! that one of your sorrows, that our one sorrow can never in this world depart from us! Well perhaps we shall never be frivolous again, eating of this everlasting wormwood."[5]

A century and a half earlier, Mather imagined that perhaps God took beloved children (laying "Wormwood on the Breasts which you have hung too much upon") to redirect parents' attention to heaven and secure their salvation. But mostly, Emerson abhorred Mather's way of thinking about and justifying God's ways. He had rebelled from Mather's explanations for why God would allow a much beloved, innocent child to die. And he was intent on finding different forms of solace. Given how ideas about God, nature and history were intertwined, that also meant, of course, that he needed new ideas about both nature and history.

Emerson's rebellion against Mather's God had commenced long before Waldo's death, and it did not at first constitute a repudiation of Christianity. Like many nineteenth-century critics of orthodox religious traditions, Emerson had originally aimed for a reformation of Christianity rather than a replacement. The previous generations' emphasis on the possibility of progress on earth (whether to recover the Garden of Eden or replace paradise altogether) had produced a conscience-ridden conflict between various orthodox dogmas and what historian Howard Murphy called "the meliorist ethical bias of the age." The increasingly pervasive idea, for example, that "the life of man on this earth both can and should be progressively improved through a sustained application of human effort and intelligence" tended to conflict, Murphy notes, with the "otherworldly-salvation motif" of orthodox Christianity. Vicarious punishment, human depravity, eternal damnation, an emphasis on God's justice over his goodness: Emerson and his friends eventually deemed all of these doctrines unjust and furthermore blamed such doctrines for complacency in the face of earthly suffering.[6]

Other Americans feared that manifestos for "human improvement" were elevating earthbound agendas above their proper status. Would not religion's spiritual claims be marginalized amid so much focus on earthly improvement? Some feared, for example, that individuals might start believing that the material earth was all that existed or mattered. Furthermore, pervasive belief in original sin and human depravity inspired distrust of the idea that humans could know so much about what constituted goodness on earth much less how to get there. In an 1825 sermon entitled "The Beloved Physician," Reverend John Marsh lamented how often physicians, lost in secondary causes, forget to look up to "the great first cause." Yes, Marsh argued, medicine was a regular science and physicians must seek to understand natural laws in the interest of alleviating the sufferings of his fellows. But he exhorted his listeners not to adore the instrument (the physician) but rather "be grateful to God for raising him up, and granting relief to beings who, for their sins, deserve nothing but indignation and wrath, tribulation and anguish." Of course, Marsh's point would have been lost on men like Emerson who rebelled against the idea that humans deserved "wrath, tribulation, and anguish" in the first place.[7]

Nineteenth-century rebellions against orthodox Christianity like Emerson's are often blamed on the discoveries of science, especially geology and evolution. Certainly, literal interpretations of the Bible's miracle-laden history seemed out of step with mechanistic science's emphasis on the uniformity of nature's laws. But many skeptics, including Emerson, initially rebelled against orthodox Christianity on moral rather than scientific grounds. The educational reformer Horace Mann, for example, "repudiated a God who gloried in the punishment of the wicked" after witnessing the despair on his mother's face when a minister preached on the terrible consequences of dying unconverted at his young brother's funeral sermon. Having abandoned orthodox beliefs about God, including original sin and a system of rewards and punishments, Mann needed a new explanation of suffering. Only then did new visions of earth and human history rush in to fill the gap left by these decisions.[8]

As a young man Emerson had adopted a version of Christianity that was among the first to lodge charges of injustice against orthodox versions of God, first in England and then on the other side of the Atlantic: Unitarianism. Unitarians emphasized God's benevolence over his "inscrutable justice," a stance that bolstered faith in man's God-given reason to discern causation rightly (initially many Unitarians still assumed men were more rational than women, although the tradition's strong impulse to question authority inspired some to doubt even that time-honored assumption). Emphasizing God's benevolence and the power of reason also encouraged Unitarians to dispense with doctrines that had traditionally been defended by faith rather than reason. Surely, Unitarians held, the creator who had allowed Newton to discern the laws of the heavens did not intend to befuddle men with mysteries that must be taken solely on faith. Indeed, the term *Unitarian* reflected adherents' denial of the Trinity, the doctrine held by Catholics and most Protestants that God is both one and triune: Father, Son, and Holy Ghost. Joseph Priestley ridiculed trinitarianism on the grounds that "three divine persons constituting one God is, strictly speaking, an absurdity, or contradiction." Priestley even dispensed with the divinity of Christ because that belief relied on a miracle, his virgin birth. But while Jefferson and Franklin expressed sympathy for Priestley's efforts, critics demanded to know in what sense Unitarian ideas constituted Christianity at all.[9]

By the early 1820s the most influential American Unitarian, William Ellery Channing, had dampened Priestley's more radical claims. Channing defended belief in both the New Testament miracles and Christ's divinity. (How would the authority of Jesus and scripture be upheld, Channing asked, if one abandoned belief in miracles?) Yet even Channing's version of Unitarianism could seem to critics heretically tolerant of new scientific ideas. After all, Unitarians' suspicion of creeds and codified doctrine made it difficult to establish some kind of rule that would limit explanations by natural law.

Although they abandoned doctrines that others believed central to Christianity, Unitarians maintained Mather's belief that God's benevolent dispensation to humanity included spiritual *and* material progress via the study of natural law. Channing preached, for example, that the fact "man has power, by arts and commerce, to multiply and spread almost indefinitely its provision for human wants" provided "eloquent testimony to the goodness of the Creator." Like both Malthus and Paley, Channing emphasized affliction as an important means of ensuring progress on earth rather than (solely) toward heaven. "The distance at which good objects are placed," he wrote, "and the obstacles that intervene, are the means by which Providence rouses, quickens, invigorates, expands, all our powers. These form the school in which our minds and hearts are trained." (Sermons like this are surely why Malthus's talk of "a struggle for existence" was easily adopted by many Christians.) Channing even argued that God's delegation of ameliorative powers to humanity undermined the arguments of those who insisted suffering disproved the existence of a benevolent God.[10]

Channing knew humans had little to show for so much confidence in their own God-given ameliorative powers. In 1840 he could rightly say, "After the study of ages, the science of medicine has not completed the catalogue of diseases; and how little can its ministrations avert their progress, or mitigate their pains!"[11] Meanwhile, he could sound a great deal like Cotton Mather as he wrestled with the great affliction of child loss. When his wife gave birth to a second daughter two years after they lost their first little girl, he warned himself against feeling happy: "Had I not learned so many lessons of this world's mutableness, I might be in danger of dreaming of a perfect joy on earth. But I do not forget where I am."[12] The problem, however, was that Unitarians like Channing abandoned many of the biblical stories that had explained "this world's mutableness" for generations. In the orthodox view, the threat of hell provided the ultimate justification of great affliction, for suffering turned undeserving humanity's attention toward heaven. But Unitarians were repudiating the fall of man, original sin, and the possibility of eternal damnation as unworthy of a good God.

As they set aside hell, the Unitarians at Harvard who trained Emerson (he was ordained a Unitarian minister in 1829) insisted that belief in a close, personal God and heaven could remain. Indeed, Harvard minister Andrew Preston Peabody argued, like Franklin, that the terrible affliction of child loss proved heaven must exist. In his 1847 sermon "Our Need of the Father," Peabody emphasized that "to a parent, above all, is this faith in the Supreme Father of unspeakable value." For, he explained, "to have a helpless being intrusted to one's care, with hosts of diseases and accidents thronging around the very gates of life . . . and then to look around upon the multitude of early graves,—who could, in view of all these things, find courage to go forward in the discharge of a parent's duties, without the assurance that the little flock have a Heavenly Shepherd, whose breath will feed their life, whose staff will guide their steps, and who, both on earth and in heaven, bears the lamps in his arms and carries them in his bosom?"[13] How, in other words, could one even bear and raise children without trust in God's providential care?

For a time, Unitarianism served as an important way station for Emerson as he tried to figure out what he believed and how he would deal with the world. Meanwhile, he had heard, and then gave, plenty of sermons on how the Book of Nature demonstrated God's benevolent governance. His education at Harvard, where Paley's *Natural Theology* was popular among even the most radical Unitarians, drove the point home: the purposeful parts of animals and plants demonstrated the existence of a benevolent God, despite the existence of pain and suffering.[14]

That demonstration received a severe blow, however, upon the death of Emerson's first wife, Ellen, from consumption (one of the leading causes of adult mortality) in 1831. The disease's name arose from the fact the body seemed slowly consumed from within (it was also called the "wasting disease"—today it is known as tuberculosis). Patients woke up in the middle of the night gagging on mouthfuls of blood. Many eventually died from asphyxiation when the lungs hemorrhaged. An incessant cough could prevent a consumptive patient from speaking for weeks at a time. Sometimes the disease would go into remission for months, if not years (Emerson was consumptive as well, but lived longer), only to return again with a more deadly grip. Ellen was just twenty years old when her lungs could draw breath no more. Three months later, delivering a sermon on Christ's blessing, "Happy are they who suffer," Emerson tried to keep the faith: "We are often made serious by domestic afflictions," he preached. "And the spiritual good that comes thence is far better than the external good that was destroyed to produce it."[15] He was going through the motions, preaching sermons on the redemptive meaning of affliction. But in private he was in the midst of a deep crisis of faith.

As Emerson sat at Ellen's bedside during her final months of life, and then as he visited her grave, he was immersed in the work of theologians calling for radical reformations in how to understand the past and the present. Based in (present-day) Germany, these theologians were intent on rescuing Christianity from irrelevance in an age that seemed to be judging God, the Bible, and the possibilities of human reason by new standards. To protect Christ's ethical message even if belief in his miracles was lost, they developed methods of "biblical criticism" or "higher criticism" to determine the authorship, date, context, and intention of the Bible's authors. Proponents of higher criticism believed that only through analyzing the Bible as a text produced by men in particular historical contexts would correct interpretations of scripture be produced. One of the most important methodological rules of higher criticism was that, in making claims about the past, one must assume that both human nature and natural laws have *always* been uniform, with no exceptions. Stories of biblical miracles, they argued, must be understood not as true accounts of what happened in the past but rather as reflections of the culture in which they were written. Only then might theologians discover the true nature of God's activity in both human and natural history.[16]

The influence of these new rules on interpretations of God's presence and activity in creation (traditionally proven via biblical miracles) was profound. For some, higher criticism destroyed belief in God altogether. For by what authority, they demanded, was any part of the Bible owed allegiance if miracles must be set aside? But for others, higher criticism inspired a rethinking of what precisely the word *miracle* means. Maybe, proposed some, natural law *is* the miracle. Maybe the proper object of man's contemplation in order to understand God is nature's *uniformity* of cause and effect. Indeed, whereas some feared that an emphasis on divine governance via natural law would turn God into a distant lawgiver, others began arguing that, in fact, God must be closer to and intertwined with creation rather than an occasional visitor.[17]

Those who found higher criticism compelling had to read scripture in new ways. As Emerson composed sermons in the months after Ellen's death, he began preaching that the passages in which Jesus cured demonic possession were in fact cases of organic disease. Christ had simply chosen to speak to the insane "in a manner conformed to their expectation" when he commanded the disease to depart.[18] Emerson was clearly trying to make sense of the suffering in God's creation by drawing on the insights of higher criticism. But as he did so, he could imagine no purpose or meaning in Ellen's death. Within a year, he could go through the motions no longer and resigned his post as a Unitarian minister. In the ensuing decade, throughout the 1830s, he helped create the American transcendentalist movement. Divinity, Emerson decided, pervaded all nature *and* humanity (this was precisely the kind of God-in-nature pantheism the mechanical philosophers had tried to crush by getting spirit *out* of nature). Then, having given up so much in the wake of Ellen's death, Emerson tried to develop a replacement for orthodox Christianity's system of heavenly rewards and hellish punishments. In an 1841 essay entitled "Compensation," he argued that there was a system of compensation in the universe "here and now," governing both animals and human beings. Immoral actions exacted a toll *on earth*. That, he said, was why one should strive to do good and be better. Concern with the preservation of individual personality in Heaven was for Emerson a selfish distraction from higher truths.[19]

Grief was supposed to be easier to bear in this transcendentalist philosophy. One need not wonder why a close, personal God permitted specific evils and afflictions if the true measure of divine goodness was at a different scale entirely. Emerson's friend Henry David Thoreau wrote of how every leaf in the forest lay down its life in its season; the law of death was the law of new life. Death, "when seen to be a law, and not an accident," Thoreau wrote, was beautiful. Thoreau did not flinch from extending this conclusion to human life: "So it is with the human plant," he wrote. "We are partial and selfish when we lament the death of an individual, unless our plaint be a paean to the departed soul, and a sigh as the wind sighs over the fields, which no shrub interprets as private grief."[20] Transcendentalists insisted that wisdom entailed transcending the individual, the particulate, the fleeting. One had to take the broader view. The cosmic view. Eventually Emerson remarried and had children. In 1838, reflecting on these worldly possessions from his new transcendentalist stance, he wrote, "If my wife, my child, my mother should be taken from me, I should still remain whole."[21]

Then, in 1842, little Waldo died. The benefits of transcendentalism had seemed clear; now the costs rushed in. Emerson could not transcend his attachment to little Waldo's life and prospects, now gone forever. He could envision no just compensation. Emerson is famous for having divinized nature, but nature had failed him: "Meanwhile the sun rises and the wind blows," he wrote, while "Nature seems to have forgotten that she has crushed her sweetest creation."[22] As Emerson searched for solace in the wake of Waldo's death, his devoutly orthodox aunt Mary Moody mercilessly placed the costs of transcendentalism before her beloved nephew. She had never agreed with Emerson's divination of nature at God's expense. What, she seemed to say, had Emerson thought nature was like in the first place?

> And art thou become like one of us? Does nature who seemed thy favored Angel to wait thy walking & musings and be the servitor of thy Muse look askance—dreamy & like as she passes the most of her children often clad in hair cloth & at best homespun silk worm stuff? And you do not like the little grave she has dug so blindly that you cannot drop your plummet into its sad chasm. Right glad am I! Abandon her! She has been covering the earth for ages with blood & war & prisons—erected her cauldrons & swept her besoms over thrones and love & wisdom & all free beautifull [sic] things. Pass her bye—go behind her stage decorations alike with her inquisitions and prostate thy higher capacities of enjoyment before Him who for some time of inscrutable purposes weilds [sic] her secret forces.[23]

For Moody, Emerson's consolations of "compensation" and "improvement" weren't good enough. Heavenly visions of progress on earth were not worth giving up heaven itself. Moody clearly preferred those who tried to make the earth better rather than those who focused solely on Christ's second coming and heaven. But she was also sure that, while people should work to abolish slavery, drunkenness, and ignorance, even a much better earth could never stand in for God's kingdom. Transformation (from an earth-bound existence to an immortal life in heaven) was the only thing worthy of God. And nature, which stole beloved children in the space of a few days for no reason, was no substitute.[24]

Ultimately, Emerson refused to "abandon nature" and adopt his aunt's belief that some "inscrutable," specially designed purpose of a personal, loving God existed behind Ellen's and Waldo's sufferings. And that meant he had to wrestle with the question of why, if the world "was made for benefit" (which he believed it must be), it is not just. Having abandoned the old explanations for the origin of great affliction and salvation, Emerson needed new explanations for both why nature was covered "with blood & war & prisons" and why, despite that fact, belief in the possibility of a better earth was warranted.[25]

Beside the Deathbed of Amiable Children

To understand the explanations Emerson chose, we must back up a bit and note that when naturalists prior to the nineteenth century argued that the perfect design of animals proved the existence of a close, personal God, they did not have to explain why God or nature allowed creatures to go extinct. "Such is the economy of nature," wrote Jefferson in 1785, "that no instance can be produced of her having permitted any race of her animals to become extinct." By the opening of the nineteenth century, however, the French naturalist Georges Cuvier had proved that entire flora and fauna had disappeared from the earth. Repeatedly.[26]

The challenges posed by the discovery of extinction for belief in a close, personal God were most famously captured by the British poet Alfred Tennyson's poem "In Memoriam." Nature seems "so careless of the single life," Tennyson writes, that God and nature seem "at strife." But surely one might "faintly trust the larger hope" for nature is, at least, careful and protective of the "type," or species. Then nature, via the fossil record, replies: "'So careful of the type?' but no. From scarped cliff and quarried stone, Nature cries, 'A thousand types are gone: I care for nothing, all shall go.'"[27] Tennyson combined the age-old question of why God allowed beloved friends, family, and children to die with the puzzle of extinction: Why would a benevolent God create such marvelous creatures, only to allow them to disappear from the face of the earth forever? Franklin had, of course, asked the same question of individual infants. But extinction expanded the scale and timeframe of the mystery to an almost unimaginable degree.

Naturalists like Charles Darwin's geology professor, Adam Sedgwick, adjusted the theory of special creation by arguing that the fossil record's constant pattern of extinction and new creations demonstrated God's continued presence in creation. But some interpreted these new discoveries quite differently. In 1844, two years after Waldo Emerson's death, an anonymously authored book entitled *Vestiges of the Natural History of Creation* appeared in England and soon crossed the Atlantic. (Although few knew it at the time, the book's author was a middle-class publisher from Scotland: Robert Chambers.) *Vestiges* argued that each species, including *Homo sapiens*, had not been independently designed by God but had developed via a purely natural process. In other words, the book proposed that species arose via a secondary cause (created by the first cause, of course, but a secondary cause all the same) called "the law of development."[28]

We have seen how, for generations, American naturalists, physicians, and ministers insisted that, although secondary causes could explain much, they could not explain the origin of species: for how could a purely mechanistic process create the wonderfully purposeful parts of animals and plants? But the author of *Vestiges*, having rebelled against the idea that God had some carefully devised purpose for letting a child die (Chambers composed the book while grieving one of his own children), decided that God's creative hand need not be involved in the creation of species either.

Vestiges was a scandalous sensation. Emphasizing the rule of natural law in the movement of planets or the formation of canyons and mountains (as astronomers and geologists were by this time doing) was one thing. Driving natural law back into the deep past as a new explanation of the origin of species was quite another. After all, proponents of the new sciences had tied the independent, special creation of species to the demonstration of the existence and attributes of a close, personal Christian God. *Vestiges*, by contrast, claimed that the creation of species via a developmental law was the more pious position. "It is the narrowest of all views of the Deity," the book proclaimed, "and characteristic of a humble class of intellect, to suppose him acting constantly in particular ways for particular occasions." Men interpreted biblical passages ("Let the earth bring forth the living creature after his kind")[29] as indicating miraculous intervention, but a truly omnipotent God would surely have created the world and *all* of its creatures via natural laws. (Notice this was a vision of creation that mapped on perfectly to higher critics' rules for explaining the human past.)[30]

Very few readers took the mechanism *Vestiges* proposed for the creation of new species seriously. (The book proposed that species advanced under the progressive influence of external conditions, namely air and light, on the developing embryo.) But *Vestiges* offered one of the first widely read evolutionary alternatives to literal readings of the Genesis creation story. Its reinterpretation of the fossil record as bound by natural law was useful to those interested in new understandings of human history that repudiated miraculous intervention. Most importantly, *Vestiges* replaced the fall of man, in which progress recovered something lost, with man's ascent. Man had lost nothing. All was a gain. In doing so, *Vestiges* offered an explanation of the origin of suffering that brought all affliction under the governance of natural laws. We see, wrote the author, the deity operating according to "fixed laws," each acting independently "according to its separate commission." As a result, those laws could only have effects that are generally beneficial, since often one law would interfere with another and inadvertently produce evil. The suffering of a boy engaged in the "lively sports proper to his age" who falls and injures his spine, for example, had been caused by two things (in the main good) coming into conflict: the love of exercise and the law of gravitation. The primary object of gravitation, *Vestiges* argued, was not to injure boys, but since natural laws are invariable, gravity must pull a boy who loses hold of a branch to the ground. "The evil is," *Vestiges* concluded, "therefore, only a casual exception from something in the main good."[31]

It is worth recalling that we saw hints of this explanation of suffering in the work of seventeenth-century proponents of the new sciences, including Robert Boyle. In *Vestiges*, the benefits of distancing God's hand from specific incidents of evil and suffering were twofold: First, God was thereby absolved of responsibility for the suffering and afflictions experienced by individuals (human mismanagement and ignorance of nature's laws were now the sole culprits). Second, human responsibility in the face of such a system seemed clear—namely, one should learn natural laws in order to prevent suffering in the future. Yes, *Vestiges* acknowledged, great suffering existed, but God had provided the means of amelioration via man's careful study of natural laws. Once, for example, physicians saw the human constitution as merely a complicated but regular process in electrochemistry, the path toward elimination of disease—"so prolific a cause of suffering to man"—became clear: to learn nature's laws and to obey them.[32]

In some ways, *Vestiges* was drawing upon a perfectly orthodox stance on the relationship between science and Christianity. In delivering a sermon on the lessons taught by sickness in the late 1830s, for example, Reverend Edward Hitchcock, professor of natural theology and geology at Amherst College, drew upon long-standing traditions of differentiating between first and secondary causes in medicine to encourage the study of natural laws while upholding belief in God's close, personal providence. Hitchcock argued that while he "would not be thought to deny that sickness is always an act of God's Providence . . . in general, He exercises that Providence and that sovereignty, as he does in respect to almost everything else, according to fixed laws: so that when disease assails us, we may be sure that there is a natural cause for it." Indeed, Hitchcock insisted, like his seventeenth-century predecessor Sydenham, that until men understood that disease and health operate according to fixed and invariable laws, they would not study those laws in order to avoid disease and preserve health.[33]

Since the seventeenth century, however, amid an increasing emphasis on secondary causes in the study of disease, proponents of the new sciences had insisted that the study of anatomy and natural history would protect faith in God's close, personal control over creation. Physicians might expand their understanding of the secondary causes through which generation, development, and reproduction were governed, but the theory of special creation demonstrated God's close, personal presence as the ultimate source of all affliction. In drawing the origin of species into the realm of secondary cause, *Vestiges* seemed to threaten this long-standing, mutually supportive relationship between Christianity, natural history, and anatomy, in which the study of nature was supposed to convince individuals of the possibility of salvation.

What, then, of heaven? *Vestiges* offered only a vague suggestion that so much benevolence in the design of the cosmos was a strong hint of "a system of Mercy and Grace behind the screen of nature, which is to make up for all casualties endured here."[34] Most of the book focused on what humans might imagine possible on earth, rather than the afterlife. Once men accepted the universal governance of natural laws, *Vestiges* declared, they could imagine that the prodigious evils caused by disease might be conquered. For "as civilization advances," the author explained, "reason acquires a greater ascendancy; the causes of the evils are seen and avoided; and disease shrinks into a comparatively narrow compass." *Vestiges* conceded that mortality tables still showed a "prodigious mortality among the young." But that concession was followed by a confident declaration that "to remedy this evil there is the sagacity of the human mind, and the sense to adopt any reformed plans which may be shown to be necessary."[35]

When Emerson read *Vestiges* in 1845, three years after Waldo's death, he found it "a good approximation to that book we have wanted so long & which so many attempts have been made to write."[36] An evolutionary vision of the deep past fit with what he wished to believe about nature, God, and human agency. (Many transcendentalists, notes historian John B. Wilson, were evolutionists long before Charles Darwin's work appeared.) Years later, he carefully transcribed a passage from Richard Owen's *Paleontology; or, A Systematic Study of Extinct Animals and Their Geological Relations* (1860) into his diary: "Paleontology teaches, as regards the various forms of life which this planet has supported, that there has been 'an advance & progress in the main.'"[37] By embracing an evolutionary interpretation of that record of life, Emerson could maintain faith in the progressive shape of history even as he dispensed with scripture as the source of that faith.[38]

Some years earlier, in 1841, Emerson had confessed that giving up the old beliefs had created a new torment—namely, "the utter uncertainty and perplexity of what we ought to do; the distrust of the value of what we do; and the distrust that the Necessity which we all at last believe in, is Fair."[39] But in embracing evolution Emerson recovered a sense of both human agency and responsibility by focusing his attention on what had arisen from the constant cycles of death to decay to life and back again over unimaginable swaths of time. Though he found the theology in *Vestiges* "civil, timid, and dull," he found the book's visions of a long-suffering past, the tribulations of the present, and promise of a higher human destiny on earth to be "just what was needed."

In a lecture called "The Spirit of the Times" delivered throughout the 1850s,

Emerson told his listeners, for example, that "the gracious lesson taught by science to this Century is that the history of nature from first to last is *melioration*, incessant advance from less to more, from rude to finer and finest organization." *This* was Emerson's reply to his aunt Mary Moody. Although the world was filled with "internecine war, a yelp of pain & a grunt of triumph," suffering ultimately "mellowed" the "whole mass" for "higher use."[40] "The whole circle of animal life," he wrote in 1860, thus "pleases at a sufficient perspective." In answer to the question of why the world was not just, Emerson eventually replied, "It was made for melioration."[41] Indeed, Emerson's good friend the physician Oliver Wendell Holmes recalled that *melioration* was one of Emerson's favorite words.[42]

Emerson could not cite clear and consistent progress in medicine when he wrote of "the gracious lessons taught by science to this Century." Although he could (and did) cite anesthetics (chloroform and ether were discovered in the 1840s) as evidence of progress in the amelioration of suffering, both had been discovered by happenstance rather than any particular insight into how nature worked. Indeed, as late as the 1860s, Emerson's friend Holmes famously argued that, aside from opium, a few medicines "which our art did not discover," wine, and "the miracle of anesthetics," the entire content of the doctor's bag *"as now used*, could be sunk to the bottom of the sea." Doing so "would be all the better for mankind," Holmes added, "and all the worse for the fishes."[43]

Holmes had been trained in Paris during the height of therapeutic nihilism, a rebellion against the violent purgatives and "heroic" treatments used by many American physicians. Therapeutic nihilists argued that physicians must stop interfering with the body's ability to heal itself. Indeed, Holmes decided that the best evidence of a benevolent purpose at work in the universe could be found not in scripture but in the vis medicatrix naturae, "the healing power of nature." For Holmes, God's goodness appeared in what every physician knew: that "a very large proportion of diseases get well of themselves, without any specific medication." Holmes even used the power of the body to fight disease to insist that belief in eternal damnation was absurd. No physician "in the perpetual presence of this great Healing Agent, who stays the bleeding of wounds, who knits the fractured bone, who expels the splinter by a gentle natural process," could believe in a God who damned![44] For evidence that progress was possible, Holmes, like Emerson, called on the "new tables of the law, placed in the hands of the geologist by the same living God who spoke from Sinai to the Israelites of old," the fossil record.[45]

For both Emerson and Holmes, faith in progress was justified by geology and the new science of paleontology rather than obvious advances in medical therapeutics. For those influenced by higher criticism, the fossil record read like a new revelation, one that mapped onto both their ethical rebellion against traditional interpretations of scripture and their hopes for the world. Having read *Vestiges of the Natural History of Creation*, the radical Unitarian minister and abolitionist Theodore Parker wrote with excitement, for example, of how the Book of Nature had finally disproved the "monstrous doctrines" of orthodox Christianity. "You find no Devil on the face of the earth to-day," Parker wrote, "no footsteps of him in the 'Old Red Sandstone,' not a track of his step amid all the 'Vestiges of the Natural History of Creation.'" The annals of human history, Parker insisted, "records no Fall of man, but rather an ascent, a continual increase in justice, philanthropy, piety and trust in God."[46] Parker then used this revised version of history to argue that true religion aspired for the Kingdom of Heaven *on earth*, believing "man as the noblest child of God; with a magnificent future before him, to be wrought out by his own toil."[47] Parker embraced the idea that God acted "providentially in Nature not by miraculous and spasmodic fits and starts, but by regular and universal laws, by constant modes of operation." Only because of that constancy, Parker insisted, could men trust that they could learn how to prevent suffering in the future.[48]

For some readers, however, *Vestiges*'s evolutionary vision of the past gave up too much. The botanist Asa Gray (who, as we will see, would christen Darwin's *On the Origin of Species* for American audiences) found the book's vague references to heaven in favor of a confident vision of progress on earth woefully unsatisfactory. "The only ground of encouragement" in such a view, Gray countered, "is, that, although the individual may suffer remedilessly, the race is going on to perfection; that, when 'man is transferred to the list of extinct forms,' some perfected form of being may succeed him, just as we have succeeded the extinct saurian and other races." *Vestiges* constantly referred to a "Divine Author," but Gray argued that its "law of development" was the old atheistic scheme of the French evolutionist Jean-Baptiste Lamarck, long since thoroughly refuted on both scientific and theological grounds. *Vestiges*, wrote Gray, "thrust the Creator out of the visible universe far back to the remotest verge of time" and confined "him there bound fast by the very laws he ordained."[49] Of course, this was precisely why Emerson found *Vestiges* a better explanation of both past and present. For Emerson, binding the Creator by natural laws was the only ethical and scientific way of making sense of the world and the best foundation for imagining that things might be done better.

In replying to his critics, the author of *Vestiges* (then still anonymous) made the connection between child loss and decisions about the nature of God's governance explicit. For he warned that we do not become sensible of truths about God and nature in hours of cheerful, active, and prosperous existence. Rather, conviction must be sought in more painful hours, when suffering arises from a misapplication of human reason to the requirements of both health and happiness. We must decide on such matters "beside the death-beds of amiable children, destroyed through ignorance of the laws of health, and hung over by parents who feel that life is nothing to them when these dear beings are no more." For it is in those moments that humans may realize that things might have been managed better and suffering ameliorated.[50]

Unitarian minister at Harvard (and a distant cousin of Emerson's) Andrew Preston Peabody agreed that "beside the death-beds of amiable children" was a place where decisions about the relationship between God's power and human agency must be made. But he came to a very different conclusion regarding what one must rely on in those moments in which physicians could do no more. Peabody warned that a profound danger lurked in claims to have solved the age-old problem of suffering via man's progressive ability to learn nature's laws: "The idea of laws of nature, omnipotent, irreversible, crushing,—of a system in the main beneficent, which yet has its hard cases and its victims,—weighs down the spirit with an iron hand." For, in the face of loss, "there always comes up the torturing question,—'Could not the issue that has taken place have been foreseen and averted, had we been more watchful and more wise?'"[51] As a minister who had attended many grieving parents, Peabody knew that such thoughts often "formed the most bitter ingredient in the cup of affliction."[52] Thoughts of one's own culpability when a loved one died, he wrote, often prevented the recognition of God as the sole author of the afflictive event. Here is the example Peabody gave:

> The child of a watchful and experienced mother is taken away by acute disease. The attack was sudden; yet the seeds of the disorder must have been lurking in the system for days or weeks previously, and there were preventive measures by which the danger might have been warded off. The mother's memory, sharpened by her grief, can now recall symptoms that might have indicated disease,—a drooping of the eyelids, or a flush of the cheek, or an unusual drowsiness, wakefulness, or peevishness; and, in remembrance of these unheeded indications, her sorrow is drugged with intense bitterness, as she reproaches herself that she had not taken alarm at the tokens of incipient illness, and administered such remedies as might then have proved effectual.[53]

To counter the torture of hindsight, Peabody argued that surely all symptoms needed the subsequent event to interpret them rightly, and thus neither science nor skill could have drawn certain conclusions before the outcome was clear. "Where knowledge cannot be," Peabody urged, "there can have been no responsibility." Indeed, he declared ignorance of future events "a part of the Divine plan; and we can hardly conceive how essentially it ministers to our happiness."[54]

To be clear, Peabody was not arguing against calling upon a physician to try and save a child who was ill. Rather, he was arguing that *Vestiges*'s answer to the problem of suffering was of no use when medicine failed. Only belief in a providential governance in which there is "no aimless suffering, no event which it is not best for us to meet and bear" could help the soul amid sorrow. "We need that faith in the Father which shall refer the trial to no second cause," he wrote, "to the uncontrolled working of no material law, but solely to the merciful purpose of one who wounds but to heal, whose very rod comforts while it chastens."[55] Only faith in the New Testament, the miracle of the resurrection, and consequent trust in the loving kindness of God and "the deathless union of those whom death has parted" could be of aid. Without this faith, Peabody argued, the unique affliction of child loss (Peabody thought no animal was subject to so much premature death) became an insolvable, even unbearable, enigma.[56]

The fact that these words were written by a Unitarian minister, who espoused a version of Christianity deemed heretical by more orthodox ministers, shows the precipice at which even some of the most unorthodox halted. Meanwhile, Peabody was right that the pressures placed on parents by attributing child loss to ignorance of nature's laws (an explanation that emphasized general, rather than special, providence) could be keen. When Fanny Longfellow lost her eighteen-month-old daughter in 1848, she relived the final days of her daughter's life over and over. She was, she wrote in her diary, "haunted by thoughts of what might have been avoided, the most pitiless of all." Amid calls for better management via better knowledge, parents, especially mothers, had to wonder: If, as one home medical guide stated, "no one can for a moment believe that the excessive and increasing infant mortality among us, is part of the established order of nature, or the systematic arrangements of Divine Providence," then who was, in fact, to blame if not themselves and physicians?[57]

Of course, as a former minister who had surely comforted parents raised in more orthodox circles, Ralph Waldo Emerson knew that blaming oneself for a child's death also took place within orthodox Christianity. He surely knew of mothers and fathers like the Methodist Episcopal bishop Leonides Hamline, who, recalling the death of his two-year-old daughter, Jane, of cholera infantum (the summer diarrhea that plagued so many infants) in 1828, explained that sad event as punishment for his own lack of faith.[58] For Emerson, the belief that God stole children away like this was worse than regret that parents or physicians might have managed the case better if only they knew more.

A Blasted Hope

Of course, knowledge is not the only thing that determines whether a child can be saved. Sometime around 1846, Dr. James McCune Smith, just beginning his practice as physician of New York's Colored Orphan Asylum, was trying to get to the asylum to care for children stricken by a measles epidemic when he was refused passage on a streetcar. For days he walked the six miles from his own practice to the asylum, until the ladies in charge of the orphanage found out what had happened and gave him funds to hire a private carriage.[59] Clearly the orphanage's patrons believed in getting a doctor to the asylum's children as often and as quickly as possible. But Smith knew that not everyone in New York thought his work worth doing.

Of both African and European ancestry, James McCune Smith had been born into slavery. New York's 1827 Emancipation Act set him free at the age of fourteen, but it was a limited kind of freedom. Smith had to leave the country to pursue his dream of becoming a physician. Columbia and Geneva College refused to admit him on account of his race. So he went to medical school in Glasgow, Scotland. There, he became the first African American to obtain a medical degree. Smith combined a devout Episcopalian Christianity with a firm commitment to mechanistic science.[60] He believed, like Malthus, Paley, and Combe, that science's rule that "like causes under like circumstances will produce like effects" provided "the basis of all our belief, all our hope—it is the very essence of that *Faith* in the stability of things, without which life would be made up of dismal, because uncertain, anticipation." Confident in that stability, "the mariner boldly launches forth on the deep; and the man of science with the same rule questions nature in her minute recesses and carries his enquiries to the very barriers of creation." This consistency of cause and effect was the only grounds, Smith insisted, for true knowledge and amelioration on earth.[61]

Smith had no doubts that the fundamental assumptions of science could be reconciled with Christianity. Christ's message, he argued, served as one of the most paramount influences that had governed "human advancement." For "without Divine Revelation," he wrote, "the human mind could never have soared to those heights of thoughts whence drop down those hallowed sentiments, which in creating the joys of home, the abeyance of a well spent present, to a glorious future life—have stimulated the human mind in its onward path."[62] Upon returning to New York in 1838, Smith spoke at the American Anti-Slavery Society about his own vision of progress. Though trained at one of the best medical schools of his time, Smith knew he could not cite advances in medical treatments as evidence that the earth could be improved. Instead, he cited the abolition of slavery in French and British colonies to demonstrate that progress was possible: "One moral victory gained," he argued, "raises the mind to an eminence whence it perceives others that must be achieved, and inspires it with new energies for the struggle."[63] Smith had decided that his role in this struggle for something better would be via devotion "to the improvement of colored *children*."[64] He helped found the New York Society for the Promotion of Education Among Colored Children and served as attending physician to the Colored Orphan Asylum from 1844 until his death in 1865. He also carried out a long, difficult fight against those who used nature's laws to argue that some children either couldn't or shouldn't be saved.

Smith was absolutely certain that, rightly interpreted, the Book of Scripture and Book of Nature could not, as testaments of the same God, contradict one another. Indeed, he believed that nature could serve as a useful adjunct to demonstrating the truths of scripture. In 1843 he quoted both Acts 17:26 and the discoveries of physiologists to demonstrate not only the unity of man but the inspired truth of scripture: "'God hath made of one blood all the nations of men,'" he wrote, "is one of those passages in scripture whose truth is inherent proof of their inspired source. The microscope has shown that globules of blood in the human species are alike, and differ from the blood of all other animals."[65]

Smith knew very well that other Americans read quite different messages from both the Book of Scripture and Book of Nature. He practiced medicine in a society in which many grouped children as white, Black, or Indian and then declared that some children died at higher rates because their bodies (or "constitutions") were different than the "constitutions" of white children. Indeed, when Smith returned from Scotland, the Colored Orphan Asylum was under the care of a physician named James MacDonald who believed that children of African descent could not by nature thrive in North America. MacDonald attributed the death of nine of the asylum's sixty-four children in a year to the "peculiar constitution and condition of the colored race."[66]

Smith knew exactly what this kind of thinking could justify. He was well aware of how explanations of mortality could be used as excuses to do nothing for these children. So he reprinted MacDonald's report in its entirety in the magazine *Colored American*, and then refuted its claims point by point. He fought back on the ground MacDonald had claimed as decisive in determining a proper attitude toward differential child mortality: science, physiology, and statistics. Using the mortality tables for New York, Smith showed that the diseases from which these and other Black children in New York had died—dropsy of the brain, convulsions, croup, whooping cough, scarlet fever, measles, chicken pox, cholera infantum, teething—actually killed far more white children. The numbers, Smith wrote, "cannot lie." If MacDonald wanted to play the game of judging children, then by the doctor's own principles it was "the colored race" that was "best fitted by nature to endure the climate of New York." Black children, Smith concluded, could survive at the same rate as white children if they were well cared for. When these children did not flourish, Smith argued, "hell-born prejudice" must be blamed, not some "peculiarity of constitution."

Smith did not have children when he wrote his response to MacDonald. But he could imagine having them, and he knew what MacDonald's kind of medical thinking might mean for any future sons and daughters of his own. As he criticized MacDonald's assumption that the asylum children's deaths were due to constitutional "peculiarities," Smith included a sarcastic appeal: "Dear, dear Doctor, if providence should ever bless us with a small family, and a hopeful little one becomes sick, we shall certainly employ you as the most *satisfactory* man in the profession. If the child recovers, you will be able to show that you have snatched it from the very jaws of death, but if you kill—No, if the patient dies, you will smother our grief in wonderment that it did not die nine times instead of one."

Smith was clearly angry at how MacDonald had harnessed science and anatomy in the name of prejudice. "Next to our Maker," Smith concluded, "do we revere Science as the clearest manifestation of his law which he has vouchsafed us. And we have hoped for much from Science . . . we have fondly dreamed that she would ever rear her head far above the buzz of popular applause, or the clash of conflicting opinion in the moral world; it is therefore almost with the anguish that springs from a blasted hope that we view this first, however flimsy, attempt to demean her to the contemptible office of ministering to public prejudices."

Smith believed that science, as God-given, had to serve—must serve—truth and justice. But it soon became apparent that MacDonald's work was just one of an avalanche of claims that science proved white Americans superior to everyone else. In 1843, Smith described this avalanche as follows:

> Learned men, in their rage for classification and from a reprehensible spirit to bend science to provoke popular prejudices, have brought the human species under the yoke of classification, and having shown to their satisfaction a diversity in the races, have placed us in the lowest rank. Now, if this were true and we were in reality such inferior beings, we would of necessity fall into this low rank in the social scale without the aid of the laws. There is no law in these states to prevent dogs and monkeys from voting in the polls. And the laws which they enact in regard to us are proof positive that our oppressors are getting more and more convinced that we are *men* like themselves; for they enact

just such laws as the experience of all History has shown it to be necessary in order to hold *men* in slavery.[67]

Sometimes Smith cited scripture, as when he called upon Acts 17:26 (God "hath made of one blood all nations of men for to dwell on all the face of the earth")[68] against this "rage for classification" and the bending of science to "provoke popular prejudices." Sometimes he used statistics, as when in 1843 he cited the fact that lower mortality rates accompanied higher standards of living. Such numbers, Smith argued, disproved claims that the "constitutions" of the "free colored inhabitants of the northern cities" could not "endure the rigors of a northern climate; and that in competition with the superior energies of white laborers, they will *'of necessity be driven to the wall.'*" Bills of mortality clearly showed, Smith argued, that it was oppressive laws and social customs that raised mortality rates, not different "constitutions." He used both new physiological discoveries, such as the fact that blood did not differ between the so-called races, and natural history, pointing out, for example, that although "animals that differ in species never procreate with one another . . . whatever differences there are between men, whether in temperament or appearance, their offspring does not therefore grow barren. In this very Republic there are some descendants of whites and blacks, who are called mulattoes; they are not sterile."[69] Both points had to be made because some naturalists and physicians were arguing so-called races were in fact different *species*.

Smith appealed to history and human nature to make sense of why, despite scripture and science, so "reprehensible a spirit" twisted both Christianity and science into defenses of slavery. Smith described belief in Africans' "natural inferiority" as "an ERROR which designing and interested men had craftily instilled into the civilized world." "Nations which term themselves civilized," Smith observed, "have one sort of faith which they hold to one another, and another sort which they entertain towards people less advanced in refinement." He declared this phenomenon, which seemed so incongruous to "our otherwise prosperous state" the "undermining influence of *caste*." Elsewhere he called the tendency of men to attribute their own group's advancement "to innate superiority of race" the "Great Idol of the Tribe." (In the seventeenth century, Francis Bacon had proposed four idols of the human mind that interfered with the perception of truth: The idol of the tribe was the tendency, once an opinion has been adopted, of drawing "all things else to support and agree with it" and of believing more readily those things one wishes to be true. Since the 1980s the phrase *confirmation bias* has been used, but the phenomena—and the warning—has existed for a very long time.) As an example of "the local manifestations of this unrepublican sentiment" of white superiority, Smith gave the following: "While 800 children, chiefly of foreign parts, are educated and taught trades at the expense of all the citizens, colored children are excluded from these privileges."[70]

Nearly a decade after sparring with MacDonald, Smith lamented "a hate deeper than I had imagined" in the "terrible majority." He continued, "I must strive humbly to draw near unto God, for renewed faith and hope and encouragement. He Reigneth over the 'raging of the waves and the madness of the people.'" Meanwhile, his faith in God's goodness inspired Smith's resolve, even amid despair and constant opposition, to continue his struggle against prejudice. Indeed, scholar John Stauffer notes that during the 1850s Smith "became more religious, and his faith enabled him to remain hopeful despite national declension."[71]

A decade after Smith composed his criticism of MacDonald's analysis of orphan mortality, the women who ran the orphanage unanimously appointed him as the institution's physician. As he took care of the asylum's children, Smith's wife Malvina Barnet gave birth to at least eleven children. In 1848 Smith wrote to abolitionist Gerrit Smith, "My wife is a fruitful vine through God's blessing, and three little souls look up to me for support and discipline and guidance: what a holy trust!"[72]

Like so many physicians, James McCune Smith often experienced the fact that medicine could do little in the wake of the worst child illnesses. Alas, his and Malvina's eldest child, Amy, died just before her sixth birthday in 1849 after a year's illness "at times painful and distressing, always obscure." Amy had borne her suffering "with child-like patience," Smith wrote to Gerrit Smith, but "it pleased God to take her home to the Company of Cherubs who continually do Praise Him. You have been afflicted in like manner and know the bitterness of it. For one thing I am deeply grateful, her mind was serene to the last, and intelligently hopeful of a Blessed Immortality." Four years later, as he prepared sketches for Frederick Douglass's antislavery newspaper, he was "sadly interrupted by the long and painful illness of one whose little chair is vacant by my hearthstone." This time it was a little boy, "whose little grave is filled on the hillside." "Again and again, as I sit by my easel," Smith wrote (he was working on a set of essays entitled "Heads of the Colored People"), "brush in hand, spirit fingers weave his golden hair upon the canvas, and those sad eyes light upon me, and spirit voices break the stillness of the night, in cadences now light and airy, now sobbing in keen agony."[73]

Smith grieved the loss of his children while wrestling with the world into which they had been born. He wrote defenses of saving all children by insisting on the role of environment in differential mortalities, and the increased capacity of all when raised in better environments. This emphasis was central to his argument that things could and must be managed better. Meanwhile, he repeatedly took up the tools of anatomy and physiology and turned them back on those using science to justify disparities. When writing of child mortality, he insisted that the question to be asked was not what was the child's race, but where did the child live? How were they treated and how did society view them? Throughout, Smith emphasized explanations based on natural, uniform cause and effect. He had faith that the Book of Scripture and Book of Nature formed a united front against prejudice and that "the great law of nature" could and must be harnessed in the name of justice and amelioration for all children. In using science to fight back against physicians who declared that the origin of Black children's mortality lay in their material constitution, Smith brought reason and science to bear on abolitionists' scriptural demand that the divine humanity of all Americans be fully recognized.

As years passed it became increasingly clear that the "blasted hope" of which Smith wrote in 1846 was not so easily revived by science. Amid increasing commitment to "better management" as a means of preventing child mortality, some Americans argued that the same criteria could not be used to explain and thus potentially reduce the mortality of all children. Faced with the rising moral and political power of abolitionists, for example, pro-slavery physicians, naturalists, and politicians continued to use differential mortality rates as evidence of "constitutional differences" between races. It was common knowledge that the mortality rates of enslaved children were higher than those of other children. (Today, scholars estimate that the mortality rate of enslaved children under the age of one was 350 per thousand, compared to 179 per thousand for white children, while the mortality rates of enslaved children between the ages of one and five averaged 201 per thousand, and 93 per thousand for white children.) As we have seen, some Americans combined the belief that these mortality rates reflected natural, constitutional differences with Malthusian laws and then argued that abolition would result in an inevitable struggle for existence that whites were "constitutionally" destined to win.[74]

Smith was not the only one to ask whether all factors had been taken into account in explaining enslaved children's mortality rates. In 1851 southern planter Thomas Affleck acknowledged that the mortality of enslaved "children is as two to one when compared with the whites, depending solely upon locality and care." Affleck pointed out that "quarters are often badly located; children allowed to be filthy; are suckled hurriedly whilst the mother is over-heated . . . not a few are over-laid by the wearied mother, who sleeps so dead a sleep as not to be aware of the injury to her infant." Clearly Affleck didn't think enslaved children's high mortality arose from either constitutional differences or nature's laws. Rather, he emphasized that the environment in which these children were born was to blame. Meanwhile, surgeon Samuel Forry diagnosed the stance that differential mortality was rooted in racial difference as part of an old tradition: "It has ever been thus, by denying an equality in the original endowments of certain families of the human race, that writers have attempted to justify the institution of perpetual servitude."[75]

Smith had hope, within a fallen world in which "it had ever been thus," that some things might be improved by appealing to scripture *and* science. He firmly believed that the Book of Nature and Book of Scripture could not be in conflict and that his beloved science must be on the side of justice. Confident that Christianity could adjust to new scientific discoveries, he did not seem bothered by rumors of a long, evolutionary past as espoused by *Vestiges* or the transcendentalists. When, dismayed by the conduct of men, he resolved in 1848 to "fling whatever I have into the cause of colored children," he wrote, "This kind of work suits me because it is very hard, and somewhat noiseless: in the series of metamorphoses, I must have had a coral insect for a millio-millio-grandfather, loving to work beneath the tide in a superstructure, that some day, when the labourer is long dead & forgotten, may rear itself above the waves & afford rest & habitation for the Creatures of his Good, Good Father of All."[76]

In comparing his life work to phenomena—coral reefs—that by this time geologists had explained as the accumulation of organisms living and dying over unimaginable time to produce a great good, Smith portrayed both his work as a physician and his writings as part of a slow advance. One could find progress, Smith insisted, in the improvements that arose from the long-forgotten struggles of those who had come before.

He had hoped to teach his own children well before releasing them on the "tide of progress, dependent upon their own well-developed resources." But in 1854, within the space of one summer, the Smiths lost three more children (Frederick, Peter, and Anna), possibly in a cholera epidemic. Again, Smith wrote to Gerrit Smith: "Oh it is sad to have no children playing round the hearthstone. I try, and may God give me the Grace to succeed, to look into other little glad eyes and listen to other little glad voices; and I try to reason myself out of the selfishness that they are not mine. Oh that meeting hereafter!"[77]

Although some criticized a focus on heaven on the grounds that it dampened activism on earth, James McCune Smith's hope of a "meeting hereafter" did no such thing. Bereft of his own children, Smith prayed for God's aid in resigning himself to care for orphaned children as though they were his own. Smith was confident that, rightly understood, *both* of God's books—the Book of Nature and the Book of Scripture—revealed all children to be "of one blood." He took on physicians like MacDonald because he saw how some Americans were using science to establish boundaries around whose children might be saved. As Smith tried to wrestle the Book of Nature away from those using science to establish boundaries around whose mortality might be lowered, others tried to wrestle scripture to the side of amelioration for all as well. For many Americans, including Smith, these were not distinct fights.

Some Great Good for Others

Harriet Beecher Stowe first learned to balance faith in both prayer and physicians with the duties of both resignation and action from the sermons of one of the most influential ministers in the country—her father, Reverend Lyman Beecher. In his autobiography, published in 1864, Lyman preached the orthodox response to bereavement when he described the death of his own child almost a half-century earlier. One little girl "was seized, when about a month old, with the hooping cough." The child's mother, Roxana Beecher, was "up night after night, taking care of the child, till she was exhausted." Lyman recorded how, when death finally came, Roxana "was so resigned that she seemed almost happy." Lyman Beecher turned the memory into a sermon on Christian grief: "I never saw such resignation to God; it was her habitual and only frame of mind; and even when she suffered most deeply, she showed an entire absence of sinister motives, and an entire acquiescence in the Divine will."[78]

Like most ministers, though he urged acceptance of God's will once the fate of the battle was certain, Lyman Beecher called upon physicians when his children fell ill. When scarlet fever struck his daughter Harriet and her little brother Frederick, the Beecher family turned to both prayer and a physician. Only after Freddy, "without much struggling, breathed his last," and God's will became clear, did resignation become the right response to what a disease might do.[79] Meanwhile, in the 1830s the religious revival known as the "Second Great Awakening" added momentum to the view that God approved of human efforts to ameliorate, rather than endure, suffering on earth. Presbyterian minister and president of Oberlin College Charles Grandison Finney argued, for example, that Christians must devote themselves to disinterested benevolence and social reform. For abolitionists, slavery must be the first great evil removed from the American landscape in order to prepare the earth for Christ's return. This did not mean abolitionists agreed on either how or when to end slavery. As president of Illinois College, Beecher tried to convince students who had organized an anti-slavery society that the time had not yet come for such radical change.

But some of those students cited Reverend Beecher's own militant insistence that sin be stamped out now, bolted from his oversight, and called for the immediate abolition of slavery in all states.[80]

Most of Lyman's own children rebelled against his orthodox views as well, even as they maintained their father's ethos of "doing good," his belief in the progressive shape of history, and faith in (a highly revised) Christianity. The eldest, Catharine, left the fold first. She repudiated her father's belief that because of Adam and Eve's disobedience all human beings were born with wicked hearts. But the missionary zeal of her father's favorite motto, "Trust in the Lord and do good," remained. She poured her energies into a campaign for educational reform, rooted in the belief that children had an innate capacity for good. She insisted that "the end for which we are made is to 'glorify God' by obedience to those laws by which 'the most happiness with the least evil' is to be secured to His vast eternal empire." Hostile reviewers called Catharine an infidel for relying on her own "*intuition* of God's justice," but Catharine recoiled from the idea that she must blame God rather than human beings for the suffering around her. Christ's life, she argued, demonstrated the "universal law" that good triumphed over evil only through self-sacrifice for others.[81] One of the actions for which Catharine is most famous was the writing of her 1829 "Circular Addressed to the Benevolent Ladies of the U. States" arguing against federal Indian removal policies. In that document, she urged Congress to continue "acknowledg[ing] these people as free and independent nations, and . . . protect[ing] them in the quiet possession of their lands." As historian Tiya Miles notes, Catharine wrote these words more than a decade after Cherokee women had themselves organized in opposition to removal, as recorded in mission documents, "in cognizance of the hardships & suffering to which it is apprehended the woman [*sic*] & children will be exposed by removal." Indeed, Miles argues that Cherokee women's campaign for justice first inspired Catherine Beecher's activism.[82]

Catharine in turn exerted a strong influence on her younger sister Harriet, who was also trying to find her own path of "doing good." Harriet eventually rebelled against their father's beliefs as well. Lyman Beecher's primary mission had always been saving souls for heaven. Given his belief in the possibility of eternal damnation, the prospect of his children dying outside the fold gave him far more anxiety than suffering on earth, including death itself. He was confident, like Mather, that God's mercy would be extended to infants and small children. But as his surviving children grew into moral, willful beings, the possibility of damnation grew. "To commit a child to the grave is trying," he wrote as he exhorted his son Edward to devote his life to Christ, "but to do it without one ray of hope concerning their future state, it seems to me, would overwhelm me beyond the power of endurance."[83]

Harriet decided that her father was wrong about what must be believed and done to be saved. She repudiated belief in eternal damnation on the grounds that, as she wrote to Catharine, "He who made me capable of such an absorbing unselfish devotion as I feel for my children so that I could willingly sacrifice my eternal salvation for theirs, he certainly did not make me capable of more disinterestedness than he has himself—He invented mothers hearts—& he certainly has the pattern in his own." For generations, belief in the importance of directing one's attention to heaven and avoiding hell had served as a primary explanation of great suffering on earth. But Harriet came to despise the burden placed on women of "good heart," trying to raise their children or struggling with bereavement, by the doctrine that "every person who does not believe certain things . . . is lost forever."[84]

Although she abandoned some of her father's beliefs, Harriet never gave up faith in heaven or her trust in God's close providence. When defending the importance of belief in a close, personal providence, she spoke of sitting at the bedside of a sick child: "Let us suppose a case. A mother sits holding her sick infant, and watching, as only a mother can, the changes of its suffering face. A deadly disease is hurrying it with irresistible force to death, and all human means and appliances bend before it as a reed before a torrent. . . . How distracting the responsibility thus thrown on the helpless, short-sighted parent. 'Have I chosen the right physician? Has he gained a right view of the case? Perhaps I am doing in my anxiety the very worst thing—perhaps he, sincerely and honestly, is pursuing a course that leads to death and not to life,' and while time is flying and a precious life is ebbing with each moment, how worthless seems human knowledge and human aid."[85]

It was amid child illness that Harriet believed that trust in God's governance was most crucial, even as she despised what some versions of Christianity (especially her father's) did to women's minds. In 1860, herself bereft of three children, Harriet wrote the following to a mother who had lost a daughter: "Be not afraid and confounded if you find no apparent religious support at first. When the heartstrings are all suddenly cut, it is, I believe, a physical impossibility to feel faith or resignation; there is a revolt of the instinctive and animal system, and though we may submit to God it is rather by a constant painful effort than by a sweet attraction." Clearly, Harriet did not believe that complete resignation in such situations was necessary, possible, or even desirable. Yes, one must believe in God's benevolent governance amid all affliction, but a good God, Harriet believed, would surely understand when the heart rebelled against certain kinds of suffering.

During the hot summer months of July 1849 Harriet and her husband Calvin Stowe lost their eighteen-month-old son Charley during a cholera epidemic. It was the second of three major cholera epidemics that had descended upon the city of Cincinnati. We now know that cholera is a bacterial disease spread by contamination and that it can be halted in its tracks by good hygiene and clean water. But no one knew that in 1849. John Snow was in the midst of studying the spread of cholera in London, but his famous argument that the disease spread from the Broad Street pump wouldn't be published for six years, and even then, doctors continued to argue about whether he was right. Meanwhile, cholera rushed through the city at breakneck speed. During just three summer months, more than nine thousand persons were said to have died of cholera within three miles of the Stowes' home. Victims died at high rates from severe dehydration following constant, painful bouts of diarrhea and vomiting. The eyes sunk in, the skin became cold and clammy, and hands and feet wrinkled. In the final stages of disease, seizures, hallucinations, and coma set in.[86]

Some made sense of the epidemic by preaching about God's wrath and the nearness of Christ's second coming. The premillennialist minister William Miller exhorted Americans to "*See! See!*—the angel with his sharp sickle is about to take the field! See yonder trembling victims fall before his pestilential breath! High and low, rich and poor, trembling and falling before the appalling grave, the dreadful cholera." But even in the midst of the epidemic, postmillennialism, which anticipated Christ's return after advancing progress on earth (rather than after increased suffering) proved more resilient and convincing to many Americans. Indeed, with each cholera epidemic physicians and ministers emphasized natural explanations more and more, all with the hope that, with God's approval, the true cause behind the disease might be discerned and human intervention one day stay the tide of death.[87]

As the town's physicians became exhausted, Harriet's own little Charley fell ill, and within days his cot became his "dying bed." "At last it is over and our dear little one is gone from us," she wrote to her husband. "He is now among the blessed. My Charley—my beautiful, loving, gladsome baby, so loving, so sweet, so full of life and hope and strength—now lies shrouded, pale and cold, in the room below." She had witnessed his "death agony, looked on his imploring face when I could not help nor soothe nor do one thing, not one, to mitigate his cruel suffering, do nothing but pray in my anguish that he might die soon."[88]

Confident of heaven, Harriet later wrote that her "only prayer to God" amid her sorrow was that "such anguish might not be suffered in vain." She hoped and believed that even such terrible suffering must be purposeful and redemptive. But rather than emphasize the role of great affliction in preparing one for heaven in the afterlife, Harriet imagined great affliction might actually prepare one to complete great earthly good. For all children.[89] Imagining increased scope for human action did *not* mean that God disappeared as an agent of individual loss, as Harriet's explanation of and response to her own bereavement shows. "There were circumstances about his death of such peculiar bitterness," she wrote, "of what seemed almost cruel suffering that I felt that I could never be consoled for it, unless the crushing of my own heart might enable me to work out some great good for others."[90]

It is important to note that Harriet did not see herself as choosing between natural processes and God's governance as she wrestled with how to make sense of Charley's death. American theologians had long since distinguished between different ways of thinking about God's activity in the world. General providence represented the usual means God chose to govern—namely through uniform, natural laws (like the universal law of gravitation). Miracles, by contrast, happened when God intentionally halted nature's laws (e.g., by breaking the uniformity of cause and effect) for some purpose. Many Protestants tended to associate belief in present-day miracles with Catholicism's system of intervening saints and miracle-healing shrines and, in rebellion against such "miracle-mongering," reserved miracles for the time of Christ's apostles. Instead, they emphasized a third category called "special providence," which existed in between natural laws and miracles (and as a result was often confused with both). Special providence was often defined as God working through natural laws to achieve a specific purpose—say, by using natural events (like the spread of disease through purely natural means) to achieve a particular providential outcome.[91] God's purposes might thus be present amid disease and suffering, yet miracles neither hoped for nor expected.

Coming from a family whose daughters were rebelling from the idea that women need "suffer and be still," Harriet imagined that suffering could be redemptive and purposeful if it led to good action on earth. This was why, as Charley lay dying, she made a resolution to fight suffering that, God willing (and she could not imagine God did not will it), could be reduced by human effort. She later wrote that it was standing beside her little boy's grave that she learned "what a poor slave mother may feel when her child is torn away from her." Certain, like James McCune Smith, that "God hath made of one blood all the nations of men," Harriet made the imaginative leap from her own loss to what other mothers felt when men, rather than God, stole their children away. This, she decided, would be her guide to action. Unlike her brothers, Harriet could not join scientific societies or become a physician or minister to "do good." But she could write and she had access to publishers. She also had several well-connected brothers with the Beecher surname (including six who could speak as ministers from the pulpit) to fight criticism and censorship.[92]

A year after Charley's death, and in the wake of the 1850 Fugitive Slave Act (the law made it a crime to shelter anyone who had escaped slavery, even in the North), Harriet Beecher Stowe wrote *Uncle Tom's Cabin*. Years later, she recorded how she wrote the novel with another infant close beside her. "I remember many a night weeping over you as you lay sleeping beside me," she wrote her son Charles, "and I thought of the slave mothers whose babes were torn from them." Stowe's influence lay in the fact that she inspired such thoughts in others, among readers that reached two million by 1857. In one scene, the main character Eliza is asked why she ran away from a kind master. She replies with a question: "Ma'am, have you ever lost a child?" The question "thrust on a new wound; for it was only a month since a darling child of the family had been laid in a grave." Met by tears and the words "I have lost a little one," Eliza tells her story, beginning with "Then you will feel for me." She had buried two children, and only the one at her knee remained. "Ma'am, they were going to take him away from me,—to *sell* him,—sell him down south, ma'am, to go all alone,—a baby that had never been away from his mother in his life! I couldn't stand it, ma'am." After the white mother retrieves the clothing of her "poor little Henry" to give "to a mother more heart-broken and sorrowful than I am," Harriet added the following lines: "And oh! mother that reads this, has there never been in your house a drawer, or a closet, the opening of which has been to you like the opening again of a little grave? Ah! happy mother that you are, if it has not been so." Harriet knew that across the nation, the latter was rare.[93] She explicitly called upon Americans' experiences of high child mortality to inspire action against laws that allowed men (rather than God) to steal children away from their parents.

For many white Americans, the tactic worked. Upon reading *Uncle Tom's Cabin*, one friend wrote to Stowe as follows: "I sat up last night until long after one o'clock reading and finishing 'Uncle Tom's Cabin.' I could not leave it any more than I could have left a dying child, nor could I restrain an almost hysterical sobbing for an hour after I laid my head upon my pillow. I thought I was a thorough-going abolitionist before, but your book has awakened so strong a feeling of indignation and of compassion that I never seem to have had any feeling on this subject until now."[94] Stowe intended for feelings of indignation and compassion to lead to moral and political action. She intended to convince Americans that talk of resignation, given the suffering caused by a system that denied thousands of Americans their own children, was sinful. God's mind and will might often be inscrutable, but human beings *could* know that God despised slavery. In the face of suffering, God's will, she argued, must demand something quite different than resignation, and furthermore, human beings were quite capable of stating this fact with confidence.

Stowe was not the first to claim such things, as Wheatley's poems show. Many abolitionists asked white readers to imagine themselves as an enslaved parent: "Let the mother gather her children about her," pleaded a circular from 1839, "and see them seized, sold, and driven away to southern markets and plantations!"[95] In a poem entitled "The Slave Mother" published in 1854, African American poet Frances Ellen Watkins Harper wrote:

> He is not hers, although she bore
> For him a mother's pains;
> He is not hers, although her blood
> Is coursing through his veins!
> He is not hers, for cruel hands
> May rudely tear apart
> The only wreath of household love
> That binds her breaking heart.[96]

Abolitionists' fundamental assumption that one mother could imagine another mother's pains and that humans, though sinners, could act against evil depended on particular interpretations of both scripture and nature. And that meant they had to counter alternative interpretations of both. Harriet explicitly fought against interpretations of the Bible and nature that pushed Black mothers and fathers outside the realm of human sympathy. When publicly criticized for her claim that pro-slavery sentiments were common within America's churches, Stowe crowed to her brother Henry: "The man has no kind of idea what he has brought upon himself." She and her brothers got to work compiling the declarations of "Synods, Presbyteries & Ecclesiastical bodies of all denominations" to prove the critic wrong: Dozens of official church declarations were actively defending slavery, in both the North and South, on the grounds that it was sanctioned by scripture. The Beechers firmly believed that, rightly interpreted, the Bible supported no such thing and that Christianity must be wrestled back to the side of abolition.[97]

Faced with calls to imagine Black suffering as their own, defenders of slavery doubled down on their claims that the Book of Nature was on their side. To do so, they called upon the mechanization and materialization of the human body to argue that Jefferson was right that "the African race" *felt* things differently. Even those who had lost their own children and were in the best position to imagine such grief in others found a justification of withholding sympathy from Black children by declaring them materially different beings. Take the Alabama physician Josiah Nott, a man who, with his wife Sarah, lost four of their children in five days to a yellow fever epidemic in September 1853.[98] None of these losses seemed to improve Nott's ability to imagine the grief felt by parents of African descent. A year after the yellow fever epidemic, in 1854, Nott published a book entitled *Types of Mankind* in which he claimed to demonstrate on scientific grounds that an impenetrable barrier existed between his own family and men, women, and children of African descent. He based his claim in part on differential mortality rates. Nott wrote, for example, that data showed that "hybrid" children died young. He did not blame the prejudice-ridden, violent environments in which these children were born. Instead, he claimed they died because their parents belonged to *different species* (lower fertility was supposed to constitute sure proof of species boundaries). By implication, there was certainly nothing anyone could do about these children's deaths, except to prevent miscegenation in the first place.[99] This attitude was in marked contrast to Nott's attitude toward what medicine might do for white children: searching for the cause of yellow fever in the interest of preventing its spread, a few years earlier Nott correctly suggested that mosquitoes played a role in transmitting yellow fever, although the hypothesis was not pursued in earnest until the 1880s.

Physicians like Nott tried to get science and the mechanization of nature on their side by classifying Africans and Europeans as different races, sometimes even different species, on what they deemed anatomical and physiological grounds. Soon, debate raged between polygenists, who held that human races arose from different original forms, either via evolution or separate, creative acts, and monogenists, who held that all races were of one blood. In 1851 the New Orleans physician Samuel A. Cartwright wrote of the South as having "two distinct races of people living in juxtaposition, in nearly equal numbers, differing widely in their anatomy and physiology, and consequently requiring a corresponding difference in their medical treatment."[100] These differences, Cartwright wrote, were so great that "the majority of naturalists" had referred the races to different species. Indeed, he argued that the South needed its own medical textbooks with close attention to the distinct anatomy, physiology, and therapeutics of each race. Drawing distinctions based on anatomy and physiology then turned into claims that differences existed in the capacity to suffer and feel pain. In 1826 Dr. Philip Tidyman had argued that Africans had a greater tolerance for pain and withstood surgery better than those of European descent. Cartwright agreed. He notoriously argued that Black men and women were subject to a hereditary disease called "Dysaethesia Aethiopis" that rendered them *insensible* to pain. (Cartwright also coined "drapetomania," the primary symptom of which was disobedience and running away.)

Whether physicians used a scientific-sounding name or not, the belief that African Americans were less sensitive to pain was held by whites throughout the North and the South. That belief meant that not everyone benefitted from the few clear advances physicians could claim in their ability to alleviate suffering. When effective anesthetics were discovered (ether in 1846 and chloroform shortly after) the Scottish physician James Young Simpson portrayed these innovations as part of divinely approved progress (including abolition) in the relief of *all* human suffering. But plenty of American physicians believed in a ladder of increasing sensibility to pain, with Black women, men, and children at the bottom, Indigenous peoples and the Irish somewhere in the middle, and white women and children at the top. This supposedly natural, law-bound hierarchy of the ability to suffer implied that rational reasons existed not to give way to misguided "sentiment" when practicing either medicine or politics. Policy and practice, some physicians argued, must not be based on sympathy for "lower" beings if races were unequal in mental, moral, or physical capacity.[101]

These claims of material difference had clear implications for how both physicians and legislators treated African American children. In the 1857 Dred Scott decision, Chief Justice Roger B. Taney had, according to physician J. H. Van Evrie, "fixed the *status* of the subordinate race *forever*." The decision's defenders appealed to both God and science by speaking of the "unchanging and unchangeable ordinances of the Eternal." Taney had even appended an essay by Cartwright entitled "Natural History of the Prognathous Species of Mankind" to the official opinion. Taney's confident conclusion that Black men were "so far inferior that they had no rights which the white man is bound to respect" was thus bound together with a taxonomy of the genus *Homo* based in part on supposed anatomical differences between different groups of children. Cartwright claimed, for example, that Black children's skulls lacked the divisions between skull plates that allowed the "largeness" of the heads of white infants to pass through "the smallness of its mother's pelvis." In a published version of the decision and appendix by John H. Van Evrie, an advertisement for Van Evrie's own book and weekly, which argued that the "specific and radical differences of the races" justified "WHITE SUPREMACY," concluded the document. Combined, these documents pushed Black men, women, and children outside the bounds of American progress by calling upon science, medicine, religion, and the law.[102]

Meanwhile, amid hints that new evolutionary visions of nature like that of *Vestiges* might get the Book of Nature back on the side of monogenists (on the grounds that all races shared a common ancestry), Nott covered all his bases:

Whether of one origin or of many, the present status of the races must be, for all practical purposes, separate. "The Caucasian, Ethiopian, Mongol, Malay and American may have been distinct creations," he wrote, "or may be mere varieties of the same species produced by external causes acting through many thousand years: but this I do believe, *that at the present day the Anglo-Saxon and the Negro races are, according to the common acceptation of the terms distinct species, and that the offspring of the two is a hybrid.*" In doing so Nott was insisting that science, reason, and observation justified pushing two million Americans and their children permanently outside the boundaries of both sympathy and the golden rule.[103]

Although he was weak and ailing from heart congestion, in August 1859 James McCune Smith put pen to paper once more to compose a rejoinder to all this talk of, in Jefferson's words, the "real distinctions which nature has made" between men. "There is no reason to infer," Smith wrote in the pages of the *Anglo-African Magazine*, "from the structure of the skeleton, that there are distinctions and permanent differences between the framework of the white and black races." Nor in the muscular system. Nor in the hair. Nor even in skin color. Naturalists and physicians may have thought that, in applying the tools of Linnean taxonomy to human beings, they were creating scientific classifications. But Smith argued that all—Jefferson, Nott, Cartwright—searched for "differences where the Almighty has stamped uniformity."[104] Any careful anatomist, Smith argued, could easily disprove Jefferson's supposedly material justifications of excluding some American children from the benefits of progress in both medicine and law. Any careful naturalist, he insisted, would find that as much variation exists within so-called races as between them. "Hence it appears," he concluded, "that the black comprises no special variety of the human race, no distinctive species of mankind, but is part and parcel of the great original stock of humanity—of the rule, and not of the exception."[105]

Two years later, as Americans went to war over how to make the earth better and for whom, copies of Charles Darwin's *On the Origin of Species by Natural Selection, or the Preservation of Favoured Races in the Struggle for Life* arrived from England. Inevitably, Americans read Darwin's works through the lens of long-standing debates over the nature of God's governance, the origin of suffering, what should (and should not) be done for children, and whose children should be favored amid the "struggle for life."

Chapter 4

NATURE GROANETH AND TRAVAILETH

WHEN DARWIN'S *On the Origin of Species* arrived on American shores in December 1859, Frederick Douglass was in the midst of his second lecture tour to Great Britain. His young daughters wrote him letters while he was away. The eldest, Rosetta, sent an account of how her little sister, Annie, was the best student in her class at a public school in Rochester, New York. Annie herself wrote as well, proudly describing her progress in German: "I expect you will have a german letter from me in a very short time." She copied out an antislavery poem that she was preparing to read aloud in school, which ended, "He whose noble heart beats warm, for all men's life and liberty, who loves alike each human form, O that's the man for me." She closed with a lament that "poor Mr. Brown" was dead, hanged on December 2 at the behest of a "hardhearted man" (probably governor of Virginia, Henry A. Wise, who refused to stay Brown's execution in the wake of the raid on Harper's Ferry. Wise also ordered Douglass's arrest.)[1]

Douglass had extended his stay in Europe to wait for things to calm down. But in March he received word that ten-year-old Annie, "the light and life of my house," had succumbed to a disease of the brain that had "baffled the doctors." Rosetta recorded that Douglass's grief was great. "I trust," she wrote to a family member after receiving his first letter, "that the next letter will evince more composure of mind." Of Annie's mother, Anna, Rosetta wrote, "She is not very well now being quite feeble though about the house." Rosetta, twenty-years-old at the time, took consolation in the belief that her little sister had "gone to Him whose love is the same for the black as the white."[2]

Of course, Douglass and his children lived in a nation that was not organized around the assumption that God's love was "the same for the black as the white." Apologists for the enslavement of 10 to 12 percent of the population defended the bondage of Black men, women, and children on constitutional, scriptural, and scientific grounds. Some claimed Africans were descendants of Canaan, the cursed son of Ham, and therefore slaves by divine decree. Others cited the claim that the apostle Paul had urged a slave to return to his master, or the fact that nowhere had Christ explicitly condemned slavery.[3] Still others used theories of "racial degeneration" from some original, created form or the separate creation of several different races (or even species) to justify "natural" hierarchies.[4] Clearly, as Americans wrestled with whose interpretations (and reinterpretations in the wake of *On the Origin of Species*) of both the Book of Scripture and the Book of Nature were correct, the stakes in debating what must be believed at the bedside of "amiable children" were high for everyone. But the stakes were not the *same* for every parent.

Those intent on either upholding or changing the world inevitably compared claims about both God and nature with their vision of what the world should be like and for whose children. Whether they called upon nature, scripture, or (most often) some powerful mix of both, clearly Americans could find what they wished in either testament. Like the Bible, evolution was used to drive calls for reform and as a new doctrine of resignation. Like the Bible, evolution was cited to justify radical change for some while upholding the status quo for others. In other words, like the Bible, evolution was used to argue diametrically opposite positions on whether, why, how, and for whom, progress in the amelioration of child suffering could and should be pursued. All of this meant that for men like Douglass the stakes in both choosing between and articulating the implications of alternative visions of nature, God, and progress were high indeed.

Work-able, Do-able Words

Frederick Douglass did not need to read Malthus to be convinced of a struggle for existence, whether that struggle arose from God's design, natural laws, or human actions. Born a slave on a plantation in Maryland in 1818, on September 3, 1838, at the age of twenty he escaped by train and boat to New York, where he married Anna Murray and was eventually able to purchase his freedom. Douglass's and countless others' escapes—ending in renewed bondage, death, or freedom—were the ultimate expressions of the belief that through struggle the earth could be better. That it was better elsewhere. And that one's children could be raised in a better world. Attempts to escape to freedom were profound repudiations of calls for submission, resignation, and quietism in the face of suffering. Having reached the North, Douglass joined the abolitionist movement, developing alliances with fellow freedmen, suffragettes, ministers, and reformers of all kinds. During his years on the abolitionist lecture platform,

Douglass had to decide how to counter the pro-slavery faction's pervasive use of both science and scripture to defend enslaving Black Americans.

Douglass challenged ministers who used scripture to justify slavery first. In one of his most popular lectures, Douglass ridiculed the slave owner's demand that the slave secure his place in heaven by following the words of the holy apostle Paul: "Servants, be obedient to your masters." The slave's soul, they said, "will live through endless happiness or misery in eternity" according to how he or she behaved in bondage. When Douglass's more sympathetic audiences (and they were not all sympathetic) did not cry or cheer, they laughed at his sarcasm. It was all so absurd: this appeal to God, heaven and hell, the salvation of souls, and the demands for resignation and submission, as men withheld the golden rule from millions of Americans on supposedly biblical grounds. For Douglass, the lesson was clear: the terrible affliction of slavery must no longer be born.[5] Resignation to suffering, in this world, was sinful.

When Douglass first entered the lecture circuit, Southern ministers drew upon the theory of special creation and natural theology to justify their vision of God's governance. He often recalled how, as a child, he was told that "'God up in the sky' had made all things, and had made black people to be slaves and white people to be masters." Furthermore, he "was told that God was good, and that He knew what was best for everybody." But given his experience and observations of the world, the last seven words "went against all my notions of goodness." On the lecture circuit, during an increasingly famous speech entitled "Slaveholder's Sermon," he mimicked the "Christian" slave holder's command to the slave: "Oh, consider the wonderful goodness of God! Look at your hand, horny hands, your strong muscular frames, and see how mercifully he has adapted you to the duties you are to fulfill . . . while to your masters, who have slender frames and long delicate fingers, he has given brilliant intellects, that they may do the *thinking*, while you do the *working*."[6] (William Paley found this use of natural theology to defend slavery abhorrent. In 1785 he argued that any government that permitted slavery—tearing apart "parents, wives, children, from their friends and companions"—rendered itself unfit "to be trusted with an empire.")[7]

Douglass pronounced that he would rather welcome infidelity, atheism, anything in preference to the gospel as preached by some American ministers. "They convert the very name of religion into an engine of tyranny and barbarous cruelty, and serve to confirm more infidels, in this age, than all the infidel writings of Thomas Paine, Voltaire, and Bolingbroke put together have done!" Such ministers had turned Christianity into "a religion for oppressors, tyrants, manstealers, and *thugs*."[8]

Distancing God's hand from evil, suffering, and affliction and placing agency increasingly within the hands of humanity may have looked like irreligion and infidelity to some, but writing these things did not mean Douglass had abandoned Christianity. He knew a man could both be a devout Christian and fight to change the world. Soon after securing his freedom, he had become good friends with none other than James McCune Smith, whom he cherished for never being "among the timid who thought me too aggressive and wished me to tone down my testimony to suit the times."[9] He surely knew of Smith's fight with physicians who declared children of color constitutionally prone to high mortality, a fight Smith firmly believed grounded in the proper interpretation of both science and scripture. As historian David Blight notes, "condemnation of religious contradiction," something at which Douglass excelled, "is not itself an antireligious prescription. In so much Christian, especially Protestant, tradition, it is precisely the opposite."[10]

Historian D. H. Dilbeck has shown how Douglass's understanding of the close relationship between certain concepts of God's governance and activism influenced how he told the story of his own life. In each version of his famous autobiographies (*Narrative of the Life of Frederick Douglass*, 1845; *My Bondage and My Freedom*, 1855, and *The Life and Times of Frederick Douglass*, 1881, 1892), Douglass changed his portrayal of God's governance in both his own life and the world in ways that unharnessed the status quo from God's close, personal governance. In the 1840s he was willing to describe an event that prepared his path toward freedom as "the first plain manifestation of that kind providence which has ever since attended me, and marked my life with so many favors," even though he knew some readers would think him "superstitious, and even egotistical" for viewing the event as "a special interposition of divine Providence." At the time, he balanced attributing events to providence with an emphasis on human agency by writing, "The Lord is good and kind, but is of the most use to those who *do* for themselves." Similarly, in 1855 he included a quote from *Hamlet*: "There's a divinity that shapes our ends, rough-hew them how we will," once again, admitting that such a view might seem "irrational by the wise, and ridiculous by the scoffer." By the time he was writing in the 1880s, however, Douglass removed all comments regarding divine providence's role from his autobiography, even when describing the same event previously attributed to "a special interposition." Instead, Douglass now emphasized that "all the prayers of Christendom cannot stop the force of a single bullet, divest arsenic of poison, or suspend any law of nature."[11]

These revisions reflected Douglass's decision that, in the face of evil and suffering, belief in special providence, miracles, and prayer provided an inadequate anchor for human experience. Prayer had not averted the cruel hand of his master. Douglass's own action in planning and executing an escape, with the aid of his friends and future wife, Anna, had done *all*.

Later, Douglass noted how, as he cast about for a means of making sense of the world, a book by the Scottish lawyer and phrenologist George Combe entitled *The Constitution of Man* "relieved [Douglass's] path of many shadows."[12] (George Combe was brother to Andrew Combe, the physician we met in the previous chapter urging that parents improve "infant management" via closer attention to natural laws). In addition to arguing that a man's character could be discerned by studying the shape of his skull, Combe also argued that humans could improve their lot by living in accordance with natural laws.[13] Indeed, Combe argued that such a system afforded "the richest and most comprehensive field imaginable for tracing the evidence of Divine power, wisdom and goodness in creation," for once humanity traced "the separate existence & operation of each natural law," the means of progress became clear.[14]

Douglass read *The Constitution of Man* sometime in the late 1830s or early 1840s. As one of his early biographers who knew Douglass wrote, the book "taught him the supremacy of law and order in nature, as well as the possibility of attaining happiness here on earth by obedience to natural laws."[15] Both Combe brothers had written at length about premature death as something that could be prevented rather than resigned to as the chastising rod of a benevolent Father. By extension, slavery was not the beneficent rod requiring resignation and hope for reward in the hereafter that its defenders claimed it to be.

It might seem strange that Douglass found comfort in the work of a man who contributed an appendix to Nott's *Types of Man*, but George Combe's version of phrenology, which divided the human brain into different, measurable sections representing "combativeness," "destructiveness," "conscientiousness," "truthfulness," and "sense of justice," was extremely malleable. In other words, this science, too, could be drawn into quite opposite campaigns. Phrenology was used both by those who defended innate, unchangeable differences between individuals and races *and* by those who believed that, through the exercise of one's undeveloped faculties, anyone's mental and moral level could be improved.[16] And while Combe's work was ridiculed as atheist or "necessarian" by its critics (and unbelievers indeed found much to like), plenty of readers read his emphasis on natural law as simply another example of humanity's increasing discovery of how God governed the world.

Although, as we will see, he was well aware of the dangers of scientists' various interpretations of the world, Douglass decided that belief in amelioration via a better understanding of natural laws provided a better foundation for progress than belief in either prayer or miracles. Meanwhile, his atheist friend Ottilie

Assing pressed him to go even further. She had come to the United States from Germany in 1852 and sought Douglass out after reading his first autobiography. Together they read the most influential works of higher criticism, including David Strauss's *The Life of Jesus* and Ludwig Feuerbach's *Essence of Christianity*. Assing told Feuerbach that reading his book had converted Douglass to atheism, but because Douglass never used that word for his beliefs, scholars have been less inclined to agree with her claim. Raised in the midst of the Revolutions of 1848 (a series of rebellions against monarchical rule in Europe), Assing was a correspondent of both Karl Marx and Friedrich Engels. All three composed radical critiques of Christianity and assumed that surely rational, enlightened individuals could not believe in God. But with or without Assing, or Strauss, or Feuerbach, Douglass clearly had reasons of his own to abandon (at the very least) orthodox Christianity. He knew better than anyone how it had been used to justify slavery by placing God's designing, benevolent hand in every last detail of creation and human experience.

Scholars have explained Douglass's apparent removal of God's presence from his autobiographies in different ways. Some argue that by the time of the Civil War Douglass saw God's direct involvement in human affairs as minimal. Others insist that Douglass's views of God did not change but that he feared that emphasizing God's role too much after emancipation might dampen individual effort to shape the destiny of the nation. But whether the changes in Douglass's narratives were strategic, heartfelt, or a mix of both, we know what kind of God, as years passed, he did not believe in. Sixteen years after Annie's death, in the summer of 1875, his daughter Rosetta's six-year-old granddaughter Allie died. Douglass urged Rosetta not to suffer from "the superstitious terrors with which priestcraft has surrounded the great and universal fact of death." Death was the common lot, he wrote, of us all; a friend, not an enemy. He would not, he explained, write of heaven: "Whatever else it may be, it is nothing that our taking thought of it can alter or improve. The best any of us can do is trust in the eternal powers which brought us into existence, and this I do, for myself and all."[17]

What did Douglass mean by "trust in the eternal powers" and why did that trust matter? Like Malthus, Chambers, and others, Douglass firmly grounded the possibility of human agency in the uniformity of natural laws. "So far as the laws of the universe have been discovered and understood," he said in a lecture in 1883, "they seem to teach that the mission of man's improvement and perfection has been wholly committed to man himself. He is to be his own savior or his own destroyer. He has neither angels to help him, nor devils to hinder him. It does not appear from the operation of these laws, nor from any trustworthy data, that divine power is ever exerted to remove any evil from the world, how great so ever it may be." Even to protect the innocent and the oppressed. "The babe and the lunatic perish alike," he said, "when they throw themselves down or by accident fall from a sufficient altitude upon sharp and flinty rocks beneath; for this is the fixed and unalterable penalty for the transgression of the laws of gravitation." The law in all directions was imperative and inexorable.[18]

In some ways such claims fell within Christian adjustments to the new sciences espoused since the seventeenth century. William Paley had conceded that an answer to the origin of evil based on divine, "general laws" was the most comprehensive answer to the origin of evil. Like Paley (and Malthus), Douglass insisted that these laws were "beneficial withal." Here is his argument as to why:

> Though it accepts no excuses, grants no prayers, heeds no tears, but visits all transgressors with cold and iron-hearted impartiality, its lessons, on this very account, are all the more easily and certainly learned. If it were not thus fixed, inflexible and immutable, it would always be a trumpet of uncertain sound, and men could never depend upon it, or hope to attain complete and perfect adjustment to its requirements; because what might be in harmony with it at one time, would be discordant at another. Or, if it could be propitiated by prayers or religious offerings, the ever shifting sands of piety or impiety would take the place of law, and men would be destitute of any standard of right, any test of obedience, or any stability of moral government.[19]

Only within a system of uniform natural laws, Douglass insisted, could humanity learn what was right. He could draw upon the recent history of science (but not medicine) as evidence for his position. Men had once thought that thunder and lightning were "visitations of divine vengeance," but the science of storms had triumphed over such superstitions. Now, "science tells us what storms are in the sky and when and where they will descend upon the continent, and nobody now thinks of praying for rain or fair weather."[20] With careful attention, humanity could study existing evils in the interest of understanding and then removing them. Naval disasters, for example, led to the perfection of naval architecture and better knowledge of ocean navigation; all depended on fashioning the minds of men more and more to the likeness of the divine mind by accumulating knowledge of natural laws. Only through a better knowledge of such laws, he argued, would progress be secured. Trusting in a consistent system of cause and effect was the only basis on which action could be successful and the only means provided by providence to fight against evil and injustice and secure an earthly salvation. Furthermore, an earthly salvation was the only kind of salvation humans could say anything about with certainty.

Douglass thus set aside confidence in both a divine purpose behind his beloved Annie's death and an afterlife where he would meet her again. He did so not because a commitment to natural laws provided a sure means of liberation (physicians like Nott had demonstrated quite the opposite, as would much of American science and medicine for generations) but because the alternatives provided neither adequate accounts of human experience nor the best guides to "what to do."

Based on all experience, Douglass agreed with Christian ministers, including Malthus and Paley, that struggle and suffering were crucial components of this natural system. In the hands of men like Lyman Beecher, the redemptive power of suffering began the century as sent by God's personal hand and remained the province of theologians. In Douglass's writings, suffering and struggle became matters of nature's laws that must be learned and harnessed in the name of progress. In 1857 (two years prior to *On the Origin of Species*) Douglass wrote, "If there is no struggle there is no progress. Those who profess to favor freedom and yet deprecate agitation, are men who want crops without plowing up the ground, they want rain without thunder and lightning. This struggle may be a moral one, or it may be a physical one, and it may be both moral and physical, but it must be a struggle. Power concedes nothing without a demand. It never did, and it never will." James McCune Smith called these Douglass's "work-able, do-able words."[21] And they could be embraced by both evangelicals, as the means provided by God to fight for heaven on earth, and by unbelievers who left God behind but kept the faith in progress via the good fight.

Was Douglass willing to drive the creative, progressive power of suffering and struggle back into deep time to explain the origin of human beings? Was he willing to embrace, as Emerson did, an evolutionary vision of the past and future as the best intellectual foundation for creating a better world for all children? To understand Douglass's response to evolution, we must watch the arrival of Charles Darwin's *On the Origin of Species* on American shores.

Grandeur in This View of Life

In his youth, Darwin believed, like most naturalists, that both scripture and science proved that the all-wise, all-benevolent and all-powerful God of Christianity had created species via independent creative acts rather than via secondary causes. Darwin had read and been, in his own words, "charmed by" Paley's argument that the purposeful parts of animals and plants could not have arisen via mechanistic, natural laws. His professors of geology, natural history, and (with one exception) comparative anatomy all taught that both natural history and human anatomy provided clear evidence of God's designing hand and the truths of Christianity. The Unitarian side of his family emphasized the benevolence of divinely designed natural laws, while the Anglican side emphasized God's close, personal presence in both nature and human lives. But both Anglicans and Unitarians believed that the study of nature was both pious and useful, on the grounds that the purposeful parts of animals and plants demonstrated God's wisdom, power, and goodness. Indeed, having recoiled from a career in medicine after witnessing two pre-anesthetic surgeries on children, Darwin's decision to join the clergy would have given him plenty of opportunities to pursue his love of natural history.

At the age of eighteen, in 1828, Darwin entered Cambridge University as preparation for a career as a minister. Three years into his studies, his natural history professors—ordained ministers all—helped him obtain a position as naturalist and "gentleman's companion" to Captain Robert Fitzroy on a surveying voyage around the globe. By the time he returned home in 1836, Darwin had witnessed much that he, trained within the theory of special creation, could not explain: mysterious patterns of geographical distribution (why did God create mockingbirds on the Galapagos Islands that were similar yet slightly different from the mockingbirds on mainland South America?); surprising patterns in the fossil record (why did God permit giant ground sloths to go extinct only to design a smaller sloth, on a similar pattern, for the same region?); indigenous plants, animals, and peoples being exterminated by creatures "designed" for habitats on the other side of the globe (why did the natives, perfectly designed for their place in nature, not vanquish the intruders?). Meanwhile, he was reading a book by a lawyer-turned-geologist Charles Lyell entitled *Principles of Geology*. There, Lyell had driven natural laws into the deep geological past by insisting that geologists must use only processes uniform in kind and degree to explain the earth's history, from the origin of mountain ranges to the complex landscapes of the British Isles. As a result (and in contrast to naturalists like Cuvier in France), Lyell used the Malthusian struggle for existence, rather than sudden catastrophes (such as continental floods), to explain extinction.

Darwin wasn't only observing rocks, fossil mammals, and Galapagos mockingbirds. He was also observing his own kind. Raised by staunch abolitionists, he witnessed the horrors of slavery in Brazil. In his account of the voyage, Darwin ridiculed those "who look tenderly at the slave-owner, and with a cold heart at the slave." They "never seem to put themselves into the position of the latter.... Picture to yourself the chance, ever hanging over you, of your wife and your little children—those objects which nature urges even the slave to call his own—being torn from you and sold like beasts to the first bidder!" Darwin wrote with revulsion of how "these deeds are done and palliated by men, who profess to love their neighbours as themselves, who believe in God, and pray that his Will be done on earth! It makes one's blood boil, yet heart tremble, to think that we Englishmen and our American descendants, with their boastful cry of liberty, have been and are so guilty."[22] Clearly Darwin did not need to read *Uncle Tom's Cabin* to be convinced that all mothers and fathers grieved alike. And because he refused to believe that humans suffered to different degrees, slavery added to his growing list of (increasingly indignant) questions for the likes of Mather, Ray, Paley, and ultimately the close, personal God he was supposed to defend from the pulpit in England.

Shortly after returning home, Darwin decided that whether guided by a first cause or not, evolution (or what he called "descent with modification") offered a better explanation of the things he had observed during the voyage, including the behavior of his own kind. Soon, his rebellion against the theory of special creation was heightened by his exposure to the same German biblical critics who had influenced Emerson, Douglass, Parker, and Stanton. These theologians had insisted that Christianity must abandon belief in miracles as a means of understanding God's governance of the world, including human history. Meanwhile, although physicians and ministers urged parents to look "beyond secondary causes" when children died, they did not usually speak of the birth and death of individuals as miracles. While the fundamental premise of European science was that the first cause, God, had created all secondary causes, naturalists and physicians agreed that God generally governed via uniform systems of cause and effect discernable by human beings. Meanwhile, recall that geologists had placed the death of species (extinction) within the realm of secondary causation as well.

Inspired by the increasingly naturalistic rules of physics and chemistry, Darwin declared it inconsistent, arbitrary, and unscientific to draw a line at the *birth* of species, especially when all agreed that varieties (which all taxonomists knew were difficult to distinguish from "good species") arose via secondary causes. Yet in searching for a purely natural explanation of the origin of species, Darwin knew that unless he could explain, via purely material, mechanical secondary cause, *how* organisms had been modified "so as to acquire that perfection of structure and coadaptation which most justly excites our admiration" then there was no point in publishing his speculations.[23] Thus, he began his search for a mechanism through which species might change while producing the extraordinary, purposeful parts of animals and plants.

Two years after his return home, in October 1838, Darwin read Malthus's *Essay on Population*. As mentioned earlier, Lyell believed that the Malthusian "struggle for existence" sufficiently explained species extinction via secondary causation. Some had even seen Malthusian laws as the means (*within* the human species) of driving mankind "upward." Indeed, as we saw earlier, some American Malthusians had argued that the struggle for existence "benevolently" eliminated "inferior" races as the species *Homo sapiens* advanced. None of these naturalists thought the "struggle for existence" could break the species boundary or create new purposeful parts. But Darwin, searching for a naturalistic means of change, turned Malthus's struggle for existence into something that (assuming an unimaginable expanse of time) could be *creative*: "As many more individuals of each species are born than can possibly survive," he wrote, "and as, consequently, there is a frequently recurring struggle for existence, it follows that any being, if it vary however slightly in any manner profitable to itself, under the complex and sometimes varying conditions of life, will have a better chance of surviving, and thus be *naturally selected*."[24]

Darwin believed that this purely natural process could both push a population across the species boundary and explain the origin of purposeful parts. In other words, he had found a secondary cause for what (to Mather, Ray, Paley, Jefferson, and so many others) had seemed to both require and thus demonstrate the active, creative power of a first cause.

Over the next twenty years, Darwin worked away at his theory. Then, a young, working-class naturalist named Alfred Russel Wallace read Malthus and Lyell too. In 1858, sick with malaria in the Malay Archipelago and reflecting on the enormous destruction of animal life that pervaded creation, Wallace asked, "Why do some die and some live?"[25] He, too, found an answer in the struggle for existence and a higher rate of survival among varieties which had some advantage in that struggle. He, too, thought that these pressures might drive varieties across the species boundary. He sent a summary of his idea to Darwin, whose friends arranged for both Wallace's manuscript and some of Darwin's early writings (thus establishing Darwin's priority) to be read at the next meeting of the Linnaean Society of London, the premier naturalists' society in England. The papers were published in the society's *Transactions* but attracted little attention. So, over the next year, Darwin composed his famous book.

Throughout *On the Origin of Species* Darwin operated on the premise that, as he wrote, the "change of species cannot be directly proved." Rather, "the doctrine must sink or swim according as it groups and explains phenomena."[26] As a result he repeatedly compared his theory of "descent with modification" with the theory of special creation espoused by his professors of geology and natural history. "How strange it is," he wrote, under the theory of special creation, "that a bird, under the form of woodpecker, should have been created to prey on insects on the ground; that upland geese, which never or rarely swim, should have been created with webbed feet." But imagine that species constantly increased in number, he continued, "with natural selection always ready to adapt the slowly varying descendants of each to any unoccupied or ill-occupied place in nature," and "these facts ceased to be strange." To support his claims, he included facts that had troubled more than just naturalists: "It may not be a logical deduction," he wrote, "but to my imagination it is far more satisfactory to look at such instincts as the young cuckoo ejecting its foster-brothers,—ants making slaves,—the larvae of ichneumonidae feeding within the live bodies of caterpillars,—not as specially endowed or created instincts, but as small consequences of one general law, leading to the advancement of all organic beings, namely, multiply, vary, let the strongest live and the weakest die."[27]

The possibility Darwin raised here—namely, that "descent with modification" and "natural selection" explained suffering as the "small consequences of one general law"—became even more explicit when Darwin finally addressed how his theory applied to human beings in his 1871 book *The Descent of Man*. In 1864, the Harvard minister Andrew Peabody, who argued against the theory of evolution posed in *Vestiges of the Natural History of Creation*, had argued, like Franklin, that the death of beloved children proved God's special care for human beings and the existence of heaven. No animal, Peabody insisted, was subject to so premature a sowing. Darwin disagreed with the premise: humans were *not* an exception, and that fact was the only means of explaining why so many children died so young. In *The Descent of Man*, within a confession that it was impossible not to regret the inevitable, evil outcomes that arose from man's rapid rate of increase, Darwin added the following line: "But as man suffers from the same physical evils as the lower animals, he has no right to expect an immunity from the evils consequent on the struggle for existence." Human beings *were* governed by the same laws as animals, including high fecundity and limited resources.[28]

Darwin did not conclude that ameliorating child suffering via better management was impossible or undesirable. He called physicians when his children were sick, and months of his life in 1851 were devoted to seeking medical treatment for his beloved daughter Annie, albeit to no avail (Annie died of a typhoid-like illness that year, at the age of ten). But Darwin did decide that evolution explained the existence of such suffering far better than the theory of special creation. Setting aside belief in a loving, divine purpose in every adaptation, he also set aside belief in a loving, divine purpose in every human affliction.

At the age of sixty-seven, in 1876, Darwin explicitly connected these two moves when he composed his autobiographical *Recollections of the Development of My Mind and Character* for his family. In a brief section called "Religious

Beliefs," Darwin included the following within a long list of reasons why "disbelief crept over [him] at a very slow rate":

> That there is much suffering in the world no one disputes. Some have attempted to explain this with reference to man by imagining that it serves for his moral improvement. But the number of men in the world is as nothing compared with that of all other sentient beings, and they often suffer greatly without any moral improvement. A being so powerful and so full of knowledge as a God who could create the universe, is to our finite minds omnipotent and omniscient, and it revolts our understanding to suppose that his benevolence is not unbounded, for what advantage can there be in the suffering of millions of the lower animals throughout the almost endless time? This very old argument from the existence of suffering against the existence of an intelligent First Cause seems to me a strong one.[29]

Darwin seems to be talking mainly of animals (i.e., other sentient beings, including the "millions of the lower animals"). But by the time he wrote this he had collapsed the boundary between humans and animals. Meanwhile, he and his wife, Emma, had also lost three of their ten children: three-week-old Mary Eleanor died in 1842, ten-year-old Annie died in 1851, and eighteen-month-old Charles died in 1858. Like Chambers, Darwin believed evolution provided a better explanation of such suffering than the theory of special creation.

Darwin applied evolutionary explanations of suffering to a range of phenomena that had puzzled him during the voyage, including differential mortality rates between indigenous and European animals in places like Australia and New Zealand. "As natural selection acts by competition," Darwin concluded, "it adapts the inhabitants of each country only in relation to the degree of perfection of their associates; so that we need feel no surprise at the inhabitants of any one country, although on the ordinary view supposed to have been specially created and adapted for that country, being beaten and supplanted by the naturalised productions from another land."[30] Design by a benevolent, all-powerful, all-wise God simply did not, in Darwin's view, make adequate sense of such patterns, especially when witnessed among God's most favored creation, human beings. The high mortality rates of Indigenous children were central to Darwin's long list of puzzles that did not make sense under the theory of special creation. He wrote, for example, of how "it was melancholy at New Zealand to hear the fine energetic natives saying, that they knew the land was doomed to pass from their children. Every one has heard of the inexplicable reduction of the population in the beautiful and healthy island of Tahiti since the date of Captain

Cook's voyages."[31] Eventually, Darwin decided that "God's benevolent design" surely could not explain such patterns.

Having put forward a purely naturalistic explanation of the patterns he observed during his travels, Darwin closed *On the Origin of Species* by referring once more to the wondrous adaptations described by generations of devout naturalists. After summarizing natural selection, his purely mechanistic explanation of the origin of those adaptations, he reminded his readers that from this "war of nature, from famine and death" the production of the "higher animals" had resulted. Then he offered a vague consolation for having dispensed with a two-hundred-year tradition of documenting God's benevolent design in the purposeful parts of animals and plants: "There is grandeur in this view of life, with its several powers, having been originally breathed into a few forms or into one; and that, whilst this planet has gone cycling on according to the fixed law of gravity, from so simple a beginning endless forms most beautiful and most wonderful have been, and are being, evolved." (In subsequent editions Darwin added "by the Creator" to "originally breathed.")[32]

Darwin's wording gave readers permission to combine his theory with the long-standing tradition within science of speaking of secondary causes as designed by the first cause. Indeed, the devout Harvard botanist Asa Gray argued that "descent with modification by natural selection" provided a helpful means of understanding scripture. Gray placed the problem of suffering at the center of the explanatory power of Darwin's theory by arguing that the war, disease, and famine that arose from a struggle for existence was precisely the kind of pervasive suffering one should expect from the fallen state of creation (for example, Romans 8:22: "For we know that the whole creation groaneth and travaileth in pain together until now").[33] Indeed, Gray insisted that Darwin had turned suffering, the great challenge of theodicies, to "creative account." Descent with modification via natural selection, Gray wrote, "explains the seeming waste as being part and parcel of a great economical process. Without the competing multitude, no struggle for life; and without this, no natural selection and survival of the fittest, no continuous adaptation to changing surroundings, no diversification and improvement, leading from lower up to higher and nobler forms."[34] (The phrase "survival of the fittest" was not Darwin's, but the British evolutionist Herbert Spencer's. Darwin adopted the phrase in the fifth edition of *Origin of Species* in 1869.)

Gray did not imagine that humans might know God's purposes in creating through so violent and destructive a process. Indeed, he blamed unbelief on the natural theology of those like Paley who insisted the world was overall a happy place and therefore God was good. Such claims provided fodder for atheists and materialists because the state of the world did not in fact map onto human concepts of either wisdom or goodness. On the other hand, criticisms of Darwin on the grounds that he had turned nature into a nasty, misery-ridden place made no sense to Gray: Darwin had simply observed what creation was like and tried to explain phenomena via secondary causes. Gray pointed out that Darwin's language allowed readers to imagine those natural laws as created by the first cause, God (indeed, Gray himself defined natural law as how God governed). Finally, Gray argued that evidence of God's goodness, which seemed to disappear if one looked at creation and human experience too closely, could be recovered in the fact that the evolutionary process was progressive: after all, it had culminated in humankind.[35]

This was suffering as redemptive on a cosmic scale, placed before the American public during a civil war that took the lives of 750,000 soldiers. Darwin's book outlining his own evolutionary theory of the origin of species crossed the Atlantic as the United States fell apart and sons brought through childhood diseases were sacrificed in the tens of thousands on the battlefields of the Civil War. Inevitably, his book would be read through the lens of how to make sense of such suffering, including whether this new portrait of the Book of Nature reinforced or undermined the Book of Scripture.

Some, including the poet Emily Dickinson, rebelled against the idea that the future justified or made sense of so much suffering, death, and war, no matter how heavenly, whether on earth or elsewhere. Such claims meant nothing to the dead, Dickinson wrote. "I wish 'twas plainer," she wrote, "the anguish in this world. I wish one could be sure the suffering had a loving side."[36] But Gray had no such doubts. Nor did he doubt that humans had a role to play in that struggle. One must fight against sin, suffering, and evil while trusting in God's providence. Gray had no children but told Darwin that he regretted that fact only because he had no son to send to war to fight for abolition.[37] In another letter, Gray wrote, "We accept our misfortunes and adversities, but mean to retrieve them, and would sink all that we have before giving up. We work hard, and persevere, and expect to come out all right, to lay the foundations of a better future, no matter if they be laid in suffering."[38] He might as well have written, *Of course* they are laid in suffering. For Gray was certain that the primary message of Christianity was that suffering, which must be expected, was redemptive and progressive after all.

Frederick Douglass was still "owned" by Hugh and Sophia Auld when Darwin returned home from his famous five-year voyage on the HMS *Beagle* in 1836. He would not secure his freedom for another two more years. By the time *Origin of Species* was published, Douglass was under suspicion for supporting

John Brown's raid on Harper's Ferry and in exile on the abolitionist lecture circuit in England and Scotland. His travels came to an abrupt end when Annie died, and he risked going home. Although by this time he had abandoned Gray's more orthodox version of Christianity, Douglass still agreed with Gray that suffering was both redemptive and progressive. He was clearly not so sure, however, that it was either correct or safe to adopt Darwin's vision of the past. Some of his heroes were quite certain that an evolutionary account of species, including human beings, was the only version of history that could secure a more just world. In 1873 one of the most influential of the higher criticism theologians, David Strauss (of whom Douglass kept a bust on his mantel),[39] tried to capture why the theory of evolution, and especially Darwin's natural, mechanistic explanation of purposeful parts, mattered so much to those trying to change the world. Philosophers and critical theologians had declared "the extermination of miracles" over and over again in vain, Strauss wrote, but "our ineffectual sentence died away, because we could neither dispense with miraculous agency, nor point to any natural force able to supply it." Always the argument that special creation—a miracle—was required for the purposeful parts of animals and plants stood in the way. In vanquishing that argument, Darwin had "opened the door by which a happier coming race will cast out miracles, never to return. Everyone who knows what miracles imply will praise him, in consequence, as one of the greatest benefactors of the human race."[40]

To Strauss, belief in miracles bolstered the corrupt political power of both churches and states. He was certain that the appeal to God's personal governance halted all political and social reform and dampened all human effort. Strauss contrasted the effect of orthodox interpretations of Genesis (including belief in the fall of man and human depravity) with the vision offered by Darwin's version of the past. Rather than being the descendants of a couple created in the image of God who had been kicked out of paradise, humanity could now be proud of having gradually worked itself up "by the continuous effort of innumerable generations from miserable, animal beginnings to its present status." For Strauss the difference in these historical narratives lay in what each inspired humans to imagine about themselves and the future: "Nothing dampens courage as much as the certainty that something we have trifled away can never be entirely regained," Strauss wrote. "But nothing raises courage as much as facing a path, of which we do not know how far and how high it will lead us yet."[41]

But although he admired Strauss, Douglass was clearly more wary of adopting Darwin's explanation of human origins so quickly. He didn't need evolution in order to damn the use of natural theology to defend the status quo. He had long since ridiculed the design argument as a defense of slavery on ethical grounds. And he knew that evolution might be a fickle friend of emancipation. In 1854 (five years before *On the Origin of Species* appeared), he had mocked evolutionists' tendency to place human groups on a ladder of "primitive" to "civilized." *All* humanity could be clearly distinguished from animals "by the possession of certain definite faculties and powers, as well as by physical organization and proportions." Douglass's targets were American ethnologists, naturalists, and physicians who argued that Africans were not, in fact, human beings. "A respectable public journal, published in Richmond, Virginia," he explained, based its "whole defence of the slave system upon a denial of the negro's manhood." The journal had conceded that if that claim was untrue, then slavery was wrong. Douglass took up the gauntlet, made his case for human unity, and then issued the following demand:

> Away with all the scientific moonshine that would connect men with monkeys; that would have the world believe that humanity, instead of resting on its own characteristic pedestal—gloriously independent—is a sort of sliding scale, making one extreme brother to the ou-rang-ou-tang, and the other to angels, and all the rest intermediates! Tried by all the usual and all the unusual tests, whether mental, moral, physical, or psychological, the negro is a *man*—Considering him as possessing knowledge, or needing knowledge, his elevation or his degradation, his virtues, or his vices—whichever road you take, you reach the same conclusion, the negro is a MAN. His good and his bad, his innocence and his guilt, his joys and his sorrows, proclaim his manhood in speech that all mankind readily understands.[42]

If anything, the reasons to be cautious in adopting evolution had only increased since Douglass wrote these lines. In 1859 Darwin had left the implications of collapsing the human-animal boundary for his own kind extremely ambiguous in *On the Origin of Species*. There, he wrote only that when the views proposed in that book were adopted "psychology will be based on a new foundation, that of the necessary acquirement of each mental power and capacity by gradation. Light will be thrown on the origin of man and his history." But plenty of writers rushed into the gap left by his silence. It isn't surprising, given the Civil War that soon raged, that the implications of Darwin's theory for long-standing debates over the origin and meaning of human "racial" variation became a primary point of contention. Throughout the 1860s and 1870s, *On the Origin of Species* was cited as proof, depending on the speaker or writer, of both abolitionist and pro-slavery views of supposed differences among America's children.[43]

In 1871 Darwin finally weighed in on the subject of human evolution by publishing an eight-hundred-page book entitled *The Descent of Man, and Selection in Relation to Sex*. Here, he explicitly addressed the debate over whether "races" shared separate or common origins in a chapter entitled "On the Races of Man." Naturalists had studied man more carefully, he wrote, than any other organic being, yet they could not agree on whether *Homo sapiens* was a single species or race. Some naturalists said there were five races, some said there were sixty-three (and everything in between). "This diversity of judgment does not prove that the races ought not to be ranked as species," Darwin offered, "but it shews that they graduate into each other, and that it is hardly possible to discover clear distinctive characters between them." Any good, cautious naturalist knew the correct conclusion to draw: "He will end by uniting all the forms which graduate into each other as a single species; for he will say to himself that he has no right to give names to objects which he cannot define." In other words, those who declared races different species, or proclaimed with confidence that there were three, five, or sixty-three human races, were quite simply very poor taxonomists.[44]

As he so often did, however, what Darwin gave with one sentence he took away with the next. One of Darwin's primary means of collapsing the boundary between animals and humans (or any other apparent gulf between species) was to establish the presence of gradations. He placed almost every supposedly unique human trait—imagination, reason, language, sense of beauty, belief in God—beneath the lens of his theory and found gradations of each in both the animal and human worlds. In doing so, he pushed some human communities closer to animals than others, constantly writing of "higher races" and "lower races."[45] Imagining gradations between "higher and lower races" in various traits became central to Darwin's argument that the differences between humans and animals were of degree rather than of kind. (Notably, there was one trait that Darwin did *not* place on a gradation: the ability to feel pain. That trait he conferred on all human beings equally, without qualification.)

Darwin was often ambiguous regarding whether these gradations, including gradations in moral sense and intellect, existed solely in the past or continued into the present. When it did look like he was talking about gradations in the present, he was also ambiguous regarding whether placement was due to nature (in which case the status of a race as higher or lower could not be changed) or nurture (in which case it was malleable and subject to education). And all of this ambiguity meant that those with interest in doing so could put a material, biological, racial, *or* cultural hierarchy right back onto a monogenetic yet divergent human species. They might also, and did, add "capacity to suffer and feel pain" to the list of traits that could be placed on a hierarchy of "sensibility" from "lower, more animal-like" races to "higher, civilized" races.[46]

Josiah Nott did not think evolution undermined his view of how to think about different human children, whether those supposed differences were due to nature or nurture. Even if Darwin was right, he wrote, and all races came from a common origin, millions of years were required for their "development." "The Freedmen's Bureau," Nott pronounced in 1866, "will not have the vitality enough to see the negro experiment through many hundred generations, and to direct the imperfect plans of Providence."[47] Nott's use of the phrase "the imperfect plans of Providence" was sarcastic. He believed that the policies of Reconstruction were based on ignorance of a natural, providential hierarchy between men and that, were it not for the interference of a "false philanthropy," the "constitutional differences" of men, women, and children of the "African race" would cause them to go extinct in North America.[48] Clearly, Douglass had good reasons to be wary of the use to which evolutionary progress via the "struggle for existence" would be put amid such fights.

The Ascent of Man

As Americans weighed the benefits of applying natural laws to the question of the origin of human beings, some ministers were quite confident that Christianity could be reconciled with both evolutionary visions of the past and long-standing traditions of working to ameliorate suffering. The most famous example of such efforts was yet another of Lyman Beecher's rebellious children: Reverend Henry Ward Beecher, who was confident that evolutionary versions of millennialist progress provided a compelling explanation of human experience and a wonderful, science-guided route to a new heaven on earth.

Henry Ward Beecher and Frederick Douglass had certain things in common. Both wished to emphasize a universe governed via natural laws, both repudiated orthodox doctrines like eternal damnation, and both were staunch, radical abolitionists. Douglass once wrote of Beecher's sermons as follows: "As a colored man and one who has felt the lash and sting of slavery, I cannot forget the powerful words of this man in the cause of justice and liberty, in the righteous denunciation of slavery."[49] They also shared the tragic bond of child loss. Beecher and his wife Eunice lost four of their eight children who survived the first month of life. George died at the age of five months in February 1846 of malaria. Katharine at the age of fifteen months in November 1847. Then twins Alfred and Arthur, born in late 1851, died on July 4, 1853, of the mumps. There had been a heat wave in New York that summer, and thousands of children had died of various ailments. Eunice described the twins' death as tinting Henry with "a shade of sadness" that never left him.[50]

Much later, Douglass wrote of Henry Beecher's ability to imagine the grief of a man whose children were taken not by God but by man. During the popular preacher's visit to the Douglass home in Rochester, New York, Douglass's little daughter Annie, "long since dead," had come into the room "and laid her little hand lovingly on my knee." "Mr. Beecher noticed the child," Douglass recalled. "I said to him, How can any man with a human heart take that child from my arms and sell her on the auction-block? I never shall forget his look at the moment. He begged me not to mention it. The thought made him sick, as if he were looking upon a tender sister being bled."[51] Douglass knew that how a fellow human being viewed one's children was the key to who would be included in new visions of heaven on earth. He obviously appreciated Beecher's feelings in those moments. But Beecher's own sermons illustrate how, in contrast to Douglass, some Americans paid less attention to how theories of evolution might be used to exclude certain children from the nation's progress-ridden future.

Henry Beecher clearly thought that he was carrying on a long tradition of combining science, Christianity, and faith in progress. While Eunice grieved openly for their children, Henry escaped to his garden, his work, and his sermons. It had always been thus, Eunice wrote, ever since the winter of 1840, when they had first experienced the death of a child, a stillborn son. After a brief note in his journal noting the twins' funeral, Henry wrote, "Began my garden. I wish to keep a little record of the progress of things." In that moment, he meant the progress of rose bushes, honeysuckle, and willow trees. But the words were an apt description of his life philosophy amid the steady beat of bereavement. Henry later recalled how, when little Georgie died after eight days of terrible suffering from malaria, "it was not for me to quail or show shrinking. So I choked my grief and turned outwardly from myself to seek occupation."[52]

Over the next decades, Henry Beecher developed new means of consolation by combining traditional Christian belief in the benefits of earthly suffering (including reunion in heaven) with grand visions of divinely governed, evolutionary progress. Historian James Moore notes that such efforts were not so much an attempt to avoid the unpleasant implications of natural selection as "a shrewd attempt at reinterpreting natural evil as preconditions for progressively greater goods." "Compelled by tooth and claw to shoulder the burden of theodicy," Beecher and others "took up a notion that was at once Christian in origin and harmonious with the spirit of the age," imposing an inevitable, upward direction to evolution that relied on long-standing postmillennialist visions of inevitable material, social, and spiritual progress.[53] Both nature and the past might be places of violence and suffering, but both Christian faith and evolutionary progress demonstrated that human effort might secure a better future, at least for some.

Evolutionary visions of progress provided a new language in which to worry about, imagine, and distribute salvation to America's children. For those rebelling against orthodox Christianity as unethical, this alternative history appeared at just the right time. Like his sisters Catharine and Harriet, Henry Ward Beecher had been raised on his father Lyman's strict view of orthodox Christianity. Then, at university, he learned about how scientists were adjusting Christianity to accommodate the discoveries of geologists. He learned from Professor Edward Hitchcock, for example, of how the discovery that the earth was millions of years old could be combined with the study of cause and effect in nature to reveal more accurate knowledge about God. He learned about phrenology and George Combe's argument that the study of mechanistic cause and effect could be applied to the study of the human mind and behavior. Then, in the summer of 1853, after the twins died, he spent time in the Berkshires with Emerson, Walt Whitman, and other transcendentalists. He soon adopted this circle's belief in the inherent goodness of human beings and the potential within society and humanity to transcend "our lower natures."[54] This and Emerson's emphasis on amelioration mapped well onto what Beecher wanted to believe about himself and what he wanted to see in the world around him.

Beecher wasn't willing, as Emerson was, to imagine Christ as an ideal human being rather than the son of God. Christ was always in Beecher's view divine. But he did adopt Emerson's view of the nature and potential of humanity. By the time he was the minister of Plymouth Church in Brooklyn, Beecher filled his sermons with passages from the New Testament and talk of Christ's love rather than the Old Testament images of God's wrath and justice for which orthodox ministers like his father were famous. "*Love* should be the *Working Principle* of religion—not blind obedience, abject submission, or cold justice," he wrote in 1843. Historian Debby Applegate notes that it is largely owing to Henry Ward Beecher that today's "Mainstream Christianity is so deeply infused with the rhetoric of Christ's love that most Americans can imagine nothing else, and have no appreciation or memory of the revolution wrought by Beecher and his peers."[55]

Fewer still know that Beecher used an evolutionary vision of nature to inspire this revolution. Mather and Bacon had found promise of a progressive shape to history in scripture. Beecher, by contrast, emphasized evidence of moral progress in abolitionism and physical progress in the shape of the fossil record. But amid their differences, each of these men agreed that through science human beings could work in concert with God's will to make a better world.

The precise shape of Beecher's vision of evolution was strongly influenced by the British philosopher Herbert Spencer, whom he read in the 1860s. Spencer had coined the phrase "survival of the fittest" in 1864 and wrote long tomes about ethics and evolutionary progress in society. Given rebellions against orthodox beliefs, Spencer was intent on securing ethics "on a scientific basis." He based his new ethical system on evolution and in doing so provided what Americans like Beecher needed as they cast aside old doctrines. In 1892 Cornell University's *Philosophical Review* described Spencer's work as revealing "the inner meaning of Nature as a whole" in which an "Immanent Deity" could be recognized, "ruling and reigning in combinations often bewildering to us,—not unfrequently causing us to shrink with sense of pain," but offering a "meliorist view of life in general" in its steady advance toward better things.[56] Despite so much suffering, in other words, better things would come. This may sound like standard Christian fare, but in Spencer's view the only thing humans could be sure of was that better things would come on *earth*. Nothing, he insisted, could be known beyond that claim.

In 1870 Beecher organized weekly, secret meetings with Edward L. Youmans, "the prophet of Spencerian science," and twenty-five Brooklyn ministers so that they could all study evolution.[57] That same decade Beecher was publicly arguing that the concept of hell and divine punishment be purged from the Plymouth Church's creed.[58] Evolution was key to justifying this move. In providing a new story of the origin of man, evolution both undermined doctrines that Beecher hated and offered an alternative explanation of human nature. Darwin's theory exterminated, for example, the old theory of sin based on the fall of man and vicarious punishment. Beecher called the latter a "repulsive, unreasonable, immoral, and demoralizing theory." "The idea that God created the race," he wrote, "and that two of them without experience were put under the temptation of the arch-fiend (or whatever the 'creature' was), and that they fell into disobedience to what they did not understand anything about, and that God not only thrust them out of the Garden of Eden, as no parent would treat a child in his own household, but that he then transmitted the corruption that was the result of disobedience through the countless ages, and spread it out and out and out, and kept on through the system of nature, mingling damnation on the right and on the left, before, and behind—I hate it, because I love God!" To Beecher, evolution proved that the direction of history had always been constantly and solely progressive. "Man never did fall," he insisted, "men never could fall because all men were born at the bottom; and if there has been any turn at all, it has been upward and onward!"[59]

Christianity had always held man to be a dual creature, composed of both animal and spirit, with "the two struggling together for supremacy." But Beecher argued that the theory of evolution gave new meaning to that belief: sin was the triumph of the animal instincts, virtue the progressive triumph of the human spirit via self-mastery, morality, and justice. Evolution thus provided an interpretation of sin, Beecher argued, that made man's duty clear: do better, deny the animal, appetites, and passions; chose morality, conscience, and reason, and contribute toward God's progressive plan for humanity. Like Chambers, Strauss, and Stanton, Beecher found in evolution an alternative explanation of human nature—its past, present, and future—for which he had been searching after rebelling against the old narratives on ethical grounds. Having abandoned belief in both a literal fall of man and the prospect of eternal damnation, he could now proclaim with confidence: "God is love. God is love. God is love. That is the first, that is the middle, and that is the last echo of the sacred word."[60]

Emphasizing God's love over his justice did not equate to a denial of the reality of suffering. Indeed, Beecher still described suffering as indicative of God's love on the grounds that affliction was creative and educative. (In this sense he sounded a lot like Mather and his father, Lyman Beecher.) But like Gray, he cited Romans 8:22—"For we know that the whole creation groaneth and travaileth in pain together until now"[61]—to argue that evolution simply provided a scientific account of what had always been evident in both scripture and human experience. He insisted that pain and sorrow were still "God's ministers, God's schoolmasters, God's police." But Beecher did not then insist (as Mather and Lyman Beecher did) that the purpose of affliction was to turn one's attention toward heaven. Instead, he drew on evolution and the struggle for existence to find in nature a clear, scientific justification of Christian faith in a progressive direction to history on earth.[62]

Like Francis Bacon more than two centuries earlier, Henry Ward Beecher implied that the realm of human agency and human responsibility in this grand progressive march was great indeed: "There was not one word of information on the face of the whole globe when men started on the voyage of the human race," he pronounced, but "now we have labeled the poisons in the jars in the apothecary's shop.... It was all to be learned, and to be learned by man's finding out himself under the inspiration of God's natural laws; for God's great material globe and its physical laws were the only schoolmaster that He sent to instruct the human race." Like the author of *Vestiges*, Beecher explained suffering in ways that called upon natural law to absolve God of the details: "The world is full of ignorance, disease, revolution, wars, pestilence, and immeasurable disasters. They spring from definite laws; but laws are cruel to those who are too ignorant to obey them."[63]

Henry knew that this evolutionary, scientific vision of progress amid suffering could be cold comfort when a child died. He once conceded that, although he could demonstrate, thanks to evolution, God's goodness with respect to mankind "generically," he doubted whether God's goodness could be proved by how nature treated individuals. Given that fact, he still urged, like Ray and Boyle before him, that evidence of "the universality of Divine sympathy and love" was best obtained from the story of Christ rather than from nature. At other times, when faced with the old question—If God is so full of love for his creation, why, then, is there so much misery?—he gave the old answers: man's reasoning was finite, and all would make sense in the hereafter.[64]

What did it mean to demonstrate God's goodness via nature with respect to mankind "generically"? Henry described the "vast waste and the perishing of unfit things" as "one of the most striking facts in the existence of this world." But like Emerson, he argued that, viewed from the right perspective, one would see that while the weak perished, "the vast universe, looked at largely, is moving onward and upward in determinate lines and directions . . . toward perfection." Individuals died, but the species—*Homo sapiens*, for example—advanced. From this point of view, he wrote, science "is but the deciphering of God's thought as revealed in the structure of this world," including the fossils that lay in the record of the rocks.[65] Speaking on how "the theory of evolution . . . throws light upon many obscure points of doctrine," Henry explained suffering as rooted in whether individuals or groups had moved either away from or toward animality.[66] Indeed, he often implied that movement toward a redeemed state did not take place in the lives of individuals but rather over eons as humanity (or was it just the "favored" races?) gradually improved. Meanwhile, he blamed the continued existence of poverty amid economic progress on the ignorance and sinfulness of the poor, who he believed were guided primarily by their "animal faculties" and thus (rightly, naturally, divinely) doomed to inferior stations in life. Thus, Henry spoke in one breath of divinely approved progress while explaining the suffering of some as the inevitable and ultimately benevolent outcome of divinely designed natural laws.

Douglass had criticized belief in individual salvation on the grounds that it directed attention away from the possibility of ameliorating earthly suffering. But Beecher's talk of "generic" salvation via evolution implied that natural laws determined whose suffering demanded action and whose did not. Ministers, naturalists, and policymakers, for example, all used language about progressive evolution and a Malthusian "struggle for existence" to justify federal policy toward Indigenous Americans. In 1867 a congressional investigation concluded that population declines were the result of disease, intemperance, war, loss of land, and "the irrepressible conflict between a superior and inferior race." As one expert witness testified during the hearings, "The causes which the Almighty originates, when in their appointed time He wills that one race of men—as in races of lower animals—shall disappear off the face of the earth and given place to another race . . . has reasons too deep to be fathomed by us. The races of the mammoths and the mastodons, and the great sloths, came and passed away: the red man of America is passing away!" This statement was an extraordinary mix of belief in the benevolence of natural laws, a (convenient) faith that God had his reasons "too deep to be fathomed," and firm certainty that—given any (presumably misguided) ameliorative effort would be up against both nature and God—nothing could be done.[67]

Not everyone claimed that God had designed the world this way. When confronted with differential mortality among Indigenous and European children, Darwin had decided that a purely naturalistic struggle for existence provided a better explanation of such patterns, and the suffering they entailed, than that the all-wise, all-powerful, and all-good God of Christianity allowed such things as part of his creation. And yet Darwin, too, implied that viewed from the right perspective such a system still produced great goods. While Darwin conceded that "moralists" might lament the extinction of "lower races," for example, he insisted that overall such a process was beneficent, for it led to "the Human race, viewed as a unit" advancing in rank.[68] Writing to William Graham in 1881, Darwin wrote that the "more civilised so-called Caucasian races have beaten the Turkish hollow in the struggle for existence." He then offered a prediction: "Looking to the world at no very distant date, what an endless number of the lower races will have been eliminated by the higher civilised races throughout the world." Darwin criticized Graham on one point only: the idea that "the existence of so-called natural laws implies purpose. I cannot see this."[69] Darwin saw natural selection, when applied to human beings, as in the main a good, albeit completely blind, force. After all, it had somehow led to the "higher races." As historians Desmond and Moore note, although Darwin refused to put a benevolent God behind either the process or the result, he assumed "an inevitability that had to be explained, not a socially sanctioned expansion that had to be questioned."[70] Some kinds of suffering, Darwin implied, must be borne in the name of progress.

As both Gray and Beecher noted with approval, Darwin's stance echoed scriptural assurances that suffering (whether of individuals or groups) would ultimately be redemptive. Confident that they could determine the nature and benevolence of natural laws, sometimes the very abolitionists who had fought at Douglass's side thought their work done after the war. Nature's laws, they declared, could now sort out the "fit" and "unfit" in a struggle for existence unfettered by the "artificial" human institution of slavery. (Abolitionism had rarely equated to egalitarianism, as many abolitionists' belief that freedmen should be "colonized" back to Africa had long since demonstrated.) Some even drew upon long-standing traditions of describing natural laws as God's means of governing to place providence behind the "survival of the fittest." In Britain, for example, Reverend Charles Kingsley, whom Darwin cited in the second edition of *On the Origin of Species* for evidence that evolution was not in conflict with Christianity, argued that natural selection proved that "the 'lowly' races were Providentially doomed and that the whites would sweep out all before them to usher in God's Kingdom."[71] Within the United States, Darwin's work was soon cited as evidence that the natural competition between races that would result from the abolition of slavery would lead to the inevitable triumph of whites over everyone else.[72] Historian John S. Haller notes that under the influence of ideas like these belief in Black extinction "became one of the most pervasive ideas in American medicine and anthropological thought during the late nineteenth century."[73] As the abolitionist lawyer Albion Winegar Tourgée lamented in 1884, many Northerners believed "the negro would disappear beneath the glare of civilization" and that "such disappearance would be a very simple and easy solution of a troublesome question."[74]

Of course, judgments regarding whose suffering was inevitable and whose might be prevented had long been made without the imprimatur of science: In choosing to enslave some children and their parents and not others. In forcing some children and their parents on deadly marches in order to clear the land for other children and parents. After all, Douglass had ridiculed orthodox Christianity's demand that he submit to suffering long before taking on scientists and physicians. Meanwhile, belief that all suffering came from God's hand had long coexisted with commitments to ameliorate suffering. Lyman Beecher's preaching against sin inspired his children and students' abolitionism, showing that activism and quietism could arise from the same soil. But clearly, in a society that placed human beings in hierarchical groups, new descriptions of nature could be harnessed, like the Bible, to justify ameliorating the suffering of some while ignoring the suffering of others.

Let Us Stand Erect

Both Henry Ward Beecher and Asa Gray insisted that Darwin had turned suffering, the great challenge of all theodicies, to "creative account." But some repudiated any God, close or distant, who created through such terrible means. The most famous "infidel" in the country (he was known as "the Great Agnostic"), Robert Ingersoll, praised Beecher for "knowing the sighs, the sorrows, and the tears that lie between a mother's arms and death's embrace" and damning "with all his heart the fanged and frightful dogma" of eternal hell.[75] But unlike Beecher, Ingersoll was not interested in replacing the God who damned with a God who designed a world full of suffering that might be ameliorated in the future. No good God, he insisted, would let things take so long, at the sacrifice of so many. "The 'Plan' of Nature I detest," Ingersoll declared. "Competition, and struggle, the survival of the strongest, of those with the sharpest claws and longest teeth. Life feeding on life with ravenous, merciless hunger—every leaf a battlefield—war everywhere."[76] He did not deny Darwin's theory; far from it. But he did deny that one could find the God of Christianity, Judaism, or Islam anywhere in such a world.

Ingersoll repudiated many of the justifications Americans had developed for believing in a progressive direction to history. He clearly preferred a vision of progress via the study of natural law over orthodox Christianity's talk of a deserved fall from grace.[77] But he despised the idea that a benevolent God had created such a suffering-filled universe and then delegated slow amelioration (for some) to humanity. He asked his listeners to consider:

> Would an infinitely wise, good and powerful God, intending to produce man, commence with the lowest possible forms of life; with the simplest organism that can be imagined, and during immeasurable periods of time, slowly and almost imperceptibly improve upon the rude beginning, until man was evolved? Would countless ages thus be wasted in the production of awkward forms, afterwards abandoned? Can the intelligence of man discover the least wisdom in covering the earth with crawling, creeping horrors, that live only upon the agonies and pangs of others? Can we see the propriety of so constructing the earth, that only an insignificant portion of its surface is capable of producing an intelligent man? Who can appreciate the mercy of so making the world that all animals devour animals; so that every mouth is a slaughterhouse, and every stomach a tomb? Is it possible to discover infinite intelligence and love in universal and eternal carnage?[78]

Ingersoll had no patience with the idea that suffering was God's means of developing humanity's moral, material, and spiritual progress, whether proponents of such views accommodated new scientific ideas or not. "What would we think of a father," Ingersoll demanded, "who should give a farm to his children, and before giving them possession should plant upon it thousands of deadly shrubs and vines; should stock it with ferocious beasts, and poisonous reptiles; should take pains to put a few swamps in the neighborhood to breed malaria;

should so arrange matters, that the ground would occasionally open and swallow a few of his darlings. . . . Suppose that this father neglected to tell his children which of the plants were deadly; that the reptiles were poisonous; failed to say anything about the earthquakes . . . would we pronounce him angel or fiend?"[79] Ingersoll's answer was clear: even Beecher's God was a fiend, not a whit better than the kind that had allowed Adam and Eve to fall because they ate from the Tree of Knowledge.

Though often accused of atheism, Ingersoll preferred the term "agnostic," defined by the British biologist Thomas Henry Huxley in 1869 as the stance that "a man shall not say he knows or believes that which he has no scientific grounds for professing to know or believe." Just as the theist could not prove that God exists, Ingersoll argued, "the Atheist cannot know that God does not exist." But he was absolutely certain that the God of Christianity could not, *must* not, exist. Child illness and death were central to Ingersoll's certainty. The historical record contains no evidence that Ingersoll and his wife Eva lost children, but they had two daughters, Eva and Maud, and he had a good, sympathetic imagination. When writing to a bereaved mother whose orthodox upbringing inspired fear that her unconverted son was in hell, Ingersoll urged her, "Take counsel of your own heart. If God exists, your heart is the best revelation of him, and your heart could never send your boy to endless pain." Ingersoll, one of his biographers wrote, "did not seek to destroy hope of a future life," but "he merely sought 'to prevent theologians from destroying this.'"[80]

Ingersoll couldn't understand why anyone would wish to believe in, much less preach about, a God that would put parents through such anguish. When he delivered an oration at a child's grave on January 8, 1882, he repeated his stance, that those standing around that little grave "with breaking hearts" need not fear. He offered not a promise of heaven but the consolation that at least we know that "the dead do not suffer." Ingersoll was the most famous infidel in the nation, but he concluded with these words: "We, too, have our religion, and it is this: 'Help for the living, hope for the dead.'"[81]

In actual fact Ingersoll didn't have much patience with those who promised heaven, either. He repudiated with evangelical passion the belief that rewards in heaven would make up for the misery on earth: "The clergy . . . balance all the real ills of this life with the expected joys of the next. We are assured that all is perfection in heaven—there the skies are cloudless—there all is serenity and peace. Here empires may be overthrown; dynasties may be extinguished in blood; millions of slaves may toil 'neath the fierce rays of the sun, and the cruel strokes of the lash; yet all is happiness in heaven. Pestilences may strew the earth with corpses of the loved; the survivors may bend above them in agony—yet the placid bosom of heaven is unruffled. Children may expire vainly asking for bread; babes may be devoured by serpents, while the gods sit smiling in the clouds."[82] Ingersoll insisted it was impossible "to harmonize all the ills, and pains, and agonies of this world with the idea that we were created by, and are watched over and protected by an infinitely wise, powerful and beneficent God, who is superior to and independent of nature."[83]

Ingersoll's greatest indictment against the belief in a God who promised rewards in the hereafter was the influence such belief had on attitudes toward earthly suffering. He was certain that a direct relationship existed between faith in a personal God and inactivity in life. "Why," he asked, "should man endeavor to thwart the designs of God?" As long as most men believed in a personal God who acted via miracles or special providence, he argued, "the world was filled with ignorance, superstition and misery." Like Strauss and Douglass, Ingersoll attacked the design argument because it provided a rational justification for belief in a close, personal God and thus, in Ingersoll's view, resignation and complacency in the face of suffering. His arguments against the design argument were damning. In a famous essay of 1872 entitled "The Gods" Ingersoll wrote, "These religious people see nothing but design everywhere." They point to the sunshine, the flowers, and the April rain, and insist on God's "personal, intelligent interference in everything." "Did it ever occur to them," Ingersoll demanded, "that a cancer is as beautiful in its development as is the reddest rose? That what they are pleased to call the adaptation of means to ends, is as apparent in the cancer as in the April rain? How beautiful the process of digestion! By what ingenious methods the blood is poisoned so that the cancer shall have food! By what wonderful contrivances the entire system of man is made to pay tribute to this divine and charming cancer!"[84]

While Ingersoll agreed with those who saw science and the uniformity of natural laws as the primary means of ameliorating suffering in the future, he refused to place God behind such a system. He was certain that when man "found that sickness was occasioned by natural causes, and could be cured by natural means," the prospect of decreasing misery increased. He believed science (and crucially, human sympathy) could indeed provide "help for the living." "From a philosophical point of view," he urged, "science is knowledge of the laws of life; of the conditions of happiness; of the facts by which we are surrounded, and the relations we sustain to men and things—by means of which, man, so to speak, subjugates nature and bends the elemental powers to his will, making blind force the servant of his brain." But he declared that humans could know nothing of why that happened to be the way of the world. "Reason, Observation and Experience—the Holy Trinity of Science—have taught us that happiness is the only good," he urged, "that the time to be happy is now, and the way to be happy is to make others so. This is enough for us. In this belief we are content to live and die. If by any possibility the existence of a power superior to, and independent of, nature shall be demonstrated, there will then be time enough to kneel. Until then, let us stand erect."[85]

As he set aside hell, original sin, the fall of man, and the bent posture of prayer, Ingersoll placed a demand before his listeners: What would they do for the world here and now? Critics may have painted freethinkers as amoral, but Ingersoll in turn declared his critics immoral for their complacency in the face of so much suffering. Whose beliefs, he demanded, led to the most ethical behavior? Whose beliefs led to help for others and a better tomorrow? To claim, as Ingersoll did, that freethinkers, agnostics, atheists, unbelievers, and infidels were more moral than Christians was a radical move, for many Americans had long wondered what incentive to goodness existed if a man did not believe in God.

Having set aside traditional justifications of morality, unbelievers like Ingersoll had to seize the moral high ground from both God and Christians to prove their goodness and worth. Historian Frank M. Turner has described how narratives of conversion from belief to unbelief followed evangelical models in which the postconversion life inspired better behavior than before. Unbelievers, Turner notes, "contended that the doctrine of atonement and hell were immoral and that their abandonment in favour of melioristic, secular moralities was good for all concerned." "Just as evangelicalism had at its inception been a faith of social action and reform," notes Turner, "so also were the new secular faiths," only science and technology alone would be the new tools of both moral and physical reform. The direction of history, notes Turner, was still postmillennialist. After all, it still "suggested a promethean vision of human beings as their own redeemers and as the redeemers of a fallen physical nature."[86] However, in emphasizing that, first, honest doubt and, second, the methods of science (rather than scripture or authority) provided the only routes to truth, agnostics like Ingersoll tried to gain moral ground over those who, they argued, took things solely on faith.

Sometimes the arguments of believers made that task quite easy. While traveling in Europe in the 1870s, Caspar Morris, the physician whom we met earlier writing about scarlet fever, described the tall chimneys of steam engines and the waste of mines as showing "that we had entered a mineral region in which the Creator had hidden the treasures of the earth to stimulate the energy and enterprise of man for their development."[87] In an age of unregulated child labor, statements like this exposed Christianity, the design argument, and the theory of special creation to easy ridicule. (Historian and philosopher of science Michael Ruse once wrote that obviously the authors of such statements had never "worked in a pit, nor did their children have to.")[88]

Ingersoll clearly saw Darwin's work as on the side of human freedom, justice, and emancipation from the confining authoritarianism of theological control of both society and the human mind. In an interview in 1899 Ingersoll spoke of Darwin as follows: "From his brain there came a flood of light. The old theories grew foolish and absurd. The temple of every science was rebuilt. That which had been called philosophy became childish superstition. The prison doors were opened and millions of convicts, of unconscious slaves, roved with joy over the fenceless fields of freedom." This, Ingersoll concluded, "is Darwin's century."[89]

Elizabeth Cady Stanton, who embraced the lessons of Andrew Combe's *Treatise on the Physiological and Moral Management of Infancy*, agreed that Darwin's theory was a crucial foundation for campaigns against an unjust status quo. She drew a tight connection between the state of society, biblical history, and the theory of special creation. To critique the subordinate status of women, for example, she organized a scandalous book called *The Woman's Bible*, published in 1895. "The real difficulty in woman's case," Stanton wrote, "is that the whole foundation of the Christian religion rests on her temptation and man's fall, hence the necessity of a Redeemer and a plan of salvation. As the chief cause of this dire calamity, woman's degradation and subordination were made a necessity." Stanton embraced Darwin's work as an alternative creation story through which she could dispense with that "necessity" entirely: "If, however, we accept the Darwinian theory, that the race has been a gradual growth from the lower to a higher form of life, and that the story of the fall is a myth, we can exonerate the snake, emancipate the woman, and reconstruct a more rational religion for the nineteenth century." Stanton urged that once individuals saw the fall of man as a figment of the human imagination, they would see suffering (including the travails of childbirth) as something to be ameliorated rather than endured.[90]

For Stanton, that "more rational religion" included both women's suffrage and full participation in civil life, including access to the knowledge that Andrew Combe convinced her, a half century earlier, might keep her children alive. She was not alone. Ernestine Rose, daughter to a Jewish rabbi and one of the first public American atheists, agreed. Rose attested in 1852 (after losing two of her own babies), "When our little ones are removed by death from our care and affection, we feel most keenly our ignorance, and long to know more about the laws of health. Woman might be physician to her self and her children. But the medical schools are closed against her; she is denied the advantages granted to men, for obtaining the knowledge of these things, more necessary if possible to her than to the other sex."[91] Physician Mary Putnam Jacobi, whom we will meet in the next chapter, also argued that only when women had access to instructions in the biological sciences would their influence "on the amelioration of human life be first made justly apparent."[92]

The abolitionist and radical minister Theodore Parker agreed with Stanton that Darwin's work was surely on the side of justice. Surely, he thought, evolution would wrest the origin of species out of the hands of those who used the Book of Nature to justify the status quo, including slavery. In Rome suffering from the final stages of consumption when he first learned of "Mr. Darwin's 'Principles of Selection in Natural History,'" Parker wrote excitedly to a friend, "It is one of the most important works the British have lately contributed to science." Darwin had dispensed with the "foolish notion of an interposition of God when a new form of lizard makes its appearance on the earth." Parker urged that "Science wants a God that is a constant force and a constant intelligence, immanent in every particle of matter" and that "the old theological idea of God is as worthless for science as it is for religion."[93] *On the Origin of Species* had provided a natural explanation of something that had previously seemed explicable only via miracle, and in doing so, in Parker's view, justified a vision of the past in which human action rather than prayer might remake the world.

Douglass counted each of these individuals—Beecher, Ingersoll, Stanton, and Parker—as either allies or good friends at various stages of his life. But in his few comments on evolution, Douglass expressed ambivalence regarding whether Americans' evolutionary visions of progress would so easily bend what Parker famously called the "long arc" of the "moral universe" toward justice for all children. Douglass agreed that orthodox Christianity's fall of man, miracles, and eternal damnation must be repudiated in favor of a universe governed by natural law. But Douglass never publicly embraced what for many had become an obvious accompaniment of that repudiation: Darwin's theory of descent with modification. In a speech on the philosophy of reform given in 1883, Douglass described his position as follows: "I do not know that I am an evolutionist, but to this extent I am one. I certainly have more patience with those who trace mankind upward from a low condition, even from the lower animals, than with those that start with him at a high point of perfection and conduct him to a level with the brutes. I have no sympathy with a theory that starts man in heaven and stops him in hell."[94]

Douglass clearly preferred a law-bound history over one filled with miracles and God's close, personal governance. The scope for human agency, reform, and progress seemed secure if humans could trust in the absolute uniformity of nature's laws. But unlike his white friends, he was not convinced that evolution, as embraced by many Americans, was an obvious ally of emancipation. Too many naturalists, physicians, anatomists, ministers, and politicians were using the idea of evolution to turn natural laws into a tool for binding Americans of color to some "primitive" stage of evolutionary progress, and keeping them there.

As visions of human evolution infiltrated American medicine, biology, and politics, Douglass had to constantly push back against those who decided nature had created the status quo. During his eighteen months as ambassador to the Republic of Haiti, for example, he argued that human actions and policies had caused Haiti's political and economic troubles, not some "natural inferiority" of the nation's peoples. At the beginning of the century, after enslaved men and women in the French colony of Saint-Domingue had wrested freedom for themselves from the French, Haiti had been founded as the second independent republic in the hemisphere. Ever since, abolitionists had cited the Haitian Revolution as evidence that enslaved peoples both hungered for freedom and were capable of self-government. In response, plenty of white Americans declared that science proved the mental and moral inferiority of Africans and their descendants.[95] That biological inferiority then became the most "scientific" explanation of Haiti's political and economic difficulties. But Douglass pointed out that surely Haiti's troubles could not be attributed to "natural inferiority" when the nation faced, first, the rampant prejudices of those who could not forgive Haiti "for being black" and, second, the staggering weight of national debt imposed by the French government, under threat of invasion, as "compensation" for its economic losses after the Haitian Revolution. (The indemnity was not paid until 1947.) Neither nature nor God, Douglass argued, had anything to do with Haiti's history or fate. The doctrine of white supremacy and the policies of the French and US governments had done all.

Douglass had first expressed disdain for the "scientific moonshine that would connect men with monkeys" five years before Darwin published *On the Origin of Species*. By the time, with the help of Asa Gray, Darwin had convinced many American naturalists and physicians of the reality of human evolution, the reasons for Douglass to be wary of such theories as a foundation for American visions of progress had only increased. He knew that science, like the Bible, could be a powerful yet fickle friend of amelioration, depending on who defined the best means of progress and who would be included in its boundaries. This historical fact would profoundly influence who benefitted and who did not when scientists finally, in the words of Unitarian minister Harold E. B. Speight, furnished "the skill which will save the child."[96]

Chapter 5

THE BENEVOLENT ARC OF HISTORY

SOMETIME IN 1853 THE PHYSICIAN and chemist John William Draper's nine-year old son, William, lay dying. For days John's sister, Elizabeth, hid the boy's favorite Protestant devotional tract. Having converted to Catholicism, she couldn't bear the thought of the boy being led toward eternal damnation in his final moments by Protestant heresies. After death silenced the little boy's pleas for the tract, Elizabeth placed the little volume on her brother's breakfast plate. Draper kicked her out of the house and they remained estranged for the rest of their lives.[1]

Like Emerson, Douglass, Stanton, and Beecher, Draper repudiated orthodox Christianity's belief in the fall of man and the prospect of eternal damnation. He famously defended that repudiation as both ethical and rational by writing a book called *History of the Conflict between Religion and Science* (1874) that contained countless stories of humanity's progressive discovery of natural laws. In doing so, Draper maintained Mather's and Bacon's millennialist faith in the possibility of progress on earth. But he moved God to so great a distance that providence was difficult to find at all. As he told his tales of scientific discovery, Draper also ignored many of the original motivations of the founders of mechanistic science. He described Newton, for example, as proving that "no arbitrary volition ever intervenes, the gigantic mechanism moving impassively in accordance with a mathematical law." This was not an accurate portrait of what Newton, who thought the universal law of gravitation demonstrated God's close, personal governance, thought he was doing. But Draper needed a version of Newton who believed in uniform, natural laws acting according to strict, mechanical rules with no exceptions.

Having decided that a universe governed by natural laws was the only means of making sense of human experience, Draper readily adopted Darwin's effort to drive this law-bound nature backward to the origin of human beings. He described the debate over Darwin's work as follows: "We are now in the midst of a controversy respecting the mode of government of the world, whether it be by incessant divine intervention, or by the operation of primordial and unchangeable law."[2] (As we have seen, Christian theologians had actually developed more subtle notions of God's governance via special providence, general providence, and miracles, but in the fight over how to explain child loss, historically nuanced discussions of the nature of God's activity often got lost.)

Draper's book struck a chord. The book went through several editions, was translated into almost every European language, and proved useful to a range of fights. His version of history appealed to atheists, agnostics, and freethinkers, of course. It also appealed to those questioning whether Christianity, as opposed to other faith traditions, should have so much power in the world. Impressed by Draper's portrayal of Islam as more supportive of science than Christianity, for example, the Ottoman scholar Ahmed Midhat Efendi translated the book into Turkish as part of his fight against Christian missionaries. Draper's appeal to a "higher concept of the Creator," his ridicule of orthodox Christianity, and his detailed account of Islamic contributions to knowledge, allowed Midhat to argue that the Islamic concept of God aligned better with science.[3]

History of the Conflict between Religion and Science was just one of many books published in the last quarter of the nineteenth century that portrayed a long history of conflict between science and (at least orthodox) Christianity. All insisted that science must triumph in that conflict in order for society to progress and suffering be ameliorated, but none could cite an increased ability to cure, as opposed to prevent, disease as evidence of their vision of history. Indeed, as a physician like Draper knew, there was very little anyone could do once a child fell ill except try not to make things worse. Draper did have evidence that mortality rates were slowly going down thanks to improved "child management" and better sanitation. But he cited those slowly and inconsistently dropping rates to argue that science "has shown medicine its true function, to prevent rather than cure disease."[4]

Within a generation, however, Draper's judgment that medicine's true function was to prevent rather than cure disease would be proved premature by the discovery of an antitoxin against one of the most dreaded of childhood diseases: diphtheria. However, it is crucial to note that the visions of progress which the discovery of antitoxin finally seemed to vindicate had been imagined long before. Meanwhile, the dual nature of these visions—imagining that science might be used to save the lives of some children while justifying the mortality of others—helps to explain why some American children benefitted from progress toward lowering mortality rates (whether via prevention or cure) more than others.

A New Revelation

The most famous "conflict" books—*A History of the Warfare of Science with Theology in Christendom* (1896)—was composed by historian and president of Cornell University, Andrew Dickson White. Like those of Darwin and other famous freethinkers, White's initial rebellion against orthodox Christianity occurred on ethical, rather than scientific, grounds (after all, when White was in college natural theology tied Christianity and science tightly together). In his autobiography, White recalled how both the Episcopal and Presbyterian ministers of his youth portrayed God as "an angry Moloch holding over the infernal fires the creatures whom he had predestined to rebel," while speaking of hell as "filled with infants not a span long." He recalled witnessing a father at the funeral of an unbaptized infant who "lost all control of himself" when the rector expressed pity the baby had not been baptized. "That is a slander on the Almighty," the father cried, "none but a devil could, for my negligence, punish this lovely little child by ages of torture. Take it back—take it back, sir; or, by the God that made us, I will take you by the neck and throw you into the street!" "The gentle rector faltered out that he did not presume to limit the mercy of God," but White decided "the doctrine had not stood the test." Soon, "under the influence of thoughts like these" he became "a religious rebel."[5]

Stories of orthodox ministers preaching about infant damnation were common in agnostics', atheists', and freethinkers' stories of their conversion to unbelief. Believers contested these caricatures of orthodox Christianity (Unitarians and Universalists didn't talk about hell for anyone). Lyman Beecher's son Edward came to his father's defense, for example, when someone claimed Lyman told the mother of an unbaptized child that it was now "a tenant of hell" and that one man had marched out of the church in protest. Edward denied his father had preached any such thing and furthermore insisted that the idea that just *one* man would have walked out on such a sermon was a slander on pious Christians.[6] But if the caricatures of orthodox ministers preaching about infant damnation were unjust, White had additional complaints. He wondered how Christ's words could be reconciled, for example, with the "virtual support" of the "whole orthodox part of the church" for slavery. That alliance, he recalled, "deepened my distrust of what was known about me as religion." For when religious men did oppose slavery, they were "generally New England Unitarians or members of other bodies rejected by the orthodox." Finally, he rebelled against orthodox ministers' tendency to insist that their version of Christianity provided the only route to salvation. As a child, White listened "Sunday after Sunday" to an Episcopal preacher who declared that Christ's promises were made to Episcopalians alone and "that those outside it had virtually no part in God's goodness; that they were probably lost." As he listened, he thought of his beloved grandmother, who was Presbyterian and therefore, according to this minister, unsaved. "My heart rose in rebellion," White wrote, for she was "the best Christian I knew, and the idea that she should be punished for saying her prayers in the Presbyterian Church was abhorrent to me." Ultimately, he decided to give up belief in orthodox Christianity—Episcopalian, Presbyterian, or otherwise—as a unique revelation.[7]

Science appears quite late in White's autobiographical account of his rebellion. White would have been twenty-seven years old when Darwin's *On the Origin of Species* was published in 1859. By that time, having rebelled against orthodox Christianity on ethical grounds, he had been searching for a new vision of humanity's past, present, and future for some time. As a young professor of history at the University of Michigan in the late 1850s, he read the histories written by "Buckle, Lecky and Draper" (each described the past as a progressive struggle between superstition and science). White recalled how these books "gave [him] a new and fruitful impulse." But "most stimulating of all was the atmosphere coming from the great thought of Darwin and Herbert Spencer."[8] The Darwinian hypothesis, he wrote, revealed "a whole new orb of thought absolutely fatal to the claims of various churches, sects, and sacred books to contain the only or the final word of God to man. The old dogma of 'the fall of man' had soon fully disappeared, and in its place there rose more and more into view the idea of the rise of man."[9] White then drew upon the history of science, as he saw it, to demonstrate what men must do to secure further material and moral progress.

White believed that the stakes were high in replacing old dogma (and the power of theologians who enforced that dogma) with "Man's Ascent" via the increasing discovery of natural laws. In an 1876 article for *Popular Science Monthly* entitled "The Warfare of Science," he imagined a particularly tragic outcome of the thirteenth-century Catholic Church's attitude toward a Franciscan friar named Roger Bacon. White claimed that the Church had imprisoned Bacon for defending the power of observation and experiment. Historians of science have demonstrated that, like many of White's tales, this was not true, but the story of Bacon's imprisonment provided a dramatic and useful vision of the past for claims White wished to make regarding the state of science and medicine in his own time.[10] "But for that interference with science," White wrote, "this nineteenth century would, without doubt, be enjoying discoveries which will not be reached before the twentieth century." He continued:

> Thousands of precious lives shall be lost in this century, tens of thousands shall suffer discomfort, privation, sickness, poverty, ignorance, for lack of discoveries and methods which, but for this mistaken religious fight against Bacon and his compeers, would now be blessing earth.
>
> In 1868 and 1869 sixty thousand children died in England and in Wales of scarlet fever; probably quite as many died in this country. Had not Bacon been hindered we should have had in our hands, by this time, the means to save two-thirds of these victims, and the same is true of typhoid, typhus, and that great class of diseases of whose physical causes Science is just beginning to get an inkling.[11]

White blamed these sixty-thousand children's deaths on the willful ignorance of natural laws that, he argued, resulted from theological control over men's minds. In doing so, he allowed readers to absolve God of so much suffering by blaming child loss on humanity's ignorance of nature's laws. And, like so many before him, White did so long *before* much evidence existed that science could help physicians save children's lives.

Though he had been compiling his stories for decades, White dedicated himself to composing his nine-hundred-page, two-volume book—*A History of the Warfare of Science with Theology in Christendom*—following the sudden loss of his wife, Mary, in 1887. The book took nearly ten years and was published in 1896. White readily admitted that he immersed himself in the lives of those who had fought "the good fight" as a means of solace. Meanwhile, while in Europe as ambassador to Russia and Germany, he found special comfort in Copenhagen's Ethnographic Museum's display of the "gradual uplifting of Scandinavian humanity from prehistoric times."[12] Thus, the progress he discerned in both evolution and the history of science offered an alternative history of man in which White found both purpose and meaning. That new history, White argued, revealed Christ's true message: "a revelation not of the Fall of Man, but of the Ascent of Man—an exposition, not of temporary dogmas and observances, but of the Eternal Law of Righteousness—the upward path for individuals and nations."[13]

In telling his own version of this grand, progressive history of the past, White urged his readers to have faith in science as the means through which humanity could achieve its own, divinely approved salvation on earth. Scientists had substituted, he wrote, "a new heaven and a new earth for the old—the reign of law for the reign of caprice, and the idea of evolution for that of creation."

Thus, the history of science and scientific discoveries constituted "a new revelation divinely inspired."[14] Geologists, for example, had convinced "the whole civilized world" that the earth was not six thousand years old and thereby infinitely increased "the knowledge of the power and goodness of God."[15] Most importantly, they had replaced the fall of man with hopes that the human condition could be ameliorated via human effort. Thanks to the discovery of evolution, White argued, men were abandoning archaic, superstitious doctrines and realizing that "the tendency of man has been from his earliest epoch, and in all parts of the world, as a rule, upward."[16] Indeed, White declared that the theologians who had loudly insisted that *Vestiges of the Natural History of Creation*'s evolutionary, law-bound vision of the past promoted atheism "ought to have put up thanksgivings" for *Vestiges* "and prayers that it might prove true."[17]

It is important to note that White used the word *theology* in his book title rather than *religion*. He explicitly stated that, while science was at war with theology, true science had no conflict with true religion. To do so, he carefully defined religion as "a Power in the universe, not ourselves, which makes for righteousness" based on the golden rule. White defined theology, by contrast, as institutionalized dogma tied to political power. White argued that, rightly understood, religion satisfied deep ethical and spiritual needs, provided the basis of social order and stability, and bolstered belief in the constancy of natural law. Theology, by contrast, relied on belief in miracles and myth to enforce belief and inspired authoritarianism, superstition, and ignorance. This distinction between religion and theology was crucial to White's argument that, rightly understood, science and religion were not in conflict.

Unlike Ingersoll, White always insisted that he was not abandoning Christianity. But he feared that if Christianity was not divested of what he called "theological," superstitious thinking (including belief in miracles), then Christ's great ethical truths—the command to love thy neighbors as thyself and the golden rule—would be swept away as science inevitably replaced the miracles and the myths.[18] White struggled with what his new understanding of providence meant for belief in immortality.[19] But he attributed his avoidance of total unbelief to the teachings of ministers like Henry Ward Beecher and Theodore Parker. These men, he wrote, prevented him from "cynicism and scoffing" and "stopped any tendency to atheism." The way they "spoke of a God in the universe gave a direction to my thinking which has never been lost."[20]

In White's version of history, scientists became the new prophets of a reformed religion. They were, he wrote in 1876, "the searchers of God's truth, as revealed in Nature." What had "the warriors of science" given to religion? Simply this: "great new foundations, great new ennobling conceptions, a great new revelation of the might of God." Indeed, at points White could sound a great deal like Ray, Mather, and Paley. Here he is, for example, describing the sixteenth-century anatomist Andreas Vesalius: "He substituted for repetition by rote of worn out theories of dead men, conscientious and reverent, searching into the works of the living God. He substituted representations of the human structure, pitiful and unreal, truthful representations, revealing the Creator's power and goodness in every line."[21]

Having lived through the Civil War, White appealed to familiar rhetoric but applied martial language to the history of science. He wrote of battle cries, strategies, combatants, champions, missiles, and weapons of "this great war" between science and theology. In White's hands (like Draper's), the history of science took on a very particular tone and legends took shape: Christopher Columbus confronted by churchmen insisting on biblical grounds that the earth is flat; Nicolaus Copernicus forced to delay his great work on the sun-centered universe until his deathbed, his work immediately condemned by the Church; the one man brave enough to openly declare the "new truth," Giordano Bruno, hunted down and burned at the stake; Galileo's telescopic proof of heliocentrism declared blasphemous by the Catholic church; the anatomist Vesalius risking charges of sacrilege and "popular fury" for digging up and dissecting bodies in order to place knowledge of human anatomy on a scientific foundation. White's stories of scientists triumphing over theologians have been passed down ever since, despite ample evidence that many of those tales (including those listed above) were woefully inaccurate.[22]

In White's view, the greatest crime of the theologians examined in his book was that they halted human effort. He called the adoption of inoculation and vaccination a triumph over opposition and "the greatest battle of science against suffering."[23] (White briefly mentioned Mather as having supported inoculation but made physicians like Boylston the heroes of the tale, gave far more lines to Reverend Massey's opposition, and didn't mention Onesimus or African knowledge at all.) When White wrote of different countries' responses to cholera, he described the contrasts as follows: Russian authorities realized the necessity of applying modern science to the problem and gave extraordinary powers to the medical and engineering professions and the police, while superstitious Russian peasants believed that "if God had wished us to drink hot water, he would have heated the Neva." According to White, the lesson of such tales was clear: progress in things like the fight against cholera directly correlated with how much freedom the politicians and theologians gave to scientists to pursue research and action unfettered by theological concerns and institutions.[24]

Deep within the first volume of *History of the Warfare of Science with Theology*

White reprinted his 1876 imaginary declaration of how many lives could have been saved had Roger Bacon not been imprisoned hundreds of years ago. Hundreds of pages later, in the second volume, he repeated the lament in the chapter on sanitation. How might things be different, White wondered again, had Bacon not been persecuted in the thirteenth century? White was confident he knew the answer: "The world to-day, at the end of the nineteenth century, would have arrived at the solution of great problems and the enjoyment of great results which will only be reached at the end of the twentieth century, and even in generations more remote. Diseases like typhoid, influenza and pulmonary consumption, scarlet fever, diphtheria, pneumonia, and *la grippe*, which now carry off so many precious lives, would have long since ceased to scourge the world."[25]

This explanation of child loss as a product of human ignorance enforced by theologians was personal. As White composed his book, two of the three children born to his second wife, Helen Magill, died as infants (in 1892 and 1896). Meanwhile, his son Fred, having graduated from Columbia College, suffered from stomach and intestinal troubles brought on by a severe attack of typhoid years earlier and was diagnosed with neurasthenia. An infectious disease had thus rendered Fred incapable of the active, useful life his father had imagined for him. To White, this was surely just one of the devastating costs of man's lack of the scientific knowledge required to fight disease.

Like Chambers, Emerson, Draper, Douglass, Stanton, and others, White believed that explaining suffering as arising solely from human ignorance provided a more useful and ethical alternative to orthodox Christianity. Indeed, White argued that "the triumph of scientific thought" led to "not only a theology but also a religious spirit more and more worthy of the goodness of God and of the destiny of man."[26] Meanwhile, White's histories defended an understanding of the relationship between providence and natural law that talked of divine benevolence but passed all earthly agency to human hands, with the lives of children at stake in the fight.

White's tales became popular amid very specific fights over the nature of God's governance, the origin of suffering, and the best routes to progress. Many of the biologists, physicians, ministers, and activists who will appear in the following pages learned their history of science from White. Indeed, some (including influential biologist and eugenicist David Starr Jordan and social reformer Florence Kelley) attended Cornell University and heard White's lectures about the "warfare between science and theology" firsthand. Clearly, White's stories resonated with many Americans' hopes. Newspapers and magazines throughout the country reviewed his book (it was also translated into French, Italian, German, Swedish, and Japanese). The *American Naturalist* reprinted the line about geologists increasing the knowledge of the power and goodness of God.[27] The nature writer John Burroughs described the book as full of tales of men who had once tried to solve the problem of God's goodness in the face of evil by talking about the devil coming to the realization that "cleanliness is a better safeguard against fever than fasting or prayer."[28]

White and his readers justified these hopes by imagining the human past as an evolutionary, progressive ascent, rather than a deserved fall and expulsion from Eden. They could not appeal to obvious medical advances at the bedside. In a footnote, White could include evidence of a decreased death rate "especially among children" in New York City in 1866. (Measured as the percentage of children that died under age five, the child mortality rate for the city had declined from 53 percent in 1867 to 47 percent in 1877 thanks to improvements in "drainage, sewerage, and ventilation of dwelling-houses, with enforced cleanliness and observance of other sanitary laws.")[29] These drops in mortality numbers were not due to new medical techniques for saving sick children. They were also difficult to stabilize. Sometimes the numbers went right back up again. But with White's history of science in hand, a lack of progress in the ability to save children's lives when they fell ill need not undermine faith that progress was possible. Indeed, White cited the absence of progress as evidence that he was right regarding the forces that prevented medical advancement. He imagined, for example, that physicians might have triumphed over disease sooner *if only* theologians had not hindered Bacon in the thirteenth century. And while at least one reviewer found the Bacon passage both unfair and unhistorical, others found the passage particularly worth reprinting.[30] It appeared in *Popular Science Monthly*, the American Medical Association's *Handbook for Speakers on Public Health*, and the *Texas Medical Journal*. Indeed, the author of the latter's "Texas' Sin of Omission—Her Sanitary Needs" cited the passage in an effort to convince legislators to pass "enlightened sanitation legislation," collect vital statistics, and fund a Public Health Department.[31]

For decades, White's tales inspired influential philanthropists to fund scientific and medical research in the hopes that he was correct about the best path to progress. Some took advice from White directly. When, in 1901, White's son Frederick died following a long struggle with the chronic effects of typhoid and "melancholia," Andrew Carnegie invited White to his castle in Scotland. Carnegie distracted White from his grief through the "long dreary summer" by asking for advice on Carnegie's plan for a national institution for scientific research.[32] Like White, Carnegie had long since rebelled against orthodox Christianity on ethical grounds. In his autobiography Carnegie wrote of how he had followed in his own father's footsteps by repudiating what "Andrew D. White ventures to call" the "Eternal Torturer" God of orthodox Presbyterianism. Having given up orthodox Christianity, Carnegie depended on White's version of the past to justify his claims that the Carnegie Institute of Washington supported "the application of knowledge to the improvement of mankind."[33] Indeed, White's grand historical narrative implied that repudiating orthodox Christianity and fighting for scientific progress were not only possible but *pious*.

Not all those who spent their days trying to make the world a better place believed that White's history was either helpful or true. The Catholic physician James Joseph Walsh was dismayed by the popularity of White's tales (both Draper and White peppered their histories with heavy doses of anti-Catholicism). Walsh pointed out that the past, including the Middle Ages, was neither as superstitious nor as ignorant as White claimed. He countered White's tales of the Catholic Church's theologians, physicians, and scientists obstructing progress with detailed historical evidence of Catholicism's support for science. He reminded his readers that the great Louis Pasteur was a devout Catholic who once wrote, "The more I study nature, the more I stand amazed at the work of the Creator. I pray while I am engaged at my work in the laboratory."[34]

Walsh clearly believed that science had advanced and could advance still further. But he insisted that both the stories and the premises of White's book were simply untrue. He repudiated White's assumption that progress depends on dispensing with belief in the close, personal God of orthodox Christianity. As both a Catholic and a physician, Walsh held that distancing God's hand from human experience, imagining earth as the only heaven, and denying man's ability to say anything about immortality cost far too much. He especially feared what loss of faith in God's close, personal providence did to parents: "The loss of children by death, particularly when there are but one or two children in a family, as is so frequent in modern times, often brings on a state of mental perturbation in which the health of mind and body, specially of women, may suffer severely. Religion, with its development of the spirit of sacrifice, whenever it is taken seriously, is the best possible sheet anchor in such cases, and the gradual diminution of religious feelings and abandonment of religious practice during the present generation have greatly multiplied the tendency to such severe breakdowns."[35] White, by contrast, believed that parents must find solace in, first, the strict uniformity of nature's laws and, second, the progress to be secured if human beings would stop praying for miracles.

Both White and Walsh believed that medicine could be improved and suffering ameliorated. Both had dedicated a good part of their adult lives to those premises. But White believed solace in science must be sufficient; Walsh that science was not enough. They were on opposite sides, not of a choice between saving lives or not saving lives, but of what must be believed when a child could not be saved. As we will see in the next chapter's examination of eugenics, Walsh also feared what men might do with science if "nature's laws" became unharnessed from the ethical guidance of Christianity's emphasis on the sanctity of the individual human life. For his part, White believed one could still find eternal truths in Christ's Sermon on the Mount and the Lord's Prayer. But why, some asked, stop there? Why maintain "love thy neighbor" as true after the rest of the scriptural history had been set aside? On what grounds did "blessed are the meek" command obedience if Genesis and Revelation were fables?

White assumed he was purifying Christianity rather than destroying it. But others set aside *The Warfare of Science*'s odes to the goodness of God and adopted his grand history as evidence that human beings must find solace and hope only in what uniform natural laws might someday allow humans to do. The philosopher Paul Carus, for example, cited White's tales as evidence of science's steady advance over the idea of an intervening God.[36] Then, inspired by the speeches of Anagarika Dharmapāla, who had traveled from Ceylon to the World's Parliament of Religions in Chicago in 1893, Carus declared that Buddhism, not Christianity, mapped best onto a world governed by natural laws.[37] Carus was not alone in his thoughts on Buddhism. A contributor to the *Freethinker* agreed that the words of Dharmapāla proved that "Buddhism is the more philosophical faith," for Gautama Buddha "taught that we suffer through our ignorance of the immutable laws of causation, and by ignoring truths of suffering, causes of suffering, cessation of suffering, and the noble way of salvation."[38] Carus also insisted that humanity must replace the hope of heaven with a different kind of hope: that what a man does, be it evil or good, lives after him so far as individuality impresses itself and influences his contemporaries. That, he insisted, was the only kind of immortality worth believing in. Thus, Carus repudiated the idea that he could or should hope to ever meet his own dead child, a son named Paul, again. The only "purpose of life and the duty of man," Carus insisted, was "activity and labor in the service of amelioration."[39]

In 1885 Carus used a new word for his beliefs: meliorism. The words *amelioration* and *melioration* (the action or process of improving something or mitigating suffering) had been used for generations (as we have seen, *melioration* was Emerson's favorite word). But melior*ist* as something one could *be*, and melior*ism* as something one could believe in, were new terms. In 1877, the British novelist Mary Ann Evans (her pen name was George Eliot) was one of the first to use the term *meliorist* to capture her position that this is neither the best nor worst of worlds, but rather a world capable of gradual improvement via human effort.[40] Two decades later the American journalist Junius Henri Browne wrote that "meliorists" looked "to secular, not to celestial, agencies for the government and regulation of this planet." Meliorism "in a religious sense," Browne wrote, openly acknowledged nothing but natural law and did not endeavor to determine what was behind that law. God, providence, the supreme being, the great first cause, the meliorist might humor any of these. But what the meliorist could not entertain was an *intervening* God who answered prayers with miracles and interference in natural law. Browne (like Malthus, Douglass, and others) argued that an intervening God (whether in the present or the past) undermined the basis on which effective meliorism depended: namely, belief in uniform natural laws that humanity might discover and through which trustworthy action could proceed.[41]

For many Americans, however, replacing hope of heaven with a purely earthly progress did not make sense of the world. Like Franklin, Elizabeth Stuart Phelps, for example, argued that only heaven could justify the "misery and waste and sin" on earth. "How else," Phelps asked, "are we going to account for the awful waste of material which goes on forever in our dark history? . . . How explain what otherwise were superfluous sacrifice and wanton cruelty?" For Phelps, the demand to give up personal immortality for a future in which the race, rather than the individual, would "liveth" was a sorry consolation to the correspondent writing on paper with "deep, black margins," who cried out, "Tell me, tell me that my lips shall touch my vanished child again!"[42] For many, hope of heaven depended in turn on belief in Christ's miraculous resurrection. Confronted by White's histories, White's friend Mary Eaton demanded to know what became of the New Testament's promise of eternal salvation if belief in its miracles must be abandoned. The main lesson of White's book, she argued, seemed to be that the Christian religion was a cunningly devised fable, the Bible a tissue of falsehoods, its divine author a myth, his son the greatest imposter the world has ever known. White might think that he was purifying Christianity of its superstitious elements in order to preserve Christian ethics in a scientific age. But Eaton insisted that he was replacing revelation with a religion evolved from *human thought*, stripped of all that is divine.[43]

It is important to note that Eaton's insistence on the reality of Christ's miracles was not a demand that one believe in miracle healings or God's miraculous intervention in the present. Many Protestants believed that miracles were limited to the time of Christ and the apostles. The key point was that those biblical miracles provided evidence of Christ's divinity and his atonement for human sins rather than evidence that one might pray for miracles in the present. Clearly, for much of the history of science, belief in Christ's miracles had not undermined a commitment to science and the uniformity of natural laws in the present. But for White, both science and effective meliorism demanded that no exceptions exist, even in the deep past, to the strict uniformity of nature's laws. A firm commitment to explaining the world, including the past, as under the governance of uniform natural laws, White insisted, was the only means of securing a history that told of man's ascent (rather than fall) and, in doing so, justified hope (and, most importantly, provided the means) of still further progress.

For some Americans, hope of neither heaven nor "progress on earth" nor some postmillennialist combination of the two offered just recompense for a world in which children suffered so much and died so often. As a young man, Samuel Clemens (otherwise known as Mark Twain) dispensed with the idea that all suffering arose from a close, personal God.[44] When his own son, Langdon, died of diphtheria at the age of nineteen months in 1872, Clemens blamed himself rather than God. (He had taken the boy on a ride in an open barouche on a "raw, cold morning" and believed that had his boy been in the care of a more careful person no harm would have come to him.)[45] Later, Clemens wrote a story in which a precocious three-year-old named Bessie asks her mother "why there is so much pain and sorrow and suffering." Bessie's mother replies with the old answers: "In His wisdom and mercy the Lord sends us these afflictions to discipline us and make us better." But what, Bessie asks, of the little boy who died of typhus? What of the "billions of little creatures ... sent into us to give us cholera, and typhoid, and lockjaw, and more than a thousand other sicknesses?" Bessie (and Clemens) give an indignant reply to the claim that all these terrible things were actually good: "It's awful cruel, mamma! And silly!"[46]

But in marked contrast to White, Clemens despised the idea that one might locate purpose and benevolence in millions of years of a progressive, evolutionary "struggle for existence" leading up to man. He found no solace in the promise that children in the future would be saved while his own son perished because humanity did not yet know enough. Clemens ridiculed codiscoverer of natural selection Alfred Russel Wallace's claim, for example, that the Age of Reptiles had been designed to prepare the earth for humanity. (All suffering, Wallace had insisted, resulted in progress in the end.) Clemens's response is dripping with satirical disdain. The poor reptiles, he wrote, "led this unsettled and irritating life for twenty-five million years, half the time afloat, half the time aground, and always wondering what it was all for, they never suspecting, of course, that it was a preparation for man.... Man has been here 32,000 years. That it took a hundred million years to prepare the world for him is proof that that is what it was done for. I suppose it is. I dunno."[47]

Clemens clearly found little comfort in the belief that, divinely approved or not, his own and millions of other children might die from human ignorance while those of the future might be saved via the slow and laborious advance of science. He wrote these words a few years after the discovery of an antitoxin that, if only it had been discovered earlier, might have saved his little boy. As we will see, that purely laboratory-based discovery would soon be touted as proof of mechanistic medicine's ability to ameliorate suffering. For some, the discovery of antitoxin vindicated more traditional arguments for why faith in progress was warranted, from postmillennialist interpretations of scripture to the moral progress represented by abolitionism and from the progressive patterns of the fossil record to the extraordinary ascent of *Homo sapiens* via evolution. For others, the evidence finally coming out of scientific laboratories that children might be snatched from death via science completely replaced the old arguments for faith in progress, and a more complex history began to be forgotten.

The Blessed Antitoxin

In his last public address, Ingersoll had included diphtheria, the disease that killed Clemens's son, in one of his long lists of challenges to belief in God: "If a good and infinitely powerful God governs this world," Ingersoll asked, "how can we account for cancers, for microbes, for diphtheria, and the thousand diseases that prey on infancy?"[48] Diphtheria also appeared in White's list of afflictions that physicians might have wiped from the face of the earth if only medieval theologians had allowed science to proceed unobstructed.

What Ingersoll and White could convey by the word *diphtheria* is hard to grasp for many Americans today, who often haven't even heard of the disease. But in 1850 diphtheria was one of the leading causes of death for both children and adults. It was known as the "strangling angel of children" due to the wing-shaped white membrane that grew into the nose, esophagus, or larynx and sometimes closed the airway entirely. One physician described losing one of his patients to this terrible illness as follows: "I recall the case of a beautiful girl of five or six years, the fourth child in a farmer's family to become the victim of diphtheria. She literally choked to death, remaining conscious till the last moment of life. . . . I watched the death of that beautiful child feeling absolutely helpless to be of any assistance." He described how "the physician's inability to cope with this scourge was one of the most devastating experiences through which a medical man could go."[49] It was also devastating for parents. Diphtheria may be the illness that the playwright George Bernard Shaw had in mind when he spoke of the "theology of women who told us that they became atheists when they sat by the cradles of their children and saw them strangled by the hand of God."[50] (Shaw may also have been thinking of the British atheist Annie Besant, who recounted losing her faith in God when witnessing her daughter suffering from whooping cough.)

During a time in which physicians could offer little more at the bedside than the confidence-building art of accurate prognosis, diphtheria terrified the very best of doctors. "On the one hand," wrote one, "very unpromising cases sometimes do well; on the other, patients whose symptoms have not been particularly alarming to the unskilled observer, very often die."[51] One founder of pediatrics, Job Lewis Smith, described diphtheria as "one of the most dreaded, one of the most fatal, and, unfortunately, one of the most common maladies of childhood."[52] Historian of medicine John Ballard Blake even suggested that the intensity of New England's first Great Awakening (when, during the 1730s and 1740s, evangelical fervor swept the land) may have been influenced by early epidemics of what was then known as the "throat distemper," "canker," and "the plague of the children."[53]

In the worst cases the membrane produced during the course of the disease could suffocate the patient. Attempts to cauterize the membrane away seemed only to open up new areas to infection. Physicians knew that the membrane incorporated itself into the mucous layer and thus could not be removed without the rupture of small blood vessels. It also re-formed within a few hours. Alcohol, bichloride mercury, steam, boracic acid, chloral hydrate, very diluted solutions of hydrochloric acid, and various patent medicines were all used to fight the disease in the 1890s.[54] But the state quarantine physician of El Paso, Texas, Ira Jefferson Bush, lamented that medicine seemed fruitless. "The bad cases die and the mild ones get well without treatment," he wrote, although "whisky in large doses is very beneficial." Performing a tracheotomy was the primary option in the case of laryngeal stenosis (closing of the airway), although many patients died from subsequent infection or cardiac paralysis. Still, Bush wrote that the relief tracheotomy gives "is great, and even though your patient ultimately dies of heart failure, you have prevented the horrors of death from suffocation."[55]

Faced with a seemingly hopeless case, a surgeon named George Buchanan performed a tracheotomy with the mother's words "for the love of God, try to save my boy!" ringing in his ears. That surgeon provided statistics for the fifty such operations he had performed: cured, nineteen; dead, thirty-one.[56] Some physicians, including Abraham Jacobi, suctioned the tracheotomy tube with their own mouths to remove secretions blocking the airway, knowingly exposing themselves to the disease in a desperate attempt to snatch children from death. Following the fatality from infection of a promising young physician who had tried to save a patient by such means, one commentator lamented, "The tracheotomy tube seems to be an ill-fated contrivance in encouraging physicians to jeopardize life to save life."[57] Such acts are emblematic of the desperation inspired by the disease.

Promises of progress seemed cruel illusions at the bedside of patients suffering from diphtheria. Jacobi, the acknowledged expert on the disease, wrote in 1880 that the outcomes of tracheotomies were getting worse over time: "Of sixty-seven tracheotomies which I published twelve years ago," he wrote in his *Treatise on Diphtheria*, "twenty per cent recovered; about two hundred tracheotomies performed by me since that time, brought down the percentage of recoveries to such a low figure that only the utter impossibility of witnessing a child's dying from asphyxia has goaded me on to the performance of tracheotomy."[58]

Three years after the publication of his classic treatise on diphtheria, Abraham and his wife Mary Putnam Jacobi (also a physician) lost their son Ernst to the disease. Abraham was one of the most influential pediatricians in the country, yet like many Americans he had lost several infants (and two wives) already. He and Mary had lost one infant earlier in their marriage. Ernst had made it to the age of seven, a full year past the age at which, as the physician Charles Meigs wrote in 1849, a child "does not surely belong to its parents," but seemed to be a loan which might be taken back at any moment if not "wisely and safely conducted up to the sixth year of its age."[59] The Jacobis had brought Ernst through this "loan" period. But at the age of seven he came down with diphtheria, and Mary and Abraham witnessed the disease slowly suffocate their son to death.

By the time they lost young Ernst to this dreadful affliction, both Abraham and Mary Jacobi were strict mechanists who repudiated the idea that "God's will" explained their loss. Abraham's parents were Jewish and Mary's Protestant, but both husband and wife had abandoned their childhood faiths in favor of navigating the world via strictly natural explanations.[60] Historian Carla Bittel notes that Mary believed "that all knowledge should be based on observable phenomena, that a focus on 'humanity' should replace metaphysics and organized religion, and that science offered solutions for social reform." Mary even defended vivisection, something male proponents of laboratory science thought an unwomanly thing to do (she also rebelled against the idea, dominant in science since the first days of the Royal Society, that the laboratory was a masculine domain).[61] Meanwhile, she had no patience for the "fantastic dogmatisms" others used to explain all of the "households desolated by untimely deaths." She had no use for prayer and no use for heaven or some distant reunion in the hereafter. "The dream of heaven," she wrote, "never has had influence except when the dreamers have been actively engaged in realizing it upon earth." Abraham agreed. He once wrote that the dualistic concepts of orthodox religion (i.e., the division of reality into material and spiritual realms) led to superstition and an interest solely in souls rather than bodies.[62] Both Mary and Abraham believed that only a focus on the material, bodily world would save children from premature, earthly demise.

Abraham Jacobi actually had good evidence that a dualist physician, even a Catholic one, could devote every waking moment to repairing and healing the bodies of children. He championed the tireless efforts of the Catholic physician Joseph O'Dwyer to develop a tube to keep a child's airway open as diphtheria ran its course. Jacobi's and O'Dwyer's beliefs about the ultimate cause and meaning of such terrible suffering could not have been more different. Though for a time O'Dwyer had believed in petitionary prayer (i.e., that one might pray to God with specific requests), he ultimately decided that "all that we can do is to say with resignation, 'Thy will be done,' and then we shall be sure that whatever happens will be for the best."[63] O'Dwyer's life work against the ravages of diphtheria provide a good hint that finding solace in "Thy will be done" did not necessarily lead to inaction, complacency, or fatalism. After years of trying, he developed an alternative to tracheotomies that avoided breaking the skin and exposing children to seemingly inevitable infection: O'Dwyer's tube. Although initially skeptical, Jacobi became one of O'Dwyer's main defenders against those who criticized his invention.

Clearly, profound differences regarding what kinds of beliefs would inspire effective action could melt away at the bedside of a suffering child. By all accounts, O'Dwyer was distraught by evidence that doctors ignored his detailed instructions when using his new invention, and he became obsessed with training physicians to use the tubes correctly in order to cause as little suffering as possible. One can imagine him demonstrating the method on a gasping child to Jacobi: a devout Catholic physician who believed firmly in providence and that "all that we can do is to say with resignation 'Thy will be done,'" standing beside an unbelieving physician of Jewish heritage who attended to his every word and movement so he could spread the word truthfully. Because lives were at stake. Both had suffered the losses that they were trying to prevent: Ernst had died just six years earlier from this terrible malady. O'Dwyer and his wife Catharine lost four of their eight sons to the "summer complaint" that took so many babies when the weather grew warm.[64] These physicians' different understandings of the presence, distance, existence, or nonexistence of God didn't matter a whit in those moments of desperately trying to intubate a child strangled by diphtheria toxins. In those moments what must be done, and what must be done better, was clear—no matter how these physicians differed on the ultimate fate of the children they could not save.

Despite their best efforts, evidence of clear progress at the bedside of children remained elusive for much of the time that O'Dwyer, Abraham Jacobi, and Mary Jacobi practiced medicine. That did not prevent them from regretting actions taken or not taken under the assumption that a different path might have saved their children. Indeed, fifteen years after their son's death, Mary Jacobi hinted at regret that she and her husband had not done everything they might have done. She feared she had trusted Abraham's expertise a bit too much: "I having naturally a great confidence in his professional opinion and knowing his great love for our little son, did not sufficiently urge the different precautions about his health, which I now think might have sufficiently increased his power of resistance when the trial came."[65] It isn't clear what she thought might have been done differently when she gently blamed Abraham (and herself) for Ernst's death. But as unbelievers, neither Mary nor Abraham could blame God. Instead, they blamed themselves and human ignorance, and took refuge in faith that things might be done better in the future.

By the time the Jacobis and O'Dwyer were practicing medicine, physicians taking stock of progress confronted an ambiguous record at the bedside of patients. Surgeons, anatomists, and physiologists had learned a great deal about the human body through adopting the assumptions, metaphors, and methods of the mechanical philosophers. The study of morbid anatomy (the correlation of symptoms with postmortem dissection) had improved physicians' ability to diagnose and develop accurate prognoses of a number of ailments, including tuberculosis. Surgeons had deadened the excruciating pain of operations (though not through any particular insight into the mechanisms through which anesthetics actually worked), and the imperatives of war had inspired heroic developments in surgery. But even in the face of improved sanitation during surgery as the result of Joseph Lister's and others' work, postsurgical infection rates were still extremely high.

We have seen how nineteenth-century ministers, scientists, and physicians who had faith in progress called upon scripture, the fossil record, the abolitionist movement, and Darwin's theory of evolution to support such faith. By the final quarter of the nineteenth century, they could also cite astonishing discoveries in the scientific laboratories of France and Germany. From the 1860s to the 1880s, experimental scientists established what precisely the vis medicatrix naturae was up against when children ran the terrible gauntlet of cholera, typhoid, septicemia, diphtheria, summer diarrhea, consumption, measles, and smallpox. Soon, germ theory (the idea that infectious disease spreads via microscopic, living germs, or bacteria, most famously espoused by Louis Pasteur and Robert Koch) seemed to provide the ultimate vindication of two centuries of hope that science might improve medicine.

As word of Pasteur's and Koch's work spread, some wondered: Did Lister's antiseptic procedure work so well in preventing infection after surgery because it was killing minute germs? Were the culprits of John Snow's claim that water spread cholera in fact what scientists now called "bacteria"? When the Hungarian physician Ignaz Semmelweis begged his colleagues to wash their hands prior to delivering babies in order to prevent childbed fever, were germs what must be washed away? Were these microscopic creatures the cause of all the terrible contagious diseases that stole children in the space of days? Soon, laboratories in France, Germany, and Japan announced a series of triumphant identifications of disease-causing bacteria. In 1874 Gerhard Armauer Hansen identified the leprosy bacillus, in 1876 Koch and Pasteur identified the cause of anthrax, in 1879 Neisser identified the agent of gonorrhea, and in 1882 Koch identified the bacillus that killed more human beings than any other disease at the time: tuberculosis.

As scientists debated germ theory throughout the 1870s and 1880s, the fact that the idea of germs mapped well onto existing ideas about sanitation heightened imperatives of cleanliness and strengthened the idea that disease could be prevented.[66] Meanwhile, the biologist Clifton F. Hodge cited the developments in the new field of bacteriology and germ theory to create a heavenly vision of what biology, physiology, and anatomy could one day do: "In a word, the faith, hope, and charity which inspire this science are to learn enough about the laws and possibilities of living Nature, to do away with all disease and premature death, and to make all life as full and perfect as these laws will permit. This is the inspiration of biology."[67]

It is important to keep in mind that when Hodge wrote these words in 1896 bacteriology labs had produced few techniques for saving lives once infection set in (as opposed to bolstering arguments for good hygiene and sanitation to prevent infection). After all, physicians could do nothing once diseases like measles, smallpox, typhoid, yellow fever, diphtheria, or scarlet fever appeared in a home except try to ameliorate the symptoms. Pasteur had used the first vaccine against rabies in 1885, but the extent to which this technique could be extended to other, more common childhood diseases was not yet clear. Thus, Hodge still had to imagine possibilities rather than write of grand triumphs. He had to admit that one half of people still died before the age of forty, almost all of diseases that were surely curable or preventable "did we but know how." And this went on, he lamented, with a standing army of one hundred thousand physicians. "It looks discouraging," Hodge conceded, "and an eminent physician has himself said that a doctor is like a man blindfolded, striking about with a club, almost as likely to hit his patient as the disease." The only hope amid such suffering,

Hodge urged, was to learn more about "the principles upon which God had deemed it wise to order the living population of the world."[68] Since the days of the Royal Society, however, natural philosophers, naturalists (by the 1870s they called themselves "scientists"), and physicians had been promising much more from experimental sciences.

The lack of new techniques at the bedside, despite all the experimental work at the "bench" (experimental laboratories), did not go unnoticed. Indeed, Hodge wrote the above lines as an argument against anti-vivisectionists who were citing the lack of medical progress in order to halt experimentation on live animals. Anti-vivisectionist sentiment had increased amid the rise of physiology (the study of organ and system function), a science that completely depended on animal experimentation.[69] It is important to note that even the great French physiologist Claude Bernard, who helped establish the function of the liver via experiments on animals, never argued that lives were being saved by such work. As we will see, within a few decades Bernard's discoveries led to lifesaving insulin treatments, but no one knew for certain that such extraordinary feats would result from Bernard's meticulous (and to critics, gruesome) experiments. In the meantime, anti-vivisectionist Dr. Albert Leffingwell made a perfectly fair point when in 1880 he wrote, "I venture to assert that, during the last quarter of a century, infliction of intense torture upon unknown myriads of sentient, living creatures, *has not resulted in the discovery of a single remedy of acknowledged and generally accepted value in the cure of a disease.*"[70]

Indeed, Hodge admitted as much when he focused his defense of vivisection not on evidence that experiments on live animals had improved therapeutics (he had none) but on the prevalence of suffering in the natural world. The American Antivivisection Society's resolution had argued that the practice opposed the intent of a beneficent, merciful creator "who wills the happiness of all his creatures." In response, Hodge described the endless suffering in nature: the "teeth and jaws, the beaks and talons, the claws and fangs" for preying on the weak, the parasitic animals and plants, the many diseases caused by microbes. Hodge was not calling God's goodness into question, for he reminded his readers that struggle and suffering had resulted in the progressive evolution of man: "The price has been great," Hodge conceded, "but the gain is priceless." The discovery of evolution, Hodge argued, had taught biologists the "divine point of view" that out of suffering a nobler form and higher life arose. "Animal life is cast into the world *as an experiment*," Hodge wrote, "often of the severest and most painful type. In this lifelong vivisection, Nature provides no ether or chloroform, nor even chloral or morphine." Then he demanded, "Taking Nature as we find it, what can man do about it?" "Numberless instances in the history of science" proved that "if man will only put forth a reasonable amount of effort, it may not be so difficult to comply with the command, 'Subdue the earth.'" "Nature is *wisely* ordered," Hodge argued, "to give *man plenty to do*, and to do this work is one of his highest duties." He then insisted that the amount of suffering in nature proved that "God clearly gives to man every sanction to cause any amount of physical pain which he may find expedient to unravel his laws."[71] As evidence of why science must not rest (or abandon the use of vivisection) in discerning the principles by which God had ordered the world, Hodge cited the fact that 49,677 children died annually of diphtheria.

None of Hodge's defenses of vivisection relied on evidence that science had yet improved physicians' ability to save lives. But within a little over a decade, Hodge and his fellow defenders of experimental medicine could cite the discovery of a cure for diphtheria. In 1883, the same year Ernst Jacobi died, the German-Swiss pathologist Edwin Klebs had identified a bacillus common to all cases of diphtheria, and a year later the German bacteriologist Friedrich Loeffler proved the bacillus caused the disease. (Germ theory required that every demonstration of a bacterial cause for a disease produce that same disease in experimental animals.) Klebs's work also meant that cases of diphtheria could be confirmed by the presence of the bacteria, ultimately allowing physicians to differentiate it from the less dangerous croup.[72] Meanwhile, the French bacteriologist Emile Roux proved that one could remove the germs yet still create the disease, thereby demonstrating that a bacteria-produced toxin caused the terrible symptoms. The discovery that exposure to diphtheria toxins conferred subsequent immunity soon led to hopes that scientists might harness that fact to produce antitoxins. Following experiments on animals, the first successful use of diphtheria antitoxin to halt the disease's life-threatening effects in humans occurred in 1891 in Berlin.

In 1896 the widely distributed magazine the *Forum* published the results of the American Pediatric Society's examination of 6,000 cases of diphtheria treated with antitoxin in the practice of 613 different physicians. Compiled by a founder of pediatrics, Luther Emmett Holt, the results provided a "decisive verdict in favor of the Antitoxin treatment." Mortality from diphtheria without antitoxin had ranged between 26 and 56 percent. The mortality rate with antitoxin treatment was 12.3 percent (the rate dropped to 4.8 percent if the antitoxin was given in the first three days, before the toxins could damage other organs). "No one feature of the cases of diphtheria treated by Antitoxin," wrote W. P. Northrup, "has excited more surprise among the physicians who have reported them, than the prompt arrest, by the timely administration of the serum, of membrane which was rapidly spreading downward below the larynx."

Northrup advised that skeptics of the treatment need watch just "one severe case of diphtheria clear up like darkness into daylight" and they would argue no more. Since the days when Lister proposed antiseptics in surgery, Northrup concluded, medicine had not made so great an advance.[73] By 1897 the American Pediatric Society was recommending antitoxin's use. "My dread of diphtheria has decreased to such proportion," wrote one physician in 1899, "as to render me very much less worried when called to see a case."[74]

Given its clear origin in laboratory research and the mechanization of nature, the influence of the addition of antitoxin to physicians' arsenal (otherwise quite empty) on perceptions of scientific progress was profound. The physician whom we witnessed earlier describing the hopelessness of pre-antitoxin days followed that tragic account with a description of what it was like to treat his own little girl with antitoxin ten years later: "To watch the choking dreadful membrane melt away within a few days, was one of the most dramatic and thrilling experiences of my professional career."[75] It was the first robust vindication of two centuries of promises that mechanistic sciences would improve the ability of physicians to save lives at the bedside. Now physicians could conquer diphtheria itself rather than waiting for nature to do its work or trying to prevent the germs from spreading. As one widely reprinted article announced, "Antitoxin is a child of pure science, and is born of logic."[76] The famous physician and pathologist William Welch emphasized that the discovery of antitoxin was "entirely the result of laboratory work. . . . In no sense was the discovery an accidental one. Every step leading to it can be traced, and every step was taken with a definite purpose and to solve a definite problem."[77] Historian Evelynn Hammonds notes that diphtheria was "the first infectious disease to be controlled by advances in scientific medicine, particularly discoveries in bacteriology and immunology."[78]

By 1914, then, defenders of experimentalism like the brain surgeon and former president of the American Medical Association William Williams Keen could cite diphtheria antitoxin as evidence that science could finally save children's lives. Keen drew upon the long tradition within American science and medicine of couching faith in progress in religious terms by explicitly attributing this discovery to God's will. He recounted a story told by one of his colleagues of standing with a young mother by the bedside of her only child: "The child, in the throes of diphtheria, was clutching at its throat and gasping vainly for breath. Suddenly the mother flung herself on the floor at the doctor's feet in an agony of tears, entreating him to save her child. But alas! it was impossible." If only the case had occurred just a few years later, Keen wrote, "when the blessed antitoxin for diphtheria had been discovered (solely by animal experimentation), this remedy would have been given early; and almost certainly within a few hours the membrane would have softened and disappeared, and that life, precious beyond rubies, might have been saved." In the dreadful days before antitoxin, Keen wrote, the only comfort physicians could give distracted mothers was that "it was God's will." "Now," Keen wrote with relief, "thank God, it is not His will." A table of mortality before and after the use of antitoxin concluded the argument. As for anti-vivisectionists, Keen wrote that "had they ever stood as in the past I have stood, knife in hand, by the bedside of a gasping, livid child struggling for breath, ready to do a tracheotomy when the surely tightening grip of diphtheria made it necessary to interfere, they would hail with delight the blessed antitoxin which has abolished the knife and enormously diminished the mortality of that curse of childhood. They would surely bless God that such a discovery as this antitoxin could be made *solely by experiments on animals*."[79] Keen's expression of gratitude to God was not opportunistic rhetoric: he was a practicing Baptist who wrote a book entitled *I Believe in God and in Evolution*.

Not every American Christian thought that visions of progress via mechanistic, experimental sciences were interpreting scriptural promises rightly. A group who called themselves Christian Scientists (founded by Mary Baker Eddy in the 1880s) refused all such interventions on the grounds that prayer and faith were the true medicines and all disease illusions premised on a mistaken belief in the existence of a material world. The Christian Science movement was a strong hint of a rebellion brewing against the compromises some Christians made to adjust to mechanistic science. In response, a physician named F. E. McCann suggested that physicians remind the faithful of the long tradition within Christianity of believing that God not only placed the drugs that belonged to the "animal, vegetable or mineral kingdom" on earth but "endowed man with a mind to find out about things put here for his welfare."[80] To parents who had chosen prayer rather than antitoxin, one doctor replied, "We do not question the power of Divine assistance, but it is our duty to make use of the material means that God furnishes us to alleviate human sufferings."[81] Whether these arguments were in earnest or not on the part of the authors (and there is no evidence they were disingenuous), these physicians' framing of science as the pious, divinely approved means of ameliorating earthly suffering was a perfectly orthodox position to take. But some Americans were beginning to wonder whether the adjustments required of orthodox belief might cost too much.

For many Americans, however, the discovery of antitoxin vindicated White's claim that the history of science revealed God's governance via natural laws that man could understand and harness in the name of amelioration. Some cited White's history as evidence of what needed to be done so that "medical science [would] in time conquer here as it [had] conquered elsewhere, giving one more bloodless victory to a suffering world." And some included White's claim that, had Roger Bacon not been hindered by theological authority, the long list of diseases "which now carry off so many precious lives, would have long ceased to scourge the world." Smallpox vaccination and diphtheria antitoxin proved what could be done. Surely even a cure for consumption and other terrible killers was near, if only science was allowed to proceed unopposed.[82]

Imagining progress depended, of course, on accepting stages during which humanity had not yet discovered the knowledge required to save lives. It required a God who delegated the task of creating a better world entirely to human beings via the laborious path of studying the natural laws through which God governed. As we have seen, both Ingersoll and Clemens preferred an "I dunno" over their fellows' confident schemes of salvation, scientific or otherwise, that claimed to know so much. They were not alone. In 1909, as diphtheria rates dropped, the Irish playwright George Bernard Shaw mocked the stance that humanity's increasing ability to fight disease reflected God's goodness: "What about the croup? It was early days when He made the croup, I guess. It was the best He could think of then; but when it turned out wrong on His hands He made you and me to fight the croup for Him."[83] Others found very little comfort indeed in the knowledge that science was finally saving some children while simultaneously being cited to withhold such salvation from their own children.

Better Far This Nameless Void

W. E. B. Du Bois was the first African American doctoral graduate from Harvard. Having completed a sociological study of Philadelphia's Seventh Ward, he moved with his wife, Nina, to Atlanta, Georgia, to take up a position as professor of history, sociology, and economics at Atlanta University. On a warm night in May 1899, Du Bois rushed through the streets of their new city trying to find a physician for his nineteen-month-old son, Burghardt, who lay in the painful throes of an intense fever.

Four years later Du Bois included an account of Burghardt's death entitled "On the Passing of the First-Born" in his 1903 book, *The Souls of Black Folk*: "He died at eventide, when the sun lay like a brooding sorrow above the western hills, veiling its face; when the winds spoke not, and the trees, the great green trees he loved, stood motionless. I saw his breath beat quicker and quicker, pause, and then his little soul leapt like a star that travels in the night and left a world of darkness in its train. The day changed not; the same tall trees peeped in at the windows, the same green grass glinted in the setting sun. Only in the chamber of death writhed the world's most piteous thing—a childless mother."[84]

Some scholars say Burghardt died of diphtheria. Others say he died of typhoid. Writing a half century later, Du Bois referred to the illness as "one of those spring intestinal infections." Du Bois's words and the length of the illness signal one of the dreaded diarrheal illnesses that took so many children each year. But whatever the cause, ten endless days gave way after Burghardt's death to nights of "dreamless terror" for Du Bois and his wife Nina.

The ideas Du Bois could draw upon to make sense of Burghardt's death show how diverse the options available to grieving Americans had become. Emerson was one of Du Bois's intellectual heroes, but they were united by more than abstract ideas. Scholar Shannon Mariotti has described how Emerson and Du Bois both faced the challenges of finding "harmony in visions of universal things" while grieving their sons. Reality suddenly came into conflict with their ideals, as their sons' deaths tested their transcendentalist stance that individuals must take comfort in the grandeur of the universe. Both men, Mariotti argues, were changed by the loss of their sons. Having given up belief in a close, personal providence, they had to determine what, as Emerson wrote, they "ought to do." Both fathers responded by tempering transcendentalism, which was supposed to transcend the material world, with increasing attention to the suffering of their fellow human beings on earth.[85]

Years later, W. E. B. Du Bois wrote that Burghardt's illness "might easily have been avoided and even stopped had we been persons of greater experience."[86] The words are poignant testimony of the pressures placed on parents by explanations of child mortality that emphasized ignorance as the cause of child loss. Andrew Preston Peabody had warned in the 1840s that emphasizing ignorance of nature's laws as the cause of child mortality would place a tremendous burden on the hearts of parents. But, as historian Richard Meckel notes, by the end of the nineteenth century public health campaigns increasingly blamed parental ignorance for stubbornly high child mortality rates.[87] This emphasis allowed the infant mortality (or "baby saving") movement to ignore the socioeconomic conditions that Du Bois himself had demonstrated to be a primary cause of differential mortality rates between racialized populations.

Attributing child mortality to ignorance was an especially high burden to place on Black mothers, who often lived in segregated communities in which generations of racist policies limited access to clean water, pure milk, physicians, and—after 1896—antitoxin. When reflecting on Burghardt's death in a memorial to Nina composed in 1950, Du Bois wrote that when their little boy died "in a sense [his] wife died too": "Never after that was she quite the same in her attitude toward life and the world. Down below was all this great ocean dark bitterness. It seemed all so unfair." Immersed in his work, Du Bois acknowledged that his and Nina's options for responding to such loss were very different: "After all Life was left and the World and I could plunge back into it as she could not. Even when our little girl came two years later, she could not altogether replace the One."[88]

In some ways, the shadows that loomed over the Du Boises' memory of their son's death echo those with which the Emersons struggled. And yet the depth of those shadows was also unimaginable to privileged New Englanders like the Emersons. In 1870 the Fourteenth Amendment declared that "no State shall make or enforce any law which shall abridge the privileges or immunities of citizens of the United States; nor shall any State deprive any person of life, liberty, or property, without due process of law; nor deny to any person within its jurisdiction the equal protection of the laws." But when Reconstruction ended in 1877, Southern legislatures instituted "Black codes" that restricted the civil rights of Black men and women throughout the South. Soon the Ku Klux Klan was terrorizing Black communities and targeting individuals who seemed to be making economic and social gains. Segregation determined access to both education and medical care. In 1896 the Supreme Court decision in *Plessy v. Ferguson* made "separate but equal" the law of the land, with a strong emphasis on "separate" (rather than "equal") in both schools and hospitals.

We have seen how, whether cited independently or more often together, Americans had used both science and scripture to justify slavery for generations. Now both science and scripture were used to justify discriminatory laws and segregation in schools, hospitals, churches, communities, and marriages. The medical establishment participated fully in upholding segregation. The American Medical Association, for example, effectively barred Black physicians from membership. (Black physicians formed their own organization, the National Medical Association, in 1895.) This exclusion allowed white physicians to classify and draw boundaries with impunity around whose children were of most concern. Seventy years after James McCune Smith had called white physicians to task for declaring Black children's deaths inevitable, some AMA physicians continued to analyze Black and mixed-race child mortality in ways that rooted justifications of segregation within the material body.

In 1910, for example, an article appeared in the *Transactions of the American Medical Association* by Mississippi physician H. M. Folkes that repeated the old claim that the progeny of miscegenation was more susceptible to infectious disease. Given this "scientific fact," Folkes praised the South for having settled the question of whether races should intermarry by maintaining segregation. He also claimed that enslaved Black children had had a higher standard of physical health than those who were now free. With manumission, Folkes wrote, children had been "removed from the control of intelligent direction" and placed in the hands of "their more or less fatalistic parents" who soon "lapsed into their African condition of irresponsibility." This, he argued, was their natural condition, one that prevented them *by nature* from paying attention to the lessons of science. Folkes claimed that Black parents were moved by emotions rather than judgment, and as a result education about nature's sanitary laws could do nothing to decrease infectious disease among Black children. He then argued that when one found lower rates of infectious disease among Black children the cause must be the race's natural power of immunity (he cited the lack of cases of diphtheria as evidence) rather than any ameliorative action taken by parents.[89]

Clearly, within a society ridden by racial, ethnic, and class prejudice, talk of progress via the study of and obedience to natural law could become a double-edged sword. It could inspire belief in progress and action to ameliorate suffering of some, while relegating entire communities to a "lower stage" and explaining their suffering as the inevitable, "beneficent" outcome of nature's laws. This meant that even when science—and two centuries of grand promises to ameliorate suffering via the study of natural law—finally seemed to be doing something extraordinary and child mortality rates began going down, not all of America's parents and children experienced those triumphs. Thus, even as science finally delivered on long-standing hopes that suffering could be ameliorated via science, those triumphs were doled out, *also* on "scientific grounds," only to certain children.

Like James McCune Smith, Du Bois fought back against such pervasive racism within both medicine and society with science. The same year Burghardt died, Du Bois had published what would become a landmark in sociology, *The Philadelphia Negro*. As the "science of society," the discipline of sociology appealed to many Americans intent on changing the world. Sociology was premised on the assumption that facts, statistics, and the careful analysis of cause and effect would be the means of a better future. The American Social Science Association, founded in 1865, insisted, for example, that public action and social reform must be based on a firm foundation of facts, including consensus-based research conducted by experts. Indeed, one of its founders, Caroline Dall (who found purely naturalistic explanations a better means of making sense of her own child's death), thought that such a vision would even allow women to participate in the sciences required for wise social reform.[90]

Americans differed profoundly regarding what precisely the lesson of sociology might be. Some, like Yale sociologist William Graham Sumner, argued that better knowledge would prevent policymakers from interfering in nature's beneficent ways. Others insisted that knowing nature better would allow humans to rationally contravene its worst effects. Thus, sociology could be harnessed to either justify the status quo by declaring it natural or undermine the status quo by declaring human beings capable of intervening in nature's cruel ways, including the groupish behavior of human beings. Du Bois was firmly in the latter camp. He harnessed the "scientific study of society" to link discrimination, inequality, and racial prejudice, or what Du Bois and Douglass called "the color line," to pervasive and systemic inequalities in socioeconomic status, education, criminal justice, and employment. He then created meticulous maps that linked these inequalities to poor health and high mortality in the Seventh Ward in Philadelphia. Du Bois illustrated how disease vectors required certain environments to thrive, and that, as a result, discrimination in both housing and the distribution of sanitation and other important components of a city's infrastructure had severe impacts on health. Du Bois's work was a pioneering examination of what sociologists now call "social determinants of health."

Recently, scholars of social epidemiology have written that if Du Bois's work had been acknowledged and acted on when written, American reformers might have dealt health disparities a severe blow. More children, in other words, would have been saved.[91] This conclusion is an unwitting echo of White's lament regarding the lives that might have been saved had Roger Bacon not been opposed by existing institutions and the powers that be. Only, in contrast to White's tale, in which scientists were the obvious heroes, claims made in the name of science and natural law rather than scripture upheld an unjust status quo in the Seventh Ward in Philadelphia and throughout the nation.

When Du Bois wrote about his little boy, he remembered Burghardt's innocence regarding the world Du Bois was trying to change. "He knew no color line, poor dear," Du Bois wrote, "and the Veil, though it shadowed him, had not yet darkened half his sun. He loved the white matron, he loved his black nurse; and in his little world walked souls alone, uncolored and unclothed." Mather had written of children as better off in heaven, where they no longer ran risk of eternal damnation. Du Bois, too, imagined Burghardt "not dead, not dead, but escaped; not bond, but free." "Well sped, my boy," he wrote, "before the world had dubbed your ambition insolence, had held your ideals unattainable, and taught you to cringe and bow. Better far this nameless void that stops my life than a sea of sorrow for you." And then, immersed in a discipline, sociology, that was entirely premised on belief in progress, Du Bois caught himself: How could he tell his son might not have borne his burden bravely? Perhaps that burden would be lighter in the future? "For surely, surely this is not the end," he wrote. Someday, surely, a mighty morning would yet dawn and "set the prisoner free," and men would ask of the workman, "not 'Is he white?' but 'Can he work?'" and of artists, "not 'Are they black?' but 'Do they know?'"[92]

Clearly that world had not yet been created. And in the meantime, Black Americans faced with the prospect or reality of child loss lived in a world in which others declared that their children mattered less on the march to progress. Du Bois's challenges to that world arose from a range of places: experience, science, and religion. He argued that the doctrines of white supremacy were destroying the radical power of Christ's message to inspire universal brotherhood and social equity (most powerfully symbolized by the Sermon on the Mount). Americans who believed in the "color line" had warped Christianity into a justification of racism, economic power, and psychological and physical oppression. Du Bois tried to get Christianity back on the side of truth and justice. In a series of prayers composed for his students at Atlanta University, he argued that true Christianity focused on the transformation of not only individuals but society and institutions. As historian Edward J. Blum notes, in those prayers Du Bois "looked to God to inspire the social action necessary to end war, poverty, and sickness" and "wielded Christian ideas and rhetoric as weapons to confront racism and economic injustice."[93]

Amid Du Bois's efforts to remind his students of Christ's message, his criticisms of American Christianity could be scathing. He famously rewrote the story of Jesus by portraying Christ as a Black man lynched by white mobs. In that rewriting, he usually left out the stories of Christ's miraculous healings and resurrection to focus on the ethics of the Sermon on the Mount and the need to rely on one's self and one's community (rather than divine intervention) for liberation. Blum notes, however, that "minimizing the supernatural elements of the Bible did not necessarily position Du Bois in the agnostic or secularist camp." Twenty years after Du Bois's death, his close friend Herbert Aptheker warned those who would dismiss his friend's religious language as mere rhetoric: Du Bois, Aptheker wrote, "never lost a certain sense of religiosity, of some possible supernatural creative force. In many respects, Du Bois' religious outlook in his last two or three decades might be classified as agnostic, but certainly not atheistic; that remained true even when he chose to join the Communist Party."[94]

Whether one maintained God's close, personal presence, relegated God's governance to a distance, or repudiated belief in God altogether, building the scientific knowledge needed to change the world could seem impossible when physicians and policymakers collected "facts" and numbers through the lens of prejudice. Dr. Rebecca J. Cole, the second African American woman to earn a medical degree in the United States (in 1867) called Du Bois to task for relying on data on mortality rates from tuberculosis that had been furnished by "inexperienced white physicians." Cole knew that physicians commonly assumed that Black individuals were naturally, *constitutionally*, predisposed to consumption. This was clearly a useful explanation to those uninterested in addressing the role of prejudice and segregation in producing unhealthy environments and inadequate education or medical care. "Let a black patient cough," Cole wrote, and white physicians "immediately have visions of tubercles." Who collected these mortality statistics, Cole asked, "but men of a class who are so warped by that strange American disorder, colorphobia, that before accepting their verdict we must be excused for saying we are not ready for the question."[95] Such data could not, Cole insisted, be trusted. In other words, Cole insisted that science and medicine were not so free of superstition as the likes of Andrew Dickson White wished to think.

Like Douglass, Du Bois knew science could be a dangerous ally. He knew that even as science was finally discovering lifesaving things like diphtheria antitoxin some Americans justified the "color line" by claiming segregation to be a biological requirement of progress. University of Virginia biologist Ivey Foreman Lewis, for example, embraced descriptions of science as representative of God's goodwill toward humanity, since science allowed men to discover natural laws and improve their condition on earth. Two of the most important of those "natural laws," according to Lewis, were that, first, hereditary determined all and, second, white racial purity led to all progress. Lewis declared the notion that "all men are brothers and therefore alike in their potentialities" a silly notion based on "sap-headed thinking." Not surprisingly, he presumed that biologists had useful things to say about the relative value of different children. His students wrote essays on the relation between biology and America's "race problem" that argued that providing Black Americans with birth control would help ensure their "extinction in a comparatively short time and then insure a white America and her place in the world."[96]

Du Bois was not alone in imagining that perhaps his son was better off dead rather than forced to live in such a world. Educator and civil rights activist Mary Church Terrell had such thoughts as well. Eight months after her marriage to Robert Heberton Terrell in 1891, she had suffered a miscarriage that almost killed her. Mourning their loss, Terrell wrote to her husband, "The Lord knows best, and I pray almost hourly for a spirit of resignation." She found only tears and irrepressible sadness until she remembered how she had learned during her pregnancy that her good friend Thomas Moss and two of his friends had been lynched. How might her grief have affected her baby? "The more I thought how my depression which was caused by the lynching of Tom Moss," she wrote, "and the horror of this awful crime might have injuriously affected my unborn child, if he had lived, the more I became reconciled to what had at first seemed a cruel fate."[97]

In her autobiography, *A Colored Woman in a White World*, Terrell appealed to a combination of belief in a merciful providence and the material links between a mother's mental state, her physiology, and the growing baby to justify her efforts at resignation: "As I was grieving over the loss of my baby boy one day," Terrell wrote, "it occurred to me that under the circumstances it might be a blessed dispensation of Providence that his precious life was not spared. The horror and resentment felt by the mother, coupled with the bitterness which filled her soul, might have seriously affected the unborn child." How many other babies, she wondered, had been damaged when their "colored mothers" were "shocked and distracted before the birth of their babies by the news that some relative or friend had been burned alive or shot to death by a mob?" In the midst of reconciling herself to this terrible loss, Terrell committed herself to changing the world in which she, or any mother, must think such things. She joined the movement against lynching, encouraged by both Frederick Douglass and Douglass's son Lewis not to give up her public life and activism for homemaking.[98]

For Terrell, home, children, and activism became inextricably linked. Given the world her children faced, she could imagine no other way of being a mother than working to change the world for her children and those of others. As a teenager she had joined the First Congregational Church and found both solace and purpose in the kind of Christianity pursued a generation earlier by evangelical abolitionists: a Christianity that focused on the reform of society's sins. She also found inspiration in the lives of the evangelicals of the past. Soon after losing her first baby, Terrell immersed herself in preparing a lecture on Harriet Beecher Stowe. Composing that address, Terrell later wrote, was "truly a labor of love.... I poured out my soul whenever I delivered it."[99] She included Stowe's words of how at the deathbed of her beloved Charley she had learned "what a poor slave mother may feel when her child is torn away from her." And she recited Stowe's prayer: "In those depths of sorrow which seemed to me immeasurable, it was my only prayer to God that such anguish might not be suffered in vain" and that the crushing of her own heart might enable her to work to some good for others. What must these words have meant to Terrell, crushed by grief of her own? Stowe, Terrell typed as she drew her address to a close, had "in the midst of trials and afflictions, whether as a young mother weeping at the grave of a child, or as an invalid struggling with disease and poverty, or as a philanthropist persecuted and despised for defending the oppressed . . . faltered not nor wavered in the work which she felt called to perform."[100]

Like Stowe, Terrell and her husband, Robert (a Harvard graduate and the second Black American to serve as a justice of the peace in Washington, DC), believed Christians were duty bound to improve the earth in the here and now rather than wait for heaven. In the late 1890s, in the wake of their second infant's death, the Terrells helped found University Park Temple, a progressive Black Congregational church. "Searching for solace in activism and faith," writes historian Alison Parker, "they hoped to create a church that would pursue 'race development' as a central part of its religious mission."[101] Terrell also became president of the National Association of Colored Women, whose motto "Lifting as We Climb" reflected many African American reformers' philosophy of "racial uplift." This work became her refuge after the loss of her second infant: "Like other mothers who have passed through this Gethsemane, I pulled myself together as best I could and went on with my work. I had to go on with it."[102]

When a third baby died in hospital two days after its birth, Terrell "literally sank down into the very depths of despair. For months," she wrote, "I could not divert my thoughts from the tragedy, however hard I tried." She had told herself that given how careful she had been during the pregnancy, it must be fate if things did not turn out well, and of fate she could not complain. But in the end, the circumstances of her baby's death only amplified her grief: "Right after its birth the baby had been placed in an improvised incubator and I was tormented by the thought that if the genuine article had been used, its little life might have been spared. I could not help feeling that some of the methods employed in caring for my baby had caused its untimely end."[103] Terrell believed that something might have been done differently. Something might have been managed better. It was a heartbreaking thought, worsened by her deep knowledge that for so many parents of color pervasive racism in healthcare (including legal segregation) compounded the shape and scale of the distressful thought amid grief that things might have been different.

In her autobiography, Terrell told a story of how when she was in the hospital after a car accident the nurses called her by her first name. "I could scarcely believe my ears," Terrell recalled. "It was the first time I had been called 'Mary' by anybody except my family and friends since I had grown to womanhood." When she asked the nurse to treat her with the "same courtesy she would accord women of other racial groups" the superintendent of the hospital angrily replied that "there is a vast difference between white women and colored women." "So far as I have been able to ascertain," Terrell wrote, "in no hospital in Washington are colored women addressed as 'Mrs.' Or 'Miss.'"[104]

The slight may seem small to those in less formal times, but for Terrell (and throughout the Jim Crow South) this refusal to extend a common courtesy was a profound reflection of the fact that Black mothers, fathers, and their children were classified outside the bounds of the golden rule. We have witnessed many parents in these pages—Wister, Franklin, Clemens, and others—who were plagued by memories of what they or a physician might have done differently. Black mothers and fathers who lost a child had to also hold "the color line" in their memories, minds, and hearts, a line that was defended by appeals to God, science, or a powerful mix of both. Terrell knew that some wondered why, in such a world, Black Americans even had children. She recalled a white friend saying, during a conversation about segregation and lynching, that she could not see "how any colored woman can make up her mind to become a mother under existing conditions in the United States. Under the circumstances, I should think a colored woman would feel that she was perpetrating a great injustice upon any helpless infant she would bring into the world." The thought, Terrell wrote, "while I could not agree with it entirely, caused me much serious reflection."[105]

In 1940, in the midst of the Second World War, the British biologist and writer H. G. Wells was asked to contribute an introduction for the autobiography in which Terrell wrote all these memories down. After reading the manuscript, Wells decided to conclude his introduction with a reminder of the grand hopes Andrew Dickson White had imagined for the twentieth century. White had envisioned a future that would form an "ideal of religion higher than that of a life devoted to grasping and grinding and griping, with a whine for mercy at the end of it . . . an ideal of science higher than that of increasing the production of iron or cotton," and a "republic greater than anything of which we can now dream." Now Wells urged Terrell's American readers to turn over her story "of the broadening streak of violence, insult and injustice in [their] country, through which she has been compelled to live her life." "If President White came back to Cornell, and happened upon this book and began to ask questions," Wells asked, "what answer would America give to him?"[106]

By the opening decades of the twentieth century, plenty of Americans would have replied, "But look at all the children saved by science!" Du Bois and Terrell knew, however, that all too often Americans with influence harnessed "nature's laws" in support of a highly racialized, industrial-capitalist status quo. Both knew that, by the turn of the century, some Americans used evolution, which had once seemed so revolutionary, as a new means of justifying old visions of who mattered most in American society.

Why Did the Children Have to Die?

Clearly, Americans were developing very different visions of how to best achieve medical, moral, and material progress. What precisely was the best course of action once one discerned nature's ways? Although child mortality rates were trending down for some communities, in some ways that fact only heightened the urgency of the problem. Why weren't they dropping further and faster? How could the terrible, continual "sacrifice of the innocents," as a contributor to *Cosmopolitan* in 1909 wrote, be halted? What was the best solution to the fact that "six million babies" had died every ten years for the past half century? How could one change the fact that "one baby in five" still died within the first month?[107] As Americans asked these questions and imagined solutions, their answers also entailed stances on the relative roles of science, education, and economic reform, whose children mattered most, and who got to say.

With a vision of history rooted in White's tales of science triumphing over superstition and ignorance, plenty of physicians, social workers, and public health experts blamed child mortality on "the modern Herod of ignorance, superstition, and halting knowledge." Immigrant (and often Catholic) mothers came in for particular criticism, such as a "buxom Irishwoman" who admonished a visiting nurse who had arrived full of advice on how to feed and care for her latest baby, with the reply, "Say, youse can't tell me nothin' about kids. I've buried nine already."[108] Of course, as Richard Meckel points out, diagnosing the main cause of child mortality as superstition and ignorance meant the obvious solution must be more science and better education (for mothers) about the lessons taught by science.

At the end of the nineteenth century, bacteriologists were the clear heroes of this version of the best means of progress. And indeed, some of the developments coming out of bacteriology labs were extraordinary. By 1909 scientists had developed a serum for cerebrospinal meningitis that cut the death rate from ninety percent to sixteen percent, discovered antitoxins for tetanus and diphtheria, and identified the bacterial cause of dysentery (the primary killer of infants). All of these discoveries made it increasingly possible to argue that epidemics could be stopped by securing clean milk and water.

Prior generations could cite postmillennialist interpretations of the Book of Revelation, the fossil record, the abolitionist movement, or the discovery of anesthetics to justify belief in progress. Now, evidence was finally mounting that children might be snatched from death by medical science. That said, developing safe and effective vaccines or antitoxins for additional diseases proved complicated, slow, and riddled with potential missteps. Pasteur had developed a vaccine for anthrax in 1881 and rabies in 1886, but these were relatively rare diseases compared to the primary causes of child mortality. When, in 1922, the US Supreme Court upheld compulsory vaccination in schools, smallpox was still the only vaccination routinely given. Efficacious vaccines against whooping cough, diphtheria, and measles weren't developed until after the Second World War. Meanwhile, preventing infection via better sanitation, launching clean milk campaigns, and fighting for better wages (demographers still debate the relative weight that should be given to each of these factors) clearly gave meliorists plenty of work to do, whether physicians possessed life-saving medical interventions at the bedside or not.

As Elizabeth Cady Stanton and Ernestine Rose had hoped, an emphasis on the role of mothers in preventing child loss increased opportunities for meliorist activism for at least some American women. When Dr. Dorothy Reed Mendenhall, who had been trained in obstetrics at Johns Hopkins University Medical School and could thus tell that an incompetent physician who delivered her first baby hadn't known what to do, she devoted her life to maternal and infant health reform. Her lifework, she wrote, was "a compensation reaction to help me bear the bitter frustration that the deaths of my babies gave me. Helping another mother have her child safely and advising countless mothers how to care for their babies was a real outlet for my grief." When Mrs. WRD, from Illinois, contacted the Children's Bureau after losing her four-month-old son, she wrote, "My baby was sacrificed thru mere ignorance. When I had to stand by and see my baby slowly starve I made up my mind I'd fight the world but what I'd find out some way to teach people more about babies." Some women expressed anger at alternative responses. Mrs. CFH wrote the Children's Bureau in 1915: "It makes me mad to hear the preachers say 'It's God's will' 'She has fulfilled her mission,' etc. Something got to be done—and done quickly."[109] These women's demands to do something seemed all the more urgent because of recent evidence that more *could* be done.

Meanwhile, the profits amassed within industrial capitalism meant that some Americans who turned to the work of "baby-saving" had fortunes to dedicate to scientific research, educating mothers, and improving access to clean milk. The choice to devote one's fortune to halting the "sacrifice of the innocents" was almost always personal. All of the most influential philanthropists of the era had lost children to infectious disease. Heiress Elizabeth Milbank lost her son Jeremiah to diphtheria, committed her fortune to public health, and built one of the most influential philanthropic organizations in the field, the Milbank Memorial Fund.[110] Mary Harriman, whose son died of diphtheria at the age of five, bankrolled biologist Charles Davenport's Eugenics Record Office (as we'll see in the next chapter, many Americans viewed eugenics as a crucial wing of the campaign against infant and child mortality). John D. Rockefeller founded the Rockefeller Institute for Medical Research after the death of a three-year-old grandson from scarlet fever in 1901 (he had lost a daughter, too, one-year-old

Alice in 1870). And the man who is credited with saving more children's lives than any other philanthropist during this era, Nathan Straus, joined the cause following the deaths of his one-year-old daughter, Sara, in 1878, and three-year-old son, Roland, in 1884.[111]

Straus attributed his children's deaths to contaminated milk, which meant he was primed to listen when Abraham Jacobi started calling on New York's wealthy to do something about the deaths of hundreds of infants each year from so-called diseases of digestion. Straus's philanthropic campaign centered on spreading the "gospel of pasteurization."[112] He wanted to provide pasteurized milk to American mothers—*all* American mothers. Straus's work shows that neither Christians nor freethinking scientists had a monopoly on visions of earthly salvation. For much of his life, Straus identified with Reform Judaism, a diverse movement that believed in science as an important means of both intellectual and social progress. (Rabbis associated with Reform called themselves "Liberalist" or "Progressionist," while their critics were Orthodox, Conservative, or Traditionalist.) Strongly influenced by the eighteenth century Haskalah (or Jewish Enlightenment), Reform Jews were intent on adapting Jewish traditions and beliefs to modernity in a way that would, they believed, allow its most important ethical imperatives to survive into the twentieth century. Like the Protestant higher critic David Friedrich Strauss, Reform Jews believed that the foundational Jewish texts must be studied according to the scientific methods of history (i.e., by abandoning the assumption that such texts were divinely written, and by eschewing miracles as explanations of events).

Reform Jews appealed to the concept of *tikkun olam* (Hebrew for "repair of the world") as an explicit injunction to ameliorate suffering and work for social justice. They also explicitly looked to modern science as the best method and guide to what precisely must be done to improve the world. When the Reform movement tried to codify its beliefs via the Pittsburgh Platform of 1885, leading rabbis expressed a commitment to foundational texts as valuable instruments of religious and moral instruction but insisted science was the best guide to knowledge of the natural world. "We hold that the modern discoveries of scientific researches in the domain of nature and history," the Platform declared, "are not antagonistic to the doctrines of Judaism, the Bible reflecting the primitive ideas of its own age, and at times clothing its conception of divine Providence and Justice dealing with men in miraculous narratives."[113] Clearly Christianity was not the only faith intent on reconciling a progressive vision of science with belief in the wisdom of God. Nor was it the only religious tradition whose members would eventually wonder whether the costs of that reconciliation were becoming too high.

For many Reform Jews, the future of Judaism was at stake in these debates. They feared that the younger generation would decide that Jewish traditions and beliefs must be set aside in order to join modern society. That threat was heightened by the fact that liberal Protestants intent on reconciling Christianity with science tended to describe Christianity as the inevitable outcome of a progressive, evolutionary revelation. In doing so they relegated Jewish beliefs to a "primitive stage" in intellectual history. Meanwhile, agnostic and atheist versions of "warfare between science and religion" narratives implied that the only rational attitude for a modern person was unbelief. Sigmund Freud and Albert Einstein were the most famous examples of secular Jews who abandoned belief in the God of Abraham on the grounds that religion was the refuge of childlike thinking (the existence of suffering influenced both men's repudiation of belief in a personal God). The Pittsburgh Platform codified reformers' attempt to counter all of these threats by distinguishing what, in their view, was essential within Jewish beliefs and what could safely be set aside.

The reformer Dr. Emil Hirsch, rabbi of one of the oldest synagogues in the Midwest, Chicago Sinai Congregation, insisted that Jewish beliefs were even more in step with meliorist values than other faiths. "Jewish idealism is meliorism," he wrote, "it flowers in the consciousness that morality is aggressive, that the moral life means resistance to evil, conquest of evil, activity in behalf of common humanity to make life more real and the world more worthy." Indeed, Hirsch insisted that both Darwin and biblical criticism had played a central role in bolstering this meliorist vision by revealing a truer relation between God and the world. "Darwin taught us," Hirsch argued, "a truer appreciation of the function of the past as a conditioning, yet stimulating preparation for the future." "Other religions speak of a paradise lost to be regained somewhere beyond the clouds," Hirsh wrote, but "Judaism points to a future to be won here, and not by one, but all humanity.... For the Jew, religion must act as a spur," rather than a balm to soothe wounded hearts.[114] The Pittsburgh Platform emphasized both a progressive revelation (the idea that humanity's relationship with God and the moral law had been revealed gradually and progressively in history, rather than all at once by Moses) and an ethic of working to ameliorate suffering on earth.

As reformers debated how best to guide American Judaism into the future, by the 1870s Straus had the will and wealth to do something extraordinary in the here and now. Hearing Abraham Jacobi speak of how the dreaded "summer diarrhea" caused almost half of the infant deaths in New York City each year, Straus, grieving the loss of his own babies, offered to help. Later, he called Jacobi a second Abraham: "The Talmud says that the whole world believed that the souls of men were perishable, and that man had no preeminence above a beast, till Abraham came and preached the doctrine of immortality.... Out of the loins of the first Abraham sprang a great people. Out of the brain and heart of this second Abraham came the great development in all the centuries of medical history—the saving of the lives of babies."[115] It was of no matter to Straus that Abraham Jacobi was an agnostic who had abandoned the Jewish faith and believed in neither immortality nor the preeminence of man above beasts. What mattered was that Jacobi had ideas for how to save babies on earth. In 1893, advised by Jacobi, the Nathan Straus Pasteurized Milk Laboratory began establishing milk depots to provide clean milk to New York City's poor. Historians agree that the influence of Straus's pasteurized milk depots was profound. Rates of typhoid fever, strep throat, scarlet fever, diphtheria, and summer diarrhea in the city plummeted wherever mothers had access to one of Straus's milk depots. In the midst of his campaign, Straus wrote, "Humanity is my kin, to save babies is my religion." It was, he hoped, a religion that "will have thousands of followers."[116]

Straus and Jacobi proceeded despite a range of critics of their particular means of amelioration. As we will see in chapter 6, some of those critics called such work "sentimental" philanthropy and argued it would lead to "race suicide" by allowing the "unfit" to survive and have children. (This is why some of Straus's fellow philanthropists insisted their wealth would be more "scientifically" directed by funneling it to the Eugenics Record Office rather than clean milk for the urban poor.) Other critics argued that depending on wealthy philanthropists and laboratory science (including technical, scientific fixes like milk pasteurization, antitoxins, and smallpox vaccination) were just convenient means of stifling criticism of a socioeconomic system that created tremendous suffering while allowing wealthy capitalists to portray themselves as the saviors of humanity. Socialists and communists argued that capitalist philanthropists, whether they poured their wealth into pure milk campaigns or eugenics, absolved themselves of having caused great suffering. Amid increasing talk of ameliorating suffering through knowledge of and obedience to natural laws, these critics wondered whether, as historian Martin Pernick writes, "economic and social injustice could force people to live in slums and work at jobs where obedience to natural laws of physiology and hygiene was impossible."[117] After all, talk of providing fresh air, pure milk, and a sanitary home assumed that mothers could stay home, maintain dirt-free homes, regularly visit a milk station, read books about hygiene, and easily apply what they learned.[118]

Given the inability of many parents to do any of these things, a minister named Walter Rauschenbusch decided much more must be done. Recalling his experiences as minister from 1886 to 1897 for the Second German Baptist

Church in New York City, adjacent to Hell's Kitchen, Rauschenbusch wrote, "Oh, the children's funerals! They gripped my heart—that was one of the things I always went away thinking about—why did the children have to die?"[119] Friends and colleagues urged Rauschenbusch to focus on saving souls rather than social questions. But he repudiated their advice and decided that the suffering in Hell's Kitchen, suffering he was asked to both console and explain, had been caused by human beings, not God. In other words, Rauschenbusch decided that the structure of the society in which these children lived (and often died) was to blame for all the funerals. He began writing books like *Christianity and the Social Crisis* that soon sold almost as well as the Bible. The books became a movement—the "Social Gospel"—aimed at directing the resources and attention of Christians toward the amelioration of conditions on earth via social and economic change.

The radical socialist and Catholic priest Father Thomas J. Hagerty agreed but decided that only a Marxist revolution (i.e., the replacement of capitalism with communism) would save America's children. Hagerty's 1902 pamphlet *Why Physicians Should be Socialists* argued that vaccines had allowed Americans to ignore the bad housing, the lack of food, the low wages, and the disease-ridden environments experienced by the poor. He declared smallpox vaccination a palliative measure that did nothing to root out the true causes of human suffering. Hagerty firmly believed that the discovery of vaccination and antitoxin meant very little when, as he argued, "social conditions play a leading part in swelling the infant death-rate." "In spite of all the science in the world," Hagerty warned, "so long as economic inequalities exist these evils will continue; and the physician's work will be palliative rather than curative and constructive."[120]

Critics of orthodox Christianity had long since lamented what they viewed as a reform-killing emphasis on resignation to God's will in the face of suffering. Hagerty, by contrast, lamented the resignation required of physicians working among America's poor and laboring class: "What," Hagerty demanded, "can the physician do in such cases but resign himself to the inevitable? He is impotent against the industrial conditions which make for ignorance and disease." One can imagine Hagerty leaving the bedsides of children, discouraged and angry: "Of what avail is all the physician's cleverness against the pitifully insufficient nourishment, the over-worked bodies, the toil-blunted brains, the cheap, half-poisoned food, and the almost necessarily unhygienic habits of the workers in sweat shops and factories? What can he do to save the child-slaves in the glassworks and cotton-mills when their weakened tissues are in the clutch of scarlet fever, meningitis, or diphtheria?"[121] To Hagerty, the best means of ameliorating so much suffering was neither vaccinations nor antitoxins but the prevention of children's tissues from becoming "weakened" in the first place. And that kind of prevention, he was sure, required a social and economic revolution. It demanded the end of capitalism via direct, radical, revolutionary action.

Hagerty eventually gave up on the Socialist Party of America (these "slow-cialists," as he called them, were too willing to compromise with the powers that be) and helped found the more radical Industrial Workers of the World (IWW). He composed the preamble of the IWW constitution, which included the words, "The working class and the employing class have nothing in common. There can be no peace so long as hunger and want are found among millions of working people and the few, who make up the employing class, have all the good things of life." Hagerty insisted that "All the good things in life" included having children who not only *lived* but who one *wished* to live. To convey what capitalist, industrialized America asked of its mothers, he included a few lines from a poem by Adelaide A. Procter entitled "Thou Art Weary":

> Hush! I cannot bear to see thee
> Stretch thy tiny hands in vain:
> Dear, I have no bread to give thee,
> Nothing, child, to ease they pain!
> When God sent thee first to bless me,
> Proud and thankful too was I;
> Now, my darling, I thy mother
> Almost long to see thee die.[122]

Where was God, Christianity, and providence in calls for revolution? The answer varied. Marx may have called religion the opium of the masses, but Hagerty did not believe that radical social change required a repudiation of faith in God. In an essay entitled "Economic Discontent and its Remedy" he insisted it was *capitalism* that must be repudiated, not God; indeed, Hagerty believed that *only then* could faith in God remain. Hagerty was also certain that, although the revolution did not require unbelief, one must act as though man is alone: no providential, inevitable process of evolution nor miracles would bring the revolution nearer. Only direct action by human hands, he insisted, would save children from the havoc of industrial capitalism.[123]

Socialists disagreed, however, on what should be changed first, how fast, and on what grounds. A young mother named Margaret Sanger agreed with Hagerty that a revolution was needed but eventually decided that giving women the ability to control the number of their children so they could feed them, capitalism or no, was the best means of preventing so much suffering. Like Hagerty, Sanger described her work as rooted in her experience of a world in which some mothers wished their children had never been born: "How often have I stood at the bedside of a woman in childbirth," she wrote, "and seen the tears flow in gladness and heard the sigh of 'Thank God' when told that the child was born dead!" As a nurse in New York City, she learned that working class women did not have access to knowledge about birth control, but they did have "emphatic views on the crime of bringing children into the world to die of hunger. They would rather risk their lives through abortion than give birth to little ones they could not feed and care for."[124]

Given physicians' failure to provide women with knowledge about birth control, Sanger believed that medical professionals, though they claimed the bedside of mothers and children as their province, were actually complicit in high child and maternal mortality rates. A tenement case during her time as a nurse became her Rubicon: Mrs. Sachs, a "small, slight Russian Jewess," had begged the physician attending her postabortion septicemia to tell her how to prevent another pregnancy. The doctor had "laughed good-naturedly" and said her husband should sleep on the roof. Mrs. Sachs soon died from a second, self-induced abortion. Sanger left the sobbing husband and three motherless children and walked the hushed streets of New York for hours. The city's "pains and griefs crowded in upon me," she wrote, "a moving picture rolled before my eyes with photographic clearness: women writhing in travail to bring forth little babies; the babies themselves naked and hungry, wrapped in newspapers to keep them from the cold . . . white coffins, black coffins, coffins, coffins interminably passing in never-ending succession. The scenes piled one upon another on another. I could bear it no longer." She quit nursing, "finished with palliatives and superficial cures," and "resolved to seek out the root of evil, to do something to change the destiny of mothers whose miseries were vast as the sky."[125]

To see such suffering as either God's will or the "just wages" of evolutionary progress was not an option for Sanger. Although from an Irish Catholic family, her father was a freethinker who admired Ingersoll and taught his children that their duty "lay not in considering what might happen to us after death, but in doing something here and now to make the lives of other human beings more decent." Sanger joined the Socialist Party in 1912 because she agreed with her father that socialism "came nearest to carrying out what Christianity was supposed to do."[126] Meanwhile, she watched her own mother go through eighteen pregnancies in twenty-two years (eleven children survived). "My first clear impression of life," Sanger wrote, "was that large families and poverty went hand in hand." She had watched her mother grieve for a four-year-old son who died of diphtheria or croup. Then she watched her die of consumption, overwork, "and the strain of too frequent child bearing."[127]

In 1798 Malthus had argued that the law of population must be understood as a benevolent design of a wise, powerful, good God. The suffering that arose from man's natural, inevitable increase, Malthus insisted, produced exertion and thus moral and material progress. Malthus offered "moral restraint" (i.e., abstinence and late marriages) as the only acceptable means of population control. But for Sanger, moral restraint simply wasn't good enough. It placed power over reproduction entirely within the fallible hands of men. Women, she insisted, needed a safer and more rational option for controlling the size of their families than abortion or infanticide, both of which she clearly opposed.

Sanger later dedicated her autobiography to "All the Pioneers of New and Better Worlds to Come." Disparate reformers, including founders of Hagerty's Industrial Workers of the World, often gathered in Sanger and her husband's living room. "Each believed he had a key to the gates of Heaven; each was trying to convert the others," she recalled of these socialist New York salons. (This could be dangerous work during and after the First World War: two regulars, Alexander Berkman and Emma Goldman, were expelled from the country during the Red Scare of 1919–1920.) But while the men shouted for fair wages and eight-hour workdays, Sanger recalled, "I was enough of a Feminist to resent the fact that woman and her requirements were not being taken into account in reconstructing this new world about which all were talking."[128] It was concern for the children that ended strikes, she insisted: "The primary reason for the failure of all labor rebellions was the hunger cries of the babies."[129]

"No Gods, No Masters" headed each issue of Sanger's pro-birth control magazine, the *Woman Rebel*, but the old, postmillennialist belief that human agency could achieve great things on earth remained. She described her free clinics as places where "scientific information" could be distributed on the grounds that, "just as man has triumphed over Nature by the use of electricity, shipbuilding, bridges, etc. so must woman triumph over the laws which have made her a child-bearing machine."[130] Meanwhile, although Sanger herself did not talk about God's approval of birth control, her more theologically inclined supporters did. Ralph Bevan wrote in the *Birth Control Review*, for example, that "Man was started in ignorance and consequent misery. Thus only could he enjoy the satisfaction of learning, utilizing, and controlling divine laws and instincts for the overcoming of human suffering."[131]

But what precisely did the revelations of science tell Americans about right action in the face of both suffering and high child mortality? In 1906 child welfare activist Florence Kelley called upon recent triumphs at the bedside to imagine great things would be accomplished in the future. Kelley described how her own childhood and youth had been "darkened by the horror of diphtheria. That is gone, forever, from human experience." What advances, she wondered, would her own son work for and witness?[132] As an alumna of Cornell University, where her "chief delight" had been attending Andrew Dickson White's lectures, Kelley had a grand history through which to justify faith in further progress.[133] White's histories of science made sense of Kelley's experiences of the world (although from a well-to-do household, all five of her sisters died in childhood). "So long as mothers did not know that children need not die," Kelley wrote, "we strove for resignation, not intelligence. A generation ago we could only vainly mourn. Today we now know that every dying child accuses the community. For knowledge is available for keeping alive and well so nearly all, that we may justly be said to sin in the light of the new day when we let any die."[134]

Initially, Kelley combined White's vision of progress via science (she founded the Cornell Social Science Club) with the calls of Engels and Marx for radical social and economic reform to ensure that all children benefited from scientific and medical progress. Later, she tried to work within the confines of American capitalism, as, for example, when campaigning for the first federal agency devoted to child health and welfare, the US Children's Bureau (founded in 1912). In all of this work, Kelley was very well aware that working within the status quo meant a constant fight against those who declared the status quo the result of biological laws. A founding member of the NAACP, she often ridiculed the idea of some "biologically-based" white supremacy. "If the dominant class in the United States believed that the Negro race was permanently inferior to the white race," she wrote, "then why should it reinforce that inferiority by attempts at statutory segregation in cities and inferior provision for education, including inadequate instruction of even the most elementary character? Why the spread of lynching from year to year? Why all the suppression applied to citizens who number only one in ten of the population?" As someone who worked with her on various attempts at social reform, Du Bois described Kelley as set "far apart from the average social worker" on account of her consistent and close attention to *whose* children benefitted from various policies and laws. She fought hard against any attempt to place different values on America's children. She opposed the Towner-Sterling bill, which would have created a federal department of education, for example, on the grounds that it had been "cunningly drafted to perpetuate the old discrimination against public education of the Negroes in the South." (The different "value" placed on America's children was literal: the Towner-Sterling bill proposed to devote "$2.98 per capita for teachers of colored children and $10.32 for white children in the fifteen states of the South and the District of Columbia.")[135]

But having set aside calls for a Marxist revolution, Kelley had to compromise with a system that placed tight boundaries around whose children mattered most. In the midst of the campaign for the Promotion of the Welfare and Hygiene of Maternity and Infancy Act of 1921 (better known as the Sheppard-Towner Act), executive secretary of the NAACP, James Wheldon Johnson, wrote to Kelley expressing concerns that under the provisions of the act states would be free to decide how to use federal funds. "Does this mean that in the South, for example," Johnson asked, "the money is to be expended for the protection of white mothers and children with the usual neglect of colored mothers? Or are there provisions to force an equitable distribution of funds among all mothers and children regardless of color?" He posed his concerns as questions, although both he and Kelley knew the answers. Kelley replied that they best secure the funds first. If they tried to make arguments about equitable distribution to mothers and children prior to the bill's passage, she was absolutely certain that Southern states would kill the act.[136]

In their fight to save children, Kelley and Du Bois were up against very different visions of the message to be read from White's insistence that science must replace superstition and ignorance. In 1924 the president of the American Medical Association, William Allen Pusey, argued that the Sheppard-Towner Act was "an unconscious endeavor to set aside the law of natural selection and to counteract nature's cruel but salutary process of eliminating the unfit." Scientific physicians, Pusey insisted, must surely oppose the legislation.[137] (As for the Children's Bureau, that, critics argued, was quite clearly a communist front, like child labor laws, "directly backed by Moscow.")[138] Indeed, to fight against Kelley's and Du Bois's visions of progress, some American meliorists decided (once again) that trying to save all children was a quite unscientific, sentimental, and superstitious thing to do.

Chapter 6

THE PROMISE AND PERILS OF SALVATION

BY THE END OF THE nineteenth century, the tradition of science brought to American shores by the likes of Mather and Franklin was finally saving children *after* they fell ill. Physicians could halt diphtheria, the "strangling angel," with a timely dose of antitoxin. Meanwhile, antisepsis procedures (including the introduction of rubber gloves at Johns Hopkins University in the 1890s) during surgery reduced infection rates. That and improved surgical techniques meant surgery had a higher chance of saving a child's life than before, although it was still dangerous for patients of any age to go under the knife (the development of antibiotics, the ultimate vindication of experimental medicine, was still nearly a generation away).

These developments clearly arose from the mechanistic, experimentalist study of nature. They were the first active (as opposed to preventative) vindication of faith that a commitment to the uniformity of nature's laws might ameliorate suffering. Ironically, however, increasing evidence that science, in combination with sanitation, was finally lowering child mortality rates inspired anxiety among some Americans that medicine might be intervening too much in nature's "benevolent" ways. Indeed, some confidently appealed to faith in progress via a law-bound "struggle for existence" to declare either that certain children need not be saved or that they not be born in the first place. This ironic trajectory within American science and medicine is critical to explaining an increasing ambivalence toward, and for some outright rebellion against, faith in science as the best route toward the amelioration of human suffering.

NATURE'S BLUNDERS

On November 12, 1915, Anna Bollinger gave birth to a boy at the German-American Hospital in Chicago. Upon examining the newborn, Chief Surgeon

Harry J. Haiselden observed various abnormalities: the baby had no neck, one ear, deformities of the shoulder and chest, and his pupils reacted slowly to light. After taking X-rays, Haiselden added to the list: premature hardening of the skull and leg bones, and a membrane blocking the lower bowel. Finally, the anus was imperforate, for tissue blocked the passage a half inch from the surface. This was the same "broken part" that had killed Cotton Mather's son and which Mather, unwilling to attribute such a terrible mechanical breakdown to God, had blamed on witchcraft.

Thanks to the technological wonder of X-rays, the study of anatomy, the discovery of anesthetics and the development of antiseptic surgery, Haiselden could have fixed the Bollinger baby's "mismade" digestive tract. Advances in science and medicine had made it possible for him to repair the imperforate anus, so that the baby could live. But Haiselden decided that to do so would be wrong. He requested Mr. and Mrs. Bollinger's permission not to perform the surgery and, having secured their consent, let the baby die. He then defended his decision in public, first in print and then in the form of a silent film entitled *The Black Stork*, by insisting that surgeons who refused to operate on such cases prevented great suffering. He had let several other "malformed" children die, he said, and other commentators admitted it was a common practice in hospitals.[1]

As news of Haiselden's inaction spread, influential meliorists and reformers split on whether he and other physicians were right to withhold lifesaving care from some children. The famous founder of Hull House (a settlement house in Chicago), Jane Addams, insisted that "under no circumstances has any human being the right to pass judgment of death for unfitness on any other human being." Catholic physician James Joseph Walsh argued that "physicians are educated to care for the health of their patients, but so far at least as I know we have no courses in our medical colleges as yet which teach how to judge when a patient's life may be of no service to the community so as to let him or her die properly. . . . Some of us know how fallacious our judgments are even with regard to the few things we know."[2] Pediatrician Abraham Jacobi agreed: "The mission of doctors," Jacobi commented, "is not to destroy life but to preserve life." Indeed, Jacobi believed that those who advocated allowing babies to die in order to relieve suffering must be speaking of someone else's suffering than the patients.[3] Others, however, came to Haiselden's defense. After she noticed the story of her own life being used to argue against Haiselden's action, Helen Keller issued a statement in which she praised him for performing a service to society and sparing "the hopeless being" from a "life of misery." She urged Americans to choose "between a fine humanity like Dr. Haiselden's and a cowardly sentimentalism."[4]

While at first glance, fixing Baby Bollinger's broken parts may have seemed the logical action amid long-standing promises of the progress of science and technology, Haiselden absolved himself of the need to do so by appealing to both God's will and nature's laws. In other words, he drew upon science as a new revelation of what, precisely, enlightened human sympathy demanded. Audiences of *The Black Stork* read the doctor's reasoning between the scenes of the silent film: "It is his will that the child die. Shall I set myself up as wiser than the Almighty? God does not want that child to live." Historian Martin Pernick notes that as a lifelong Methodist, Haiselden viewed science as the foundation of a truer religion in which nature's laws provided ethical guidance. Haiselden's rhetoric, notes Pernick, "reflected the survival of nineteenth-century natural theology, the belief that the laws discovered in nature and the laws revealed in religion each manifested the goodness and authority of their common creator."[5]

Of course, many atheists and agnostics also declared nature a good guide to whose children mattered most, while setting aside God as the designer of natural processes. Nature's ways were ultimately benevolent, in this view, because they had led to evolutionary progress. The director of the Carnegie Institution's Eugenics Record Office, biologist Charles Davenport, weighed in on the Haiselden case by insisting that the lesson in nature's laws and the means of evolutionary progress were clear: "Shortsighted are they who would unduly restrict the operation of what is one of Nature's greatest racial blessings—death." As commentaries on Haiselden's decision flooded newspapers, some turned the fact that his decision was supposedly rooted in reason and science rather than "sentiment" into a matter of national salvation. Haiselden, wrote one supporter, "performed the highest moral duty in permitting the natural death of this child," while the mother served "as an example of true action in parental duty, even patriotism" by preventing suffering in both the near and distant future.[6]

As ministers, activists, physicians, public health experts, biologists, and prominent public figures debated whether Haiselden was right to withhold lifesaving surgery from Baby Bollinger's "mismade machine" in the name of some higher, "eugenic" good, very little was recorded about what Anna Bollinger and her husband thought. We know they had three other children, and we know they named the little boy who died Allan. In news accounts, Anna was quoted as saying, "It is one of nature's blunders, and I am willing that nature should correct its error by my baby's death." The *Texas Medical Journal* offered the following commentary on Anna's words: "These were the words uttered by a mother of a hopelessly defective baby whose life might have been saved by an operation, and they mark the change in our regard for human sympathy, in our deeper understanding of social responsibility, and in eugenic progress."[7] We do have one hint from the Bollinger baby story that the cost, to some, of these new forms of both sacrifice and resignation in the name of "the greater good" were high indeed. On July 27, 1918, a brief note appeared in the *Chicago Examiner* that Anna Bollinger had died "broken-hearted." The paper noted that ever since her baby's death "she had declined in health. Her husband declared that grief had killed her."[8]

The debate over Haiselden's actions shows that even as scientific advances (finally) allowed surgeons to save an increasing number of children, some cited evolution as evidence that these newfound powers might actually interfere with nature's progressive ways. In 1883 Darwin's cousin, Francis Galton, had coined the word *eugenics* to capture the possibility of guiding human populations toward further progress now that the principles of natural selection had been discovered. The word *eugenics* wasn't needed for the ideas Galton defended to exist. Prior to reading Galton's new word for "the science of improving stock," Henry Ward Beecher had asked, "Do you not believe that in the coming time there is to be such knowledge of heredity as shall lead men to wise selections, and that the world which has learned how to breed sheep for better wool, horses for better speed, and oxen for better beef, will not by and by have it dawn on their minds that it is worth while to breed better men too?" Such a vision fit with Beecher's belief, which he justified by appealing to evolution, the fossil record, and recent human history, that a commitment to "the new heaven and the new earth" required only that one "take the larger view" and recognize the progress produced by "the gradual struggle which has been running on through thousands of years."[9]

The relationship between ameliorating individual suffering and the prospect of future evolutionary progress of the "human race" as a whole was up for debate early on. Darwin himself had expressed ambivalence regarding both the possibility and rightness of guiding human evolution by judging whose children should be saved. He was not, for example, sure how to assess certain kinds of medical and social progress, given his belief that "man's present rank" had arisen from a struggle for existence. Writing on the question of whether natural selection still operated within "civilized" societies, Darwin wrote, "We build asylums for the imbecile, the maimed, and the sick; we institute poor-laws; and our medical men exert their utmost skill to save the life of every one to the last moment. There is reason to believe that vaccination has preserved thousands, who from a weak constitution would formerly have succumbed to small-pox. Thus the weak members of civilised societies propagate their kind." No one, he wrote, who had attended to the breeding of domestic animals "will doubt that this must be highly injurious to the race of man." This looks like a defense of eugenics. But Darwin had quickly added that, given that humanity's moral sense depends on the instinct of sympathy, refusing "the aid which we feel impelled to give to the helpless," even "if so urged by hard reason," could not be done without "deterioration in the noblest part of our nature." Ultimately, Darwin assumed that commitments to the higher values of sympathy and individual liberty would prevent the application of the "principles of the barnyard" to human beings.[10]

Darwin was willing to consider the possibility, once heredity was better understood, that man "might by selection do something not only for the bodily constitution and frame of his offspring, but for their intellectual and moral qualities." Both sexes, he wrote as an example, "ought to refrain from marriage if they are in any marked degree inferior in body or mind."[11] But he explicitly repudiated the idea that physicians should be making decisions about whose children should be allowed to live and whose children should be allowed to die. As William Jennings Bryan noted during his campaign against teaching evolution in American schools, some of Darwin's followers were "more hardened."[12]

In a society quite willing to rank children according to levels of "fitness," the supposed dilemmas that might accompany medical progress had been hinted at before much could actually be done to save children. Oliver Wendell Holmes, who trained white American doctors for four decades at Harvard, noted in a lecture in 1860, "Hydrocephalus, tabes mesenterica, and other similar maladies, are natural agencies which cut off the children of races that are sinking below the decent minimum which nature has established as the condition of viability, before they reach the age of reproduction. They are really not so much diseases, as manifestations of congenital incapacity for life; the race would be ruined if art (surgery) could ever learn always to preserve the individuals subject to them."[13] (Tabes mesenterica is an old term for tuberculosis that infects the lymph glands.) Note that Holmes's claim depended not only on belief that race progress occurred via the elimination of the "unfit," but confidence that a physician could say, first, what constitutes fitness and, second, which diseases were hereditary.

The point may have seemed less pressing at the time because "art" (i.e., medical knowledge and technology) could not in fact do very much. But as surgeons learned how to fix "broken parts" of the "infant machine" and physicians concluded that sanitation, vaccinations, and antitoxins were finally saving children's lives, debates over whether scientific interventions might increase suffering (generally via some kind of degeneration of "the race") in the long run commenced. In a popular book entitled *The Spiritual Interpretation of Nature*, Scottish obstetrician James Young Simpson, whom we met earlier citing the discovery of anesthetics as evidence of progress, commented on the decreasing death rates in England and Wales as follows: "The reduction was mainly due, however, to a conspicuous fall in the infantile death-rate. Under the improved conditions it is the less fit forms that benefit more than the fit," reducing "racial efficiency." Simpson appealed to the animal world as evidence that death at a young age was necessary to maintain species at an average strength. While he insisted that he did not support withholding humanitarian effort to individuals, surely, he argued, the "unfit" thus saved should not be allowed to reproduce. "The very humanitarianism of some of the tendencies embodied in present-day legislation," Simpson concluded, "conceals very real dangers to racial progress. The probability is that while immediately reducing suffering, they will in reality increase it for the generations to come."[14]

The eugenics movement isn't often described as part of the fight against child mortality or suffering. The facts that, first, the movement depended on those in power judging the relative worth of different human beings and, second, eugenic thinking ultimately caused tremendous suffering make it difficult to imagine that the *amelioration* of suffering may have played a role in motivating proponents. After all, eugenicists presumed to know what constituted suffering (for all time and for everyone) in ways that today inspire quick dismissal for their often racist, ableist, and classist assumptions. But placing eugenics within the context of the effort to prevent suffering via science illuminates one of the reasons the movement had such broad appeal. For many, eugenics seemed a logical and benevolent addition to a whole gamut of science-based reforms aimed at lowering infant and child mortality, from educating women about breastfeeding to providing pure milk and better ventilation in tenements.

Eugenics fit particularly well within Andrew Dickson White's vision of the history of science as a story of the triumph of reason (here represented by the sciences of evolution and genetics) over superstition. After all, that history was premised on the assumption that new, scientific knowledge must be harnessed for progress and amelioration. As the applied wing of the science of heredity, eugenics thus fit long-standing American narratives of reducing suffering via science. As one of White's most prominent students, the biologist and president of Stanford University, David Starr Jordan, explained, while genetics "is the science of birth, development, and heredity among living organisms," eugenics is "the science and art which treats of conditions under which a human being may be well born."[15] To its defenders, eugenics was simply the logical outcome of both an evolutionary understanding of human beings and a commitment to progress. (This does not mean that proponents agreed on which policies were truly "eugenic.") Indeed, faith in scientific expertise as the best means of progress and reform, anxiety regarding the implications of lowering infant mortality rates for the health of "the race," and biologists' interest in doing their part to "save the world" (especially after the war) inspired colleges across the nation (350 by 1928) to offer courses on eugenics. Such courses were viewed as a requisite part of up-to-date biology curricula well into the 1940s.[16]

Of course, proponents of eugenics also assumed that those in power (or, for socialists, those whom they wished were in power) could rightly judge what constitutes suffering, who should say, and whose future children mattered most. Such thinking was premised on the assumption that science could sort children according to fitness and that lack of fitness increased pain and suffering (for individuals, parents, society, and the future). Galton, for example, turned Darwin's warning about the danger to human sympathy (if the principles of the barnyard were applied to human beings) on its head by positing eugenics as a kinder, more sympathetic replacement of nature's ways: "Man is gifted with pity and other kindly feelings; he has also the power of preventing many kinds of suffering," Galton wrote. "I conceive it to fall well within his province to replace Natural Selection by other processes that are more merciful and not less effective. This is precisely the aim of Eugenics."[17]

The fact that some feared that the grand campaigns for saving babies might easily become *dys*genic and therefore increase suffering over time (if, that is, such campaigns were not accompanied by more eugenic mate choice) is evident in caveats expressed by Galton's protégé, the British statistician Karl Pearson. In putting forward his own conclusion that due to natural selection "a heavy mortality leaves behind it a stronger population," Pearson felt obliged to point out that "to assert the existence of this selection and measure its intensity must be distinguished from advocacy of a high infant mortality as a factor in racial efficiency. This reminder is the more needful as there are not wanting those who assert that demonstrating the existence of natural selection in man is identical to decrying all efforts to reduce the infantile deathrate."[18] The influential California-eugenicist (and founder of marriage counseling) Paul Popenoe repeated Pearson's warning in his popular textbook *Applied Eugenics*. But Popenoe also insisted that if "save the babies" campaigns were unaccompanied by, at the very least, wiser mate choice, disaster would surely result.

Popenoe was the editor of the *Journal of Heredity*, the main publication of the American Genetic Association, and brought the authority of the science of genetics to his claims. "If the infant mortality problem is to be solved on the basis of knowledge and reason," Popenoe wrote, "it must be recognized that sanitation and hygiene can not take the place of eugenics any more than eugenics can dispense with sanitation and hygiene. It must be recognized that the death-rate in childhood is largely selective, and that the most effective way to cut it down is to endow the children with better constitutions. This can not be done solely by any euthenic campaign; it can not be done by swatting the fly, abolishing the midwife, sterilizing the milk, nor by any of the other panaceas sometimes proposed." (Euthenics was the study of how to improve the environments in which children were raised. Some saw euthenics as working in concert with eugenics; others saw euthenics as an alternative to eugenics.)[19]

Popenoe believed that "if the baby who would otherwise have died in the first months is brought to adult life and reproduction, it means in many cases the dissemination of another strain of weak heredity, which natural selection would have cut off ruthlessly in the interests of race betterment. In so far, then, as the infant mortality movement is not futile it is, from a strictly biological viewpoint, often detrimental to the future of the race." Popenoe gave a "unqualifiedly, no!" reply to the questions, "Do we then discourage all attempts to save the babies? Do we leave them all to natural selection? Do we adopt the 'better dead' gospel?" And he repeated Darwin's stance that the sacrifice of "the finer human feelings" that accompanied such a course "would be a greater loss to the race than is the eugenic loss from the perpetuation of weak strains of heredity."[20] But clearly some readers might calculate the costs to "finer human feelings" differently when deciding on what constituted right eugenic action.

Indeed, some confidently translated fears that lower infant and child mortality rates might be "dysgenic" into apocalyptic visions of the future. Science writer Albert Wiggam warned, "You raise great milk funds first, for feeding babies born to lives of feebleness, second, born from mothers too weak by nature to suckle their own offspring, and third, from parents one or both of whom are too feeble mentally to provide food for their children." Yes, Wiggam wrote, "brute nature slays its thousands, but in the end your hand-to-mouth charity will slay its tens of thousands" if not accompanied by eugenics.[21] He was certain that unless Americans learned to have faith in "natural salvation" rather than supernatural salvation, a "biological hell" would result. "These warnings," he wrote, "at first should make you tremble, they should secondly make you pray, and they should thirdly fill you with the militant faith of a new evangel."[22]

Appeals to Americans' faith in the progressive revelation of God's laws peppered Wiggam's vision of what both politicians and the public must believe and do in the face of the "warnings of the biologist." He insisted that, just as Christ had added the golden rule and the Sermon on the Mount to the Ten Commandments, "in our day, instead of using tables of stone, burning bushes, prophecies and dreams to reveal His will, He has given men the microscope, the spectroscope, the telescope, the chemist's test tube and the statistician's curve in order to enable men to make their own revelations."[23] According to Wiggam, eugenics represented "a method ordained of God and seated in natural law for securing better parents for our children, in order that they may be born more richly endowed, mentally, morally, and physically for the human struggle."[24]

Eugenic salvation schemes like Wiggam's completely altered the meaning to be found in the mortality rates of America's children. In this vision of progress, children died not because God wished to direct parents' attention to heaven, or because God wished to inspire human effort to improve medicine, but because those children were unfit for survival. "The weaker members of primitive New England," wrote Marshall Dawson in the book *Nineteenth Century Evolution and After*, "that were in the main swept away by Nature, uninterfered with by either preventative medicine or modern plumbing, while the selected stock which survived propagated large families."[25] Franklin's little "infant machines" died as a result solely of nature's great, impersonal, yet benevolent laws. Some declared those laws designed by God. Some did not. But clearly eugenicists thought they could find guidance regarding what must be done or not done in how (they thought) nature operated.

Some, including a contributor to the *American Journal of Public Health*, pointed out that Darwin's proof that "the law of natural selection operates in man just the same as in the lower animal and vegetable kingdoms," did not thereby prove that "a high infant mortality is favored as a means of improving the race."[26] But, having adopted evolutionary explanations of the origin and meaning of suffering, many American biologists, pediatricians, and public health experts held that campaigns to lower infant mortality must be accompanied by "better breeding." In doing so they embraced warnings that, as Wiggam wrote, if man cured tuberculosis, insanity, pneumonia, and other diseases and did "nothing else [they would] again wreck the very race [they] have saved."[27]

What eugenicists meant by the term *race* varied. When American Museum of Natural History trustee Madison Grant entitled a book *The Passing of the Great Race* (1916) or Lothrop Stoddard published *The Rising Tide of Color: The Threat Against White World-Supremacy* (1920), it was very clear that by "race" Grant and Stoddard meant Europeans. When Harvard geneticist Edward M. East wrote in *Mankind at the Crossroads* (1923) that the main point at issue with eugenics was "who shall inherit the earth," he tried to console his white readers by insisting that Stoddard's predictions of "colored" populations increasing so much faster than whites were wrong.[28] University of Wisconsin sociologist Edward Alsworth Ross (from whom President Theodore Roosevelt learned the phrase "race suicide") warned that Europe's colonies would soon be "filled with the children of the brown and yellow races" if white Europeans didn't fill them with their own children.[29] Historians have demonstrated how these racist versions of eugenics culminated in federal anti-immigration laws, various state anti-miscegenation laws, and the legalized, compulsory sterilization of women of color in states like California and North Carolina.

Other proponents of "race hygiene" meant the "human race" as a whole. These eugenicists described the proper goal of eugenics to be the reduction of suffering, disease, and mortality for all humanity. Family physician W. Grant Hague, for example, described the primary aim of eugenics to be the lowering of infant and child mortality rates of all communities. He bemoaned the terrible results of man's inability to prevent loss of life, including "the wasted years, the hopeless prayers, and the anguish of those who fight a battle predestined to failure." Hague's 1913 guidebook, *The Eugenic Mother and Baby*, included sanitation, pure food and milk programs, and the Food and Drug Administration within the purview of eugenic reform. But he also argued that a key means of preventing both mortality and suffering was assuring that the right babies were born. For Hague, that meant that both science and public health must be accompanied by better, scientific mate choice to avoid the propagation of hereditary disease.[30]

As eugenic thinking garnered funding and support (pervading state fairs, biology curricula and textbooks, and medical fields), clearly what proponents meant by the term *eugenics*, and the policies recommended, varied a great deal. Meanwhile, Hague's book is a good example of how eugenic thinking pervaded the new fields of both public health and pediatric medicine. In 1911 the American Association for the Study and Prevention of Infant Mortality thought it perfectly consistent and right to adopt a resolution urging states to prevent the "unfit" from breeding. Pediatricians like Luther Emmett Holt, who served as president of the Association, wrote of unwise mating as a cause of child mortality alongside passionate laments regarding how many healthy babies died of disease. In this world, it made perfect sense for a notice (in the 1913 issue of the *American Journal of Public Health*) of a book on Straus's campaign for pure milk to be immediately followed by a review of *Eugenics in the United States*. That review concluded that "Galton's dream that eugenics will become the religion of the future is finding realization in America." "May his prophecy come true!" commented William L. Holt, who authored the review.[31]

Hague explicitly tied eugenics to Americans' long tradition of recovering God's goodness by explaining suffering as rooted in humanity's ignorance of nature's laws. In the face of so much suffering, Hague wrote, individuals doubted "the justice of the Omnipotent Mind who created us and left us seemingly alone—derelicts in the eddies of eternity." For Hague, such doubt entailed an error of scale and perspective. "The truth," he urged, "is that the scheme of the universe is unalterable, we are but part of the whole and must share in the evolution of the process." In adjusting to natural law, in accepting "that the fit should survive is the genetic law of nature," eugenics offered a way—"the only way"—to "cure the ills of the world." (Hague's list of individuals who, in the interest of eugenics, should not mate included "those who are deaf, dumb, blind, epileptic, feeble-minded, insane, criminal, consumptive, cancerous, haemophilic, syphilitic, or drunkards, and those known to be victims of disease of any other special type.") Eugenics would, Hague insisted, be the final culmination of the "patient progress" illustrated by history and culminating in a true "eugenic aftertime."[32]

Primed by long-standing, postmillennialist visions of earthly progress, eugenic thinking thrived across the political spectrum and (almost) the entire gamut of stances on religion. Americans on (almost) all sides of debates over social reform agreed that the "science of better breeding" had important things to say about how American society should be organized and different children viewed.[33] Wiggam, for example, thought capitalism did a fine job of sorting out the "fit" and "unfit," and explicitly used eugenics to argue against socialist reforms. Indeed, in 1924 the British writer John Langdon-Davies identified prominent American eugenicists' real target to be socialism. Eugenicists, Langdon-Davies argued, used biology as a "scientific figleaf" to whitewash race and class prejudice and justify ignoring existing social problems and an unjust environment.[34] By contrast, socialists, while they argued against eugenics under a capitalist system, expressed confidence that within socialist societies women could choose their mates more wisely and thus more eugenic babies would be born.[35]

Unambiguous opposition to eugenic thinking tended to be expressed primarily by physicians and social reformers who were Catholic. Walsh, the Catholic physician who criticized White's "warfare between science and theology" tales, warned, for example, that biologists often forgot "how much harm they may do while all the time meaning to do so much good." He urged that "we have an immense number of social ills that are much more responsible for criminality, tendencies to insanity and the birth of weakling children than heredity is. These social ills need to be reformed. In correcting these we are applying the principles of justice and not violating human rights. We have no nice off-hand cures for social, any more than physical evils. What we need is co-operation in the use of all the natural means at hand for the alleviation of social complaints, not a ready-made panacea for social ills that will divert attention from the real problem."[36] In his famous 1922 critique, *Eugenics and Other Evils*, the English writer and philosopher G. K. Chesterton called into question the assumption that anyone could safely judge either the worth of another human being or what should be done with those deemed worth "less." Whom, exactly, Chesterton demanded, did eugenicists trust "when they say this or that ought to be done?" The only answer he could determine from eugenicists' writings was that "the individual Eugenist means himself, and nobody else."[37]

But usually even critics of the more racist and classist brands of eugenics found it difficult to repudiate eugenic thinking entirely. The Columbia University anthropologist Franz Boas was adamant that most eugenicists shirked their first duty of determining empirically and without bias whether a particular trait of interest—say, poverty—was hereditary or not. But Boas conceded that "the humanitarian idea of the conquest of suffering" made eugenics attractive. He then defined the "proper field of eugenics" as "the attempt to suppress those defective classes whose deficiencies can be proved by rigid methods to be due to hereditary causes, and to prevent unions that will unavoidably lead to the birth of disease-stricken progeny."[38] In other words, Boas expressed agreement with the assumption that genetics, as the science of natural laws governing heredity, could and should prevent certain kinds of suffering *if* the science was sound.

Americans' long tradition of looking to science as the means of reducing suffering explains why eugenic thinking was (and continues to be) so hard to give up, even for those trying to create more inclusive visions of American progress. Margaret Sanger was clearly wary of the middle class, anti-socialist bias of many American eugenicists. Too often, she believed, proponents of "better breeding" ignored the need for better environments and attributed all social problems to biology. She pointed out the uselessness of the terms "fit" and "unfit," for who would decide who fell into each category? But she also knew eugenicists were the only ones openly discussing contraception. Granted, influential biologists like Charles Davenport mentioned birth control only to ridicule it: Davenport despised campaigns for "voluntary" birth control on the ground that "the less thrifty and the less foreseeing—the proletariat, or proliferators" wouldn't use it, and the "fit" (which he equated to white, Protestant, and middle-class) would.[39] Sanger countered such fears by insisting birth control was a necessary tool for truly ameliorative eugenics.[40]

W. E. B. Du Bois, who supported Sanger's effort to expand Black women's access to birth control, also appealed to eugenic ideas, especially when countering the racist eugenics of Madison Grant and Lothrop Stoddard.[41] Du Bois pointed out, for example, that "unfit" and "inferior" individuals existed in every race. No race, Du Bois argued, had a monopoly on fitness or superiority. Similarly, the Black (and Catholic) biologist Dr. Thomas Wyatt Turner, who taught eugenics at Howard University, argued that the "best blacks" were every bit as "fit" as the "best whites."[42] Present-day readers are often surprised to learn that articles espousing eugenic ideas and goals appeared in the NAACP's publication, the *Crisis*, throughout the 1920s. But it is crucial to note that, given the tight relationship between eugenics and public health, giving Black families access to the scientific knowledge required to produce and raise "better babies" was a matter of justice, akin to ensuring access to diphtheria antitoxin, clean water, and pure milk.

Of course, Du Bois was very aware that science could be a dangerous ally when pursued in a society that racialized human beings. He knew that "biological laws" were often used to justify ignoring the mortality of some children. As a result, he faced the constant challenge of embracing science in the name of progress while repudiating its use to justify oppression. Du Bois eventually abandoned his early faith that science provided the best means of fighting suffering, at least as practiced in the United States. Later in life, he decided that only extraordinary social change would ensure that scientific and medical progress reached everyone and increasingly looked to political action and a redistribution of economic and social power as the most effective means of ensuring that science could one day save *all* children.[43]

Du Bois was not alone in his rebellion against claims that science provided the best means of American progress. As we will see, the grounds on which many white Americans began to question the power of both science and scientists were quite different from those that inspired Du Bois. But as the twentieth century dawned clearly many Americans were questioning the visions of progress to which the mechanistic sciences had become so tightly linked.

The Costs of Scientific Consolation

In February 1914 Christian evangelist Billy Sunday delivered a fiery sermon in which he denounced the earth-centered, scientific salvation schemes that had developed by the twentieth century. People were dissatisfied with science as "panaceas for their heartaches," he declared. He placed the new visions offered by transcendentalists, evolutionists, modernist Christians, eugenicists, and geologists before his listeners: "Go to that dying man. Tell him to pluck up courage for the future. Use your transcendental phraseology upon him. Tell him he ought to be confident in the great- to-be, the everlasting now and the eternal what-is-it and where-is-it! Go to that widowed soul! Tell her it was a geological necessity that her husband died, just as in the course of evolution it was necessary that the megatherium had to pass out of existence!"

Sunday then challenged scientists to take their "scientific consolation" into a room where a mother had lost her child: "Try your doctrine of the survival of the fittest. Tell her that her child died because it was not worth as much as the other one!" His message to those offering various versions of "scientific consolations" was firm: "When you have gotten through with your scientific, philosophical, psychological, eugenic, social service, evolution, protoplasm and fortuitous concourse of atoms, if she is not crazed by it, I will go to her and, after one-half hour of prayer and the reading of the Scripture promises, the tears will be wiped away and the house from cellar to garret will be filled with calmness like a California sunset!"[44]

Ingersoll may have insisted orthodox Christians must be mad to accept a doctrine in which one could never know whether a child was destined for heaven or hell. But for Sunday, belief that the world arose from some "fortuitous concourse of atoms" (i.e., solely via chance) was far worse. Sunday countered that in the face of tremendous affliction only a return to the fundamentals of Christianity could prevent madness, save souls, and heal the world. Sunday was not alone. In the opening decades of the twentieth century "fundamentalist" ministers urged Americans to turn away from the nineteenth century's compromises with science and return to the "fundamentals of the faith": faith that the Bible is the inspired word of God and authoritative throughout, belief in the miraculous virgin birth of Christ and Christ's substitutionary atonement for human sin on the cross, faith in the bodily resurrection of Jesus, belief in rewards and punishment in the hereafter, and faith in the promise of Christ's second coming. Even if this meant a return to belief in the "ever-lasting conscious suffering of the lost."[45]

In deciding that Americans had taken meliorism either in the wrong direction or too far, fundamentalists repudiated postmillennialists' stance that suffering arose entirely from human ignorance. Instead, fundamentalists embraced a strongly premillennialist interpretation of the Book of Revelation, the final book of the New Testament. They believed that Christ's second coming and one-thousand-year reign would occur after a time of *increased* suffering. This interpretation in turn depended on a more literal reading of the Old Testament, especially the fall of man, as an explanation of the origin of suffering in the first place.

Fundamentalists found evidence of their vision of the past, present, and future in the suffering caused by the war of 1914–1918, the devastating influenza pandemic of 1918, the Russian Revolution of 1917, and the Jazz Age collapse of Victorian values. As historian Paul Conkin notes, to fundamentalists "the great heresy tied to postmillennialism was an impious expectation that humans could build the kingdom by their own efforts."[46] Premillennialism did not mean believers either ignored human suffering, repudiated medicine, or accepted disease as "God's will." One must still follow Christ's example of aiding those in need. (To take just two examples of how medical progress could take place within premillennialist contexts, the first successful infant heart transplant and the development of proton therapy for cancer were developed by strongly premillennialist, Seventh Day Adventist physicians at Loma Linda University in California.) Rather, premillennialists believed that the scale at which progress might be imagined must be adjusted. They focused on ameliorating individual suffering rather than imagining that human beings might shift the amount of suffering overall downward through purely human efforts. In doing so, they repudiated postmillennialist visions (whether those of modernist Christians, deists, or unbelievers) of slowly decreasing suffering at a population scale over long swaths of time. Premillennialists believed that, in the process of passing so much agency to human beings and emphasizing grand progress on earth, Christianity's true goal—winning souls for heaven—had been lost. They decided, in other words, that the cost of seeing God's governance as solely evident in nature's laws had become too high. And the returns within the mustard gas–filled trenches of Europe woefully inadequate.

Americans who decided that driving natural laws into the deep past was costing too much found a champion in former secretary of state William Jennings Bryan. In some ways Bryan was a strange hero for those who declared grand visions of progress via human action misguided and impious. Biographer Michael Kazin notes that Bryan's personality and politics "made him a tacit ally of the liberal optimists for whom Jesus was a benevolent figure and hell an anachronistic abstraction."[47] Indeed, Bryan never took a clear stand on premillennialism versus postmillennialism. He once insisted that more attention must be focused on getting people to believe in the first coming than spending a great deal of time talking about the second. (Historian Edward Larson described Bryan's views as a "distinctive combination of left-wing politics and right-wing religion.")[48] When critics said Christianity led to madness because of the prospect of damnation, it was not a fair characterization of Bryan's beliefs. Like the Beecher children, he talked a lot about God's love, the potential goodness of human nature (rather than human depravity), and the possibility of creating a better earth. He supported economic reform, women's suffrage, pacifism, missions, and prohibition. Bryan firmly believed, however, that success in all these things depended on belief in God's close, personal governance. He also was certain that the previous century's increasing commitment to the absolute uniformity of nature's laws (whether via higher criticism, the theory of evolution, or both) inevitably led to atheism.

It is important to note that neither Sunday nor Bryan repudiated medical science or social reform. Sunday ridiculed American Christians who went to church on Sunday, paid starvation wages, and employed children in their factories. They made donations to hospitals from wealth massed through "child-labor which crushes and kills and maims more children in one year than the hospital will heal in twenty." He once admonished his audiences for taking medicine for granted: "Did you ever thank God," he demanded, "for the doctors and nurses and hospitals? For the surgeon who comes with scalpel to save your life or relieve your sufferings?"[49] Bryan accepted even more of scientists' claims than Sunday. He accepted an old earth, for example, and, in arguing that the universe cannot be the result of chance, wrote of the "reign of law, universal and eternal" which "compels belief in a Law Giver." He, too, believed in the good certain kinds of science might do. "What of vaccination and the labours of Pasteur?" he asked, "Who will estimate the value of the service rendered by the man who gave us a remedy for typhoid?"[50]

Thus, this was not a debate over whether progress was possible. Rather, it was a debate over what beliefs provided the best foundation of effective progress. Indeed, well aware that these men were also trying to change the world, Wiggam wrote that he thought it "passing strange" that Bryan and Sunday had not lent their immense power to eugenics.[51] Ultimately, Wiggam decided that belief in heaven undermined progress on earth. Promise of heavenly salvation, he wrote, demanded men "grunt and sweat under a weary load of life" in the hopes another world would right the ills of this one, while science, by contrast, was lighting the world with a different faith—namely, "that this world, too, can be made clean and sane and happy."[52] Wiggam conceded that the new eugenic religion was a painful philosophy for unprepared minds, for it offered no everlasting resting place for the individual: no personal God, no immortality, no heaven. A man must be content, he urged, with only the solace of a good fight and hope that individual suffering (of others) would be reduced in the future.

The most famous "modernist" minister in the country, Reverend Shailer Mathews of Chicago, pressed back against caricatures of Christianity like Wiggam's. "Only the historically illiterate," Mathews wrote, "can think of Christianity as being concerned solely with *post mortem* salvation."[53] Clearly, however, Bryan and Sunday differed from both Mathews and Wiggam regarding how ameliorative action should be understood relative to God's governance. Mathews argued that true amelioration via the study of natural laws demanded that strict uniformity be assumed *throughout time*, both in the interest of consistency and a true understanding of human nature. Bryan, by contrast, believed that both a correct understanding of history and the best foundation for progress depended on faith in the miraculous creation of human beings.

Despite their differences, Sunday and Bryan shared a deep belief in the existence of a benevolent, personal God who guaranteed purpose and meaning to individual suffering (including the direct connection of that suffering to the promise of heaven). As a result, they believed that the costs of driving the uniformity of natural laws into deep time and the origin of human beings had been much too high. In 1904 Bryan spoke of his fear that "we shall lose the consciousness of God's presence in our daily life, if we must accept the theory that through all the ages no spiritual force has touched the life of man and shaped the destiny of nations."[54] As Sunday explained, "If you begin to limit God, then there is no God."[55] Sunday firmly believed that the fall of man, scripture's promises of salvation, and a reunion in heaven were the only things that could make sense of human experience and provide solace for the human heart. Even if it meant one must believe that God personally stole beloved children away.

Bryan's sights were first set on evolution in earnest when, on April 6, 1917, the United States entered the Great War. After two and a half years' effort to keep the country out of the conflict as President Woodrow Wilson's Secretary of State, Bryan resigned in protest. Soon, boys who had in their youth been saved by antitoxin or access to clean water and pure milk were blown to smithereens by scientific and technological "progress" in the trenches of Europe. The virulent influenza that broke out in 1918 took even more lives. Bryan didn't talk about all this suffering as a signal of Christ's second coming. But he did try to discern where human beings had both thought and imagined wrong, to cause so much suffering.

He was not alone. Even secular historians, trained by John Draper and Andrew Dickson White to see history as progressive, found their optimistic visions useless after the war. As Clarence Walworth Alvord wrote, the old tales of progress suddenly seemed "false, cruelly false. Our edifice was perhaps pretty but it had fallen at the touch of reality, so chaotic, so unmoral." Groping for explanations of so much destruction, Alvord wrote that he could only describe himself and his colleagues as previously "drunk on belief in inevitable progress."[56] Popular tales of an age-old conflict between science and religion suddenly took on a different cast after 1914. As the historian Lynn Thorndike wrote, "We, who invent poisonous and deadly gases to slaughter mankind wholesale, hold up our hands in horror at the more discriminating activities of the Holy Inquisition, which as a matter of fact very seldom persecuted any one for scientific views."[57] The past, so ridiculed for its superstitious ways by Draper and White, didn't look quite so inferior anymore. The twentieth century had within the short space of four years fallen far.

Some recovered a progressive narrative by seeing the war as "the war to end all wars." Others described the war as a much-needed and ultimately benevolent solution to the "degeneration and feminization" of men that resulted from civilizations, or the natural result of an inevitable "struggle for existence" between nations. Meanwhile, Bryan, out of his job as secretary of state and unable to sit still for long, read a book by the biologist Vernon Kellogg that described how German military officers had defended the war as the inevitable and beneficent manifestation of natural processes: the struggle for existence and the survival of the fittest. Bryan added this to his list of other "natural tendencies of Darwinism": industrial capitalists' use of Darwinism to justify cutthroat business practices, communists' citation of a Darwinian struggle to explain conflicts between labor and capital, and eugenicists' use of evolution to argue against social reform. He decided that all of these things—German militarism, class warfare, industrial competition, eugenics—were the "ripened fruit of Darwinism." And a tree, he insisted, "is known by its fruit."[58]

Ultimately, Bryan built his objections to Darwinism on evidence of what had become of the world around him once people believed certain things about the origin of human beings. "If hatred is the law of man's development; that is, if man has reached his present perfection by a cruel law under which the strong kill off the weak," he wrote, "then, if there is any logic that can bind the human mind, we must turn backward toward the brute if we dare to substitute the law of love for the law of hate."[59] This was the logic, Bryan argued, used to sanction the German army's invasion of Belgium in 1914 and which led to the loss over the next four years of tens of millions of lives.

Biologists Vernon Kellogg, Peter Kropotkin, Charles Darwin, and others had long insisted that evolution sanctioned no such thing. Each argued that one could find messages of mutual aid, sympathy, humanitarianism, even pacifism, in evolution. Henry Ward Beecher had conceded that if the grand law that the strong prevail was all that existed, there would be reason to believe that evolution was in direct conflict with Christian experience and faith. But Beecher found a different "message of evolution" in the parental instinct to protect the helpless and the weak. Here, he insisted, the "love-power comes in," providing a restraint on force. No longer must the "weak go to the wall," Beecher wrote, but man could study the laws of the world and harness them in the name of countering the law of force with the law of love, selflessness, protection, and mercy.[60]

Bryan countered such claims with evidence that emphasizing some evolved "love-power" was not, in fact, what men were doing with Darwin's ideas. Indeed, he argued that Darwinism halted all effort to prevent suffering. "As hope deferred maketh the heart sick," he wrote, "so the doctrine of Darwin benumbs altruistic effort by prolonging indefinitely the time needed for reforms." (He had a point: Beecher's good friend Edward L. Youmans, founder of the magazine *Popular Science*, insisted "there was no use in trying to fight evils of which he himself is as conscious as anyone, as to get rid of them is a matter of thousands of years.")[61] Evolutionists' only program for reform, Bryan warned, "is scientific breeding, a system under which a few supposedly superior intellects, self-appointed, would direct the mating and the movements of the mass of mankind—an impossible system!" Bryan thus believed he had plenty of evidence—in German militarism, eugenics, and evolutionary defenses of class warfare—that using evolution as the basis of explaining the world caused more suffering rather than less.

We have examined the work of many Americans who argued that explaining the world as governed by uniform, natural laws made better sense of the world and provided a securer foundation for human action. But Bryan, while he agreed that humans could and must work to improve the human condition, was certain that such efforts must be grounded in Christianity. From the war to eugenics, he took the present and recent past as the best evidence of what might otherwise go so terribly wrong. His certainty that effective reforms depended on belief in a close, personal God meant that he saw the future progress of the nation and world at stake in any fight over the nature of God's governance. Belief in Christ's redemptive power over individuals, he argued, could cleanse the heart instantly and turn the world from sin to righteousness. It was through the change of many hearts, not better breeding, that a new world would be born, he insisted. "It is this fact that inspires all who labor for man's betterment" and why Christians prayed "Thy kingdom come, Thy will be done in earth as it is in heaven."[62] He urged that only in "looking heavenward" could humanity "find inspiration in his lineage; looking about him he is impelled to kindness by a sense of kinship which binds him to his brothers. Mighty problems demand his attention; a world's destiny is to be determined by him."[63]

Bryan's driving assumption that "delight in doing good" depended on belief in God explains his dismayed response to surveys showing that "*more than one-half* of the prominent scientists of the United States, those teaching Biology, Psychology, Geology and History especially, have discarded belief in a personal God and in personal immortality."[64] By graduation, 40–45 percent of male undergraduates surveyed had abandoned the cardinal principles of Christianity, including Christ's divinity and salvific power. For Bryan, who believed that only a commitment to Christian ideals could ensure that America's youth returned from college "prepared to lead the altruistic work that the world so sorely needs," this was a national disaster.

Bryan knew that evolutionary explanations of the origin of species were cited as the foundation of liberal and modernist interpretations of scripture that dispensed with the fall of man, Christ's miraculous atonement for human sins, and Christ's resurrection. Meanwhile, by early 1921 Bryan had also learned that little consensus existed *among biologists* regarding Darwin's theory of the origin of species. (At the time natural selection was competing with at least five other mechanisms of evolutionary change.)[65] Why then, Bryan wondered, was Darwinism being taught in American schools as though it was proven? Given the ties Bryan drew between Darwinism and the state of the world, what must be done to secure moral and material progress seemed clear: Recover a sense of God's close providence in the lives of individuals, the nation, and the world. In early 1921, with eugenics, the war, and industrial competition as evidence of the "natural tendencies of Darwinism," Bryan launched a campaign to outlaw the teaching of human evolution "as true" in public schools.

Part of Bryan's argument against teaching evolution in taxpayer-funded schools centered on the claim that Darwin's "hypothesis" was not, in fact, science but an alternative religion (he could thus insist that "real neutrality" on religion did not exist in American schools). He argued, for example, that Darwinism provided the foundation of ethical systems that had a "natural tendency" to replace Christianity. To make the implications of Darwinism for both meliorism and medicine clear, Bryan often cited a passage from *The Descent of Man* in which Darwin expressed ambivalence regarding poor laws, asylums, medicine, and smallpox vaccination. Such things resulted in "the weak members of civilized societies" propagating their kind, Darwin wrote, and then added, "No one who has attended to the breeding of domestic animals will doubt that this must be highly injurious to the race of man." Bryan pounced: "Can you imagine anything more brutal?" he asked. "Medicine is one of the greatest of the sciences and its chief object is to save life and strengthen the weak. That, Darwin complains, interferes with 'the survival of the fittest.' If he complains of vaccination, what would he say of the more recent discovery of remedies for typhoid fever, yellow fever and the black plague? And what would he think of saving weak babies by pasteurizing milk and of the efforts to find a specific for tuberculosis and cancer? Can such a barbarous doctrine be sound?"[66]

To his credit, Bryan admitted something not all proponents of eugenics did when citing this passage from *The Descent of Man*—namely, that in the very next sentence Darwin's heart rebelled, in Bryan's words, "against the 'hard reason' upon which his heartless hypothesis is built." For Darwin had immediately warned that to check our sympathy at the urging of hard reason, and thus "intentionally neglect the weak and the helpless," would destroy human sympathy, "the noblest part of our nature." But no matter what Darwin himself thought, the writings of his followers proved, Bryan argued, that Darwinism was "directly antagonistic to Christianity, which boasts of its eleemosynary [charitable] institutions and of the care it bestows on the weak and the helpless." This was what came, Bryan warned, of driving natural law into the deep past to explain human origins, breaking the animal-man boundary, and abandoning belief that man was made in God's image.

Meanwhile, Americans who were certain that changing the world (and healing the world after the war) depended on a strict adherence to science and rationalism watched the rise of both Bryan and Sunday's brands of Christianity with disbelief. They thought the fight against orthodox schemes of salvation, rooted in the fall of man and (for fundamentalists) belief in the prospect of (deserved) eternal damnation had been fought long ago. The physician William J. Robinson, a proponent of birth control who edited the works of Abraham Jacobi, attended Billy Sunday's sermons and could not contain his disgust: "Not only to permit but to encourage the public preaching of punishment by hell fire in the second decade of the twentieth century—to what lower depths can we descend?" It was almost enough to make Robinson, an ardent freethinker and reformer, throw in the towel entirely: "Is a world which approves of Billy Sunday worth saving, worth fighting for?"[67]

In the wake of Bryan's campaign against evolution, leading biologists doubled down on the claim that evolution and belief in God must be reconciled in the name of both material and spiritual progress.[68] Biologists Henry Fairfield Osborn, Charles Davenport, and Edwin Grant Conklin publicly countered Bryan by insisting that evolution was "one of the most potent of the great influences for good that have thus far entered into human experience."[69] The American Medical Association passed a resolution "that any restrictions of the proper study of scientific fact in regularly established scientific institutions be considered inimical to the progress of science and to the public welfare."[70] Meanwhile, modernist and liberal ministers (i.e., those influenced by higher criticism and willing to adjust Christian belief to evolution) tried to help ward off Bryan's accusations that evolution destroyed moral and material progress.

Shailer Mathews, for example, recruited thirteen scholars for a volume called *Contributions of Science to Religion* (published in 1924) to get God back on the side of science and demonstrate that science was on the side of God. Mathews's God administered the universe via natural law, a concept with which his scientific contributors were obviously quite comfortable. Eugene Davenport, author of the article on recent advances in agriculture, confidently concluded, for example, that "whoever soberly considers what science has achieved for agriculture in the short space of a half century, can but render thanks to Almighty God for His revelation of the laws of nature, and he will face the future with confidence unlimited and with gratitude unbounded."[71] The contributor on medicine, John M. Dodson of the American Medical Association, argued that the fact that infant mortality had been cut in half in most American cities demonstrated that more lives could be saved by science. As for whether recent advances in medical sciences had developed anything that should disturb religious faith, Dodson gave a confident "not at all." "Many phenomena of nature," he wrote, "thought by the ignorant to be due to some supernatural power have been found to be explainable by natural laws, but the conception of a world governed by immutable laws and of a race of men able to discover and interpret them is much more wonderful and inspiring than that of a world where science is unknown and in which the intervention of a special Providence must be constantly invoked."[72] Progress became impossible, Dodson declared, the moment supernatural explanations were invoked for what science could not yet explain.

In the midst of this fight, defenders of evolution often forgot the complicated distinctions Christians had long since drawn between miracles, special providence, and general providence. As we have seen, theism had rarely equated to a simplistic reliance on miracles (after all, most of the founders of mechanistic sciences, from Newton to Ray, had been devout theists). Both sides also tended to forget the ethical dilemmas that had inspired law-bound visions of nature, including evolution, in the first place. Most defenders of evolution forgot, for example, that the desire to remove God's hand from human suffering had inspired an emphasis on natural laws long before progress in the amelioration of that suffering was proved possible. And most opponents of evolution forgot how ethical rebellions against orthodox explanations of suffering had created a need for new explanations of both natural and human history. Only a few commentators tried to fight Bryan's and Sunday's claims by reminding Americans why the reformations and rebellions against orthodox doctrines had happened in the first place. In 1921 Reverend H. A. Delano of the First Baptist Church of Evanston, Illinois, advised readers of the *Illinois Medical Journal* that surely the unbelieving physician mustn't be blamed for their skepticism: "He has moved through hospitals of pain and suffering supreme; witnessed the horrors of an inferno upon battlefields of blood; invaded alleys rank with filth and tenement houses malodorous and sickening; seen humanity swarm and struggle, spawn and die; beheld the birth of monstrosities appalling; seen the iron-handed, inevitable relentless trend of heredity . . . and yet men wonder that he is often a materialist, a doubter of humanity and a relentless foe of religious shams, follies and crimes." Delano hoped for a day "when the doctor shall find the preacher sometimes attributing the death of a child to green apples rather than providence. . . . When we shall have taught people that disobedience to the laws of nature is a crime against the law of God, then I know there shall be fewer skeptics among our earnest and learned physicians."[73] To Delano, adjusting Christianity to the discovery of natural laws would sustain, rather than destroy, faith.

Meanwhile, for some, the fight against Bryan's effort to halt the application of natural laws to the origin of human beings was personal. Charles Evans Hughes, who had served as both secretary of state and on the Supreme Court, warned of dire consequences for medical progress if Bryan's campaign against teaching evolution succeeded. Preventive medicine, Hughes wrote, had saved "countless lives" and put "an end to indescribable agonies of human beings." Yet legislatures were now intent on hampering "scientific investigations through which alone the scourges of disease now beyond remedy may come under control."[74] Some readers would have known that Hughes's daughter Elizabeth had been one of the first children in the world saved by the entirely laboratory-based discovery of insulin as a treatment for type 1 diabetes. For reasons unknown at the time, diabetic patients did not produce insulin, the hormone that regulates the amount of glucose in the blood. Unable to break down glucose for energy, the body begins to break down fat in high amounts, producing acidic ketones and a potentially deadly condition: ketoacidosis. One might stave off ketoacidosis with a "starvation treatment," but the strict regulation of sugars left patients malnourished and vulnerable to infectious disease. Elizabeth had been on this strict diet for three years and weighed forty-five pounds at the age of fifteen when her mother, Antoinette Hughes, heard of two scientists at the University of Toronto who had isolated the hormone insulin via a long series of animal experiments. A course of treatment beginning in August 1922 (and for the rest of her life) saved Elizabeth. She lived nearly sixty more years, graduated from Barnard College, and had three children.[75]

By the time Mathews was organizing *Contributions of Science to Religion* in response to Bryan's campaign, the author of the contribution on medicine, Dodson, could add insulin treatment to diphtheria antitoxin as examples of "the application of the methods of scientific, exact experiment to the problems of medicine."[76] For Dodson, binding science to strictly naturalistic, mechanistic explanations had clearly (and finally) resulted in the development of lifesaving treatments. Despite the sobering lessons of the war, White's *A History of the Warfare between Science and Theology in Christendom* still provided the arc of history needed for these defenses of law-bound, mechanistic science. After all, White gave scientists' triumphs over superstition as evidence that God's goodness, although difficult to find in individual lives, could be discerned in the progressive revelation of God's governance via natural law. In his contribution to Mathews's volume, the sociologist Ellsworth Faris described White's history as the story of why "war, poverty, and crime which were formerly defended, apologized for and even conceived as a part of the divine plan, appear to our modern eyes as problems to be solved, as challenges to the technique of control which scientific men persistently seek."[77] Reviewing Mathews's volume, the Scottish zoologist James Young Simpson singled out the final sentence as "magnificently typical of the growing present-day point of view."[78]

In reality, contributors to Mathews's volume differed regarding what precisely the "present-day point of view" commanded. The author of the chapter on eugenics, biologist Charles Davenport, thought that "anybody who visits an institution for the feeble-minded and sees the hundreds of 'children' who are bedridden for life" would be impressed by society's failure "to make use of one of the most valuable means that nature has provided for purifying the race"—namely, natural death. Instead, physicians—who should have known their duty better—prevented the congenitally deformed and senseless from dying. Davenport believed society was inspired by a "perverted instinct" to value life rather than death, yet the latter was nature's primary means of purifying the race.[79] By sharp contrast, the biologist who composed the chapter on scientific methods, William Ritter, despised Davenport's brand of eugenics and thought it would "establish an aristocracy more heartless and insolent than anything the world has ever seen."[80] But although they disagreed about what the science of heredity commanded in the wake of suffering, both Davenport and Ritter agreed that the barrier between humans and animals must be broken in order for humans to truly understand themselves and secure progress.

In the midst of these debates, a few biologists explicitly acknowledged that the "the old problem of suffering" was at stake in the fight. In a 1922 book entitled *The Human Direction of Evolution* Princeton University biologist Edwin Grant Conklin insisted that only "from the standpoint of nature as a whole" could the benevolent purpose of great affliction be revealed. Struggle and death, Conklin argued, "are factors in a great world movement, in an infinite process of evolution in which the 'whole creation groaneth and travaileth in pain' ... waiting for the manifestation of the sons of God." Take that perspective, he urged, and "the religion of evolution is thus at one with the religion of revelation," for "the past and present tendencies of evolution justify the highest hopes for the future and inspire faith in the final culmination of this great law in 'one far-off divine event, / To which the whole creation moves'" (the quotations are from Romans 8:22 and the final lines of Alfred Tennyson's *In Memoriam*). Such a religion, he argued, "prays 'Thy kingdom come, thy will be done on *earth*.'" Only from this perspective, he argued, could evil, unfitness, and disharmony be both understood and transformed into challenges to be alleviated. "Disease, suffering, and death," Conklin wrote, "are challenges to man of the most insistent and persistent sort to find out their causes and to eliminate or control them. Millions of human beings suffered and died from tuberculosis, plague, cholera, typhoid, yellow fever, malaria, syphilis, cancer, and other diseases before remedies for some of these were found, and millions more will suffer and die before they are eliminated—but does any far-seeing person doubt that this will ultimately be achieved?"[81]

To justify these hopes, Conklin appealed to scripture, faith in God's sovereignty (manifested solely through natural laws), and the Christian command to give up one's own selfish will in obedience to God's. But in stark contrast to Sunday and Bryan, Conklin held that a search for egocentric, individual salvation represented a lower form of religion. The highest type of religion, he argued, followed Christ's example, forgot self entirely, and inspired service and sacrifice for the good of others. Clearly, however, not all Americans were so confident in either the possibility or the benefits of such human-driven means of salvation. Indeed, some were quite certain that giving up old versions of salvation was not worth what biologists like Conklin asked them to believe instead.

Visions of Progress on Trial

In the summer of 1925, Americans' diverse visions of the best means of progress culminated in the first trial over teaching evolution in US public schools. Inspired by Bryan's anti-evolution campaign, state legislators in Tennessee had passed a law that made it illegal to teach "any theory that denies the Story of the Divine Creation of man as taught in the Bible, and to teach instead that man had descended from a lower order of animals" in public schools, including universities. The American Civil Liberties Union offered to pay the expenses of any teacher willing to challenge the law, and John T. Scopes, a young high school teacher from Dayton, Tennessee, volunteered. WGN radio station rented AT&T cables from Chicago to Dayton to broadcast the trial across the nation (a first for American radio). As a result, Americans listened to very different visions of progress over the nation's radio waves for a week in July.[82]

Historians have demonstrated that the "Scopes Monkey Trial" was about many things, from the right of taxpayers to choose what their children learned in schools to the right of educators to ensure an up-to-date curriculum and be given academic freedom. The trial was also a debate over what constitutes true Christianity, the rights of the minority versus the majority in a democratic state, and the proper constitutional relationship between church and state. One thing the trial was not about, from the perspective of almost all involved (with the exception of Clarence Darrow), was a conflict between science and religion. Most individuals on both sides of the aisle in the Dayton courtroom believed in God and in science. (Fundamentalist Christians wanted the anti-evolution laws on the books; liberal and modernist Christians wanted them repealed.) All sides believed that suffering might be ameliorated via science. Upon close inspection, proponents and opponents of the anti-evolution law actually differed among themselves regarding the limits of scientific explanations and the best means of that amelioration. But amid variation, each side was united against the other side on at least one point: Both the defense and the prosecution saw enormous peril in the schemes of salvation of those on the opposite side of the courtroom aisle. In the weeks leading up to the trial, Bryan said things like "We must win if the world is to be saved," while his primary antagonist during the trial, Clarence Darrow, insisted Bryan "would block enlightenment with law."[83]

The trial promised to be high drama early on, especially when Darrow, the most famous defense lawyer in the country and an agnostic materialist, signed on to help Scopes's defense team. A year earlier Darrow had used strict materialist interpretations of human nature and behavior to convince a judge that two teenagers, Nathan Leopold and Richard Loeb, who had murdered a school fellow should be imprisoned for life rather than hanged. This was the ultimate application of European science's emphasis on mechanizing nature: Darrow argued that man is a machine whose behavior and drives were determined by how well his machine was running. For Darrow, this conclusion provided the only route toward a more just world: "What we are depends on heredity and environment," he argued, "and we can control neither. As a result, I never condemn, never judge."[84] He also drew a direct connection between strict materialism and progress in medicine. In a speech against the existence of a spiritual soul, Darrow spoke of how physicians searched for the cause of disease: "If he finds it, what is it? In every single instance it is a mechanistic cause, every one, and no doctor ever finds a cause until he finds a mechanistic cause."[85]

For Darrow, matter and mechanism were not a means of explaining how God governed the world; matter and mechanism were all that exists. This was not, of course, the conclusion that seventeenth-century mechanical philosophers like Sydenham, Boyle, and Ray had hoped for when they mechanized nature. They thought imagining animals and plants as machines proved the existence, benevolence, and wisdom of God (after all, machines could not make themselves!). Bryan believed that Darwin, in proposing that purposeful parts and species could arise without divine intervention, had destroyed the only effective path toward a better earth: belief in God. But for Darrow, evolution represented the logical extension of a purely materialist worldview that he believed crucial to building a more tolerant, just world.

Bryan, by contrast, was certain that he was on the side of truth, amelioration, and progress for all Americans. During a speech on the fifth day of the trial, he got a lot of mileage out of one diagram in the textbook through which, by teaching, Scopes had broken the law: George W. Hunter's *Civic Biology*. This widely used biology textbook (examined in the next section) contained an illustration of the "Tree of Life" with the number of species in the various classes of the animal kingdom. "And then we have mammals," Bryan thundered, "3,500, and there is a little circle, and man is in the circle. Find him, find man. There is that book! There is the book they were teaching your children that man was a mammal and so indistinguishable among the mammals that they leave him there with thirty-four hundred and ninety-nine other mammals. Including elephants? Talk about putting Daniel in the lion's den?" In his undelivered closing statement, Bryan added the words, "What shall we say of the intelligence, not to say religion, of those who are so particular to distinguish between fishes and reptiles and birds, but put a man with an immortal soul in the same circle with the wolf, the hyena, and the skunk? What must be the impression made upon children by such a degradation of man?"[86]

We have seen, however, that moving human beings into the same circle as mammals had solved a profound problem for many Americans. In the wake of rebellions against orthodox belief in a God who demanded that parents resign themselves to their children's absence and focus on the possibility of reunion in heaven, collapsing the animal-man boundary offered several compensations: an alternative origin story, a completely progressive history (with no fall of man or depravity of man), and an alternative explanation of why humanity was not exempt from premature death. For Americans who adopted evolution, *hope* and *benevolence* (God's, nature's, or some powerful combination of the two) could be found in the progress read from both the evolutionary record and the advance of human knowledge, whether lives had actually been saved by that advance or not. From this perspective, Bryan's campaign to make the origin of human beings an exception would halt all material, social, moral, and religious progress.

It would also return American Christianity to a version of God against which many had rebelled on ethical grounds. Indeed, the man responsible for ensuring the trial that challenged the Butler Act happened in Dayton was inspired by what fundamentalist Christianity demanded bereaved parents believe. George Rappleyea was working in Dayton as a mining engineer when he attended the funeral of a six-year-old boy killed in a railroad accident. Rappleyea later recalled the boy's mother lamenting, "Oh, if I only knew that he was with Jesus! If only I knew that!" Rappleyea was appalled when the preacher replied, "The ways of the Lord are His. You know and everybody here knows that this boy had never been baptized. He had never confessed Christ. There can be no doubt but at this moment, he is in the flames of Hell." Rappleyea claimed to have rebuked the preacher for "torturing" the grieving mother by demanding, "If your conscience won't let you think of anything to say that will bring a little comfort to that poor creature, shut up."[87] A few days later, Rappleyea heard that "this same bunch, the Fundamentalists" had passed an anti-evolution law and made up his mind to show fundamentalists "to the world." When he saw an advertisement posted by the American Civil Liberties Union offering to defend any teacher willing to challenge the law, Rappleyea helped town boosters convince John T. Scopes to volunteer. Defending evolution, in other words, would undermine fundamentalists' vision not just of the past but of the best means of salvation.

With White's narrative in hand (and with the exception of Clarence Darrow), the lawyers, scientific experts, and ministers who travelled to Dayton to defend Scopes explicitly avoided portraying the fight over teaching evolution as a battle between science and religion. Instead, they emphasized that many Americans believed that a more "enlightened religion" was in complete harmony with science. To back up this claim, defense lawyer Arthur Garfield Hays told reporters to read White's *A History of the Warfare of Science with Theology*, quoted from the book in his legal arguments, and even handed out copies of the book to people he met in Dayton.[88] Immersed in White's vision of what progress demanded, Scopes's defenders emphasized that giving up on the ultimate application of nature's mechanical uniformity—the theory of evolution—would equate to giving up on both science and medical progress.

By the 1920s, the truly lifesaving discoveries of antitoxins and insulin treatment meant that the defense could appeal to more than postmillennialist interpretations of Christianity, the fossil record, or White's tales of scientists' advancing knowledge of nature's laws. Three centuries after Francis Bacon had first promised that new, experimental methods could serve for "the relief of man's estate," mechanistic medicine had given physicians the power to save children from deadly diseases. For those who assumed that these developments had depended on a commitment to the uniformity of nature's laws, the Tennessee law's assumption that Darwin's theory of evolution was irreligious was not only wrong but dangerous. A number of scientists, ministers, and a rabbi came to Dayton to testify that evolution was not in conflict with belief in God. Ultimately, the prosecution successfully argued that presenting such arguments before the jury would place the law, rather than Scopes, on trial, and that the jury's only task was to decide whether Scopes had broken that law. The judge did, however, permit the defense team's experts to read their statements into the record (with the jury excused from the courtroom) for the purpose of filing an appeal to a higher court. In giving his statement, Harvard geologist and Baptist Christian Kirtley Fletcher Mather, a descendant of Cotton Mather, read the words of Henry Ward Beecher claiming that God revealed himself in the Book of Nature. Kirtley Mather then argued that rejecting the theory of evolution as revealed by scientists would impede "the physical progress of mankind."[89] Zoologist Maynard Metcalf was even more pointed: "God's growing revelation of Himself to the human soul cannot be realized," Metcalf insisted, "without recognition of the evolutionary method he has chosen." And that included developing "any proper grasp of the facts of structure or function of living bodies as involved in medicine."[90]

After Scopes was convicted (as both sides expected he would be), biologists and their supporters doubled down on old arguments that the study of natural laws provided the best guides to improving the world. In his posttrial writings, Mather continued to emphasize the cost to future progress if Americans refused to believe in the strict uniformity of natural laws. Mather explicitly addressed the relation between this assumption and explanations of suffering in his 1928 book, *Science in Search of God*. God's administration of the universe via natural law, Mather argued, answered age-old questions: How could God, who is all-powerful and in direct control of affairs, "at the same time, be all-wise and all-loving? Why does he permit the suffering and sin, the unhappiness and distress, which is so obviously a part of the life which we know?" Mather's reply echoes stances we witnessed in the writings of Malthus, Douglass, and Paley: In showing the administration of the universe to be orderly, science had demonstrated not only that "God is a God of Law" but that only a God of law is "worthy of our trust. Only if he operates consistently can we hope to discover his nature and his purposes."[91]

Mather conceded Bryan's point that an emphasis on God's governance via natural laws led inevitably to the question, "If God is a God of Law, operating always and everywhere in the same way, how can he also be a God of Love, helping you and me in our time of need?" He then gave what was in fact an old answer, visible in the earliest foundations of modern science: God's governance via natural law and the consequent rational intelligibility of the universe provided the best signal of "governance of a benevolent creator." Better, indeed, than all the biblical miracles. For based on the careful examination of "the rational chain of connecting cause with effect," great things—indeed, great miracles—were being accomplished. "Consider," Mather urged, "for example, the establishment of a medical service which will within a few years rid India of leprosy. Something in the universe has worked through the minds of technically trained men so that they have been led to a discovery of the cause of that dread disease and thereby have learned how it may be prevented.... A greater work is being accomplished than the healing of ten lepers among the hundreds on the shores of Galilee."[92]

Unitarian minister John H. Deitrich agreed. He cited White's history as having proved that the fundamentalist's return to an orthodox Christian interpretation of the universe, with its belief in the fall of man, human depravity, and miracles, "stands in the way of the real salvation of the world." The clear lesson of White's history, Dietrich insisted, was that the biblical history must be abandoned "so that we may address ourselves to the real causes of human sorrow and human pain." End ignorance of nature's laws, "and the world can be saved."[93] Both Mather and Dietrich called upon an old tradition within American science of seeing the world as designed in such a way that skill could be slowly improved via the study of natural laws. Each believed progress, not scripture, to be the best evidence of God's goodness. Dietrich's fellow Unitarian Reverend Harold E. B. Speight even believed that the prospect of progress could console the human heart amid child loss: "When we are now at the end of our tether in the crises of our need," he urged, "when once prayer was all that a man had, we can turn to the doctor and surgeon and nurses to furnish the skill which will save the child."[94]

Of course, some parents had little reason to trust that any of these visions of salvation, whether anti-evolutionist or evolutionist, would include their children. Any observer who knew their history would have had a difficult time determining on which side of the courtroom in Dayton stood the best allies for ameliorating the suffering of Black children. Some African American observers of the trial, in the tradition of Douglass, did insist that science provided the best means of intellectual emancipation and material and social progress for *all* Americans. Indeed, historian Jeffrey Moran notes that, in marked contrast to some of the white biologists filing briefs for the defense, Black scholars who commented on the trial linked anti-evolutionism with the South's legalized destruction of Black lives. Black professionals like William Pickens argued, for example, that anti-evolutionism arose from the South's stance that God had created some human beings different from others. Tennessee anti-evolutionists feared Darwinism, explained a writer in the *Chicago Defender*, because evolution implied "that the entire human race is supposed to have started from a common origin. Admit that premise, and they will have to admit that there is no fundamental difference between themselves and the race they pretend to despise."[95] (These commentators tended not to mention the historical use of Darwinism to justify white supremacy, perhaps because they assumed a truly enlightened science would dispense with such biased claims.)

On the other hand, plenty of African American ministers were wary of what Scopes's defenders asked Americans to believe. Reverend A. B. Callis insisted that "there couldn't be any relation between man and monkey. A monkey has no soul, therefore has no salvation. But man has both a soul and a salvation." Given strong traditions within African American Christianity of emphasizing God's care for the poor and oppressed, it wasn't clear what Black Americans would gain by binding God by natural law. In any case, Moran notes that antievolutionist activism was muted because Black parents did not have to worry about their children even learning about evolution. Given the dismal state of support for "colored" schools (segregated by decree of the state of Tennessee's constitution), few of their children attended high school, where evolution appeared in textbooks. Furthermore, so many other forces threatened the well-being of their children, including pervasive racism within American medicine, that Black mothers and fathers had more tangible things to worry about.[96]

Ironically, it was Scopes's most famous defender, the strict agnostic Darrow, who highlighted the ambiguous legacies of Americans' centuries-old fascination with science as the best means of progress. Although obviously willing to defend the theory of evolution, Darrow agreed with Bryan that one of the main means through which biologists argued they could ameliorate future suffering—eugenics—offered a poor panacea for America's ills. Darrow clearly despised prominent biologists' talk of the "eugenic lessons" of nature's laws. Asked by a reporter for the *Washington Post* what he thought of Dr. Haiselden's actions, Darrow had glibly replied, "Chloroform unfit children. Show them the same mercy that is shown beasts that are no longer fit to live."[97] Although Darrow's words have been cited as evidence that he supported eugenics, nothing could be further from the truth. The reply was classic Darrowian sarcasm. Darrow had no sympathy with those who would judge and sort children on biological grounds, unless it be to extend *more* sympathy and tolerance to "broken machines." Nowhere in his voluminous writings did he defend eugenics, and Darrow was never one to shy away from speaking if he believed in something.

We know what Darrow really thought about eugenics because a year after the Scopes trial he composed a withering critique of "eugenist" claims that "doom hangs over the human race." He ridiculed the "cries in the night of 'race suicide,' 'the rising tide of color,' 'the race dying out at the top,' and 'torrents of degenerate and defective protoplasm.'" "Amongst the schemes for remolding society," he wrote, "this is the most senseless and impudent that has ever been put forward by irresponsible fanatics to plague a long-suffering race." He clearly despised eugenicists' confident claims that the human race could be made "better." "Do we even know," he demanded, "what we mean by the word?" Ultimately, he declared eugenicists' claim that evolution is a beneficent process a religious idea rather than science.[98]

In marked contrast to Bryan, Darrow did not conclude that Americans' misguided use of evolution meant that Darwin must be wrong about how nature worked. But he was skeptical of *any* confident campaign to reform society (whether rooted in science, religion, or both) because those campaigns were made by fallible human beings. Given the structure of American society, Darrow knew very well whose children would be considered unworthy and who would be bracketed outside the category of concern when Americans talked of saving "the race." A few months after the drama in Dayton, he was in Detroit defending a Black physician, Dr. Ossian Sweets, who was charged with murder after a white mob attacked his family. Darrow tried that case by emphasizing the learned race prejudice of every member of the all-white jury. "Would this case be in this court," he demanded, "if these defendants were not black?"[99]

In contrast to other critics of eugenics (both in the past and the present), Darrow did not focus his critique on eugenicists' poor knowledge of heredity. As a result, he never implied that if only geneticists knew more, Americans might safely make decisions about which children should be saved. Although he, too, had been raised on Draper's and White's histories and adopted their criticisms of theologians in full, Darrow saw something different in the past than constant progress via science. "The history of the race," he wrote, "shows endless examples of the pain and suffering that men have inflicted upon each other by their cocksureness and their meddling." Scientists and their grand campaigns were no exception. Of eugenics, he wrote, "I, for one, am alarmed at the conceit and sureness of the advocates of this new dream. I shudder at their ruthlessness in meddling with life. I resent their egoistic and stern righteousness. I shrink from their judgment of their fellows. Every one who passes judgment necessarily assumes that he is right. It seems to me that man can bring comfort and happiness out of life only by tolerance, kindness and sympathy, all of which seem to find no place in the eugenists' creed."[100]

Darrow clearly preferred evolutionists' explanation of the origin of suffering over Sunday's and Bryan's. But he also knew that even a cursory glance at American history showed that it wasn't necessarily clear that science as practiced in the United States would be the best ally for ameliorating the suffering of all of America's children.

There Was No Doctor Then

Eight years before Americans' debates over the best means of progress were broadcast over the radio during the Scopes trial, on May 24, 1917, Harriette hayalča? Shelton's sister Ruth died of tuberculosis. Harriette and Ruth were students at the Tulalip Indian Boarding School on the Tulalip reservation forty miles north of Seattle, when the school sent Ruth home. When permission came for Harriette to go home as well, she knew her sister must be dying. She stayed by Ruth's bedside, with her parents, for weeks. A lifetime later, as Harriette Shelton Dover, she spoke of her sister's death: "I can only say to die of tuberculosis is a dreadful, painful death. I don't think many people see that kind of death anymore. I think there is medication to help the pain, and there are things the doctors can do now. There was no doctor then for my sister. I used to kneel down by her bed and hold her hand when she was suffering such pain, and she always had such a high fever. She would pull at my hand. The suffering she went through came at every hour or two hours. It lasted, maybe, an hour. It seemed the pain wracked her whole body, not only her chest, but also her stomach. She cried out. She screamed."[101] Dover's cousin Marguerite died two weeks after Ruth, also of tuberculosis. "Another Indian girl and two Indian boys" died on the reservation that summer as well.[102]

The federal Bureau of Indian Affairs reservation system inadvertently served as a proving ground for various stances in Americans' debates over the origin and meaning of suffering, whose suffering mattered most, and the means and boundaries of amelioration. As a result, Harriette and her family lived debates over the best means of American progress and whose children mattered most as they played out in the shifting policies of the Bureau of Indian Affairs, on the Tulalip reservation, and in its schools and churches.

Each September when school started Dover would ask about some of her schoolmates only to be told they had died. Meanwhile, babies whom she had held in her arms as "they moaned with the awful pain" died of spinal meningitis. Sometimes children died of measles. "There was no doctor for all those deaths," Dover recalled. "Of course, you might say no doctor would be able to cure tuberculosis anyway," but things, she insisted, were different for her people. Whites had freedom of movement, which for those who could afford it meant they could pursue the main treatment for tuberculosis: a change of air at a sanitarium. But even if they had the means to travel, those living on the Tulalip reservation had to obtain permission from the reservation agent to leave.[103]

Dover's sister died of the greatest killer of the nineteenth century but at a time when mortality rates from tuberculosis had declined in much of the United States. In the 1830s consumption was responsible for one of every four deaths. By the 1880s that number was one of eight.[104] These numbers did not go down for everyone. Historian Elizabeth James notes that during the 1890s tuberculosis "posed the same threat to American Indians . . . as smallpox and other viruses had during previous decades and centuries."[105] By 1913, the estimated rate of tuberculosis among whites was 12.1 percent, while among American Indians it was 35.4 percent.[106] On the Yakama reservation in eastern Washington, deaths from tuberculosis remained high or *increased* between 1910 and 1940, even as they declined in other populations. (The Yakama death rate from tuberculosis was seven times higher than that of Black Americans and twenty-three times that of white Americans.)[107]

By the 1920s a consensus existed within organizations like the National Tuberculosis Association that improper or insufficient diet was an important factor in the spread of tuberculosis, but that consensus had little influence on the policies of the Bureau of Indian Affairs. Reservations concentrated Indigenous communities within small areas in order to turn individuals into farmers and eventually (with the allotment system) into private property owners. Rules against leaving the reservation while white populations expanded around its borders cut off traditional food supplies and restricted seasonal movements. Dover recalled that for years the Tulalip reservation was plagued by a lack of clean water, infrastructure, and roads. Sometimes settlers shot at men who tried to canoe to long-standing hunting grounds like Whidbey Island. Meanwhile, a growing white population meant increased exposure to disease carriers, including consumptives who migrated west in search of "better air." Entrance into the wage economy, including logging and salmon canneries, required living in close quarters and purchasing food from white merchants. All of these changes increased the risk of infectious diseases on reservations. And all of this happened while scientists identified the germs and sanitarians improved the ability to provide clean water and better sanitation elsewhere.

The treaties between the US Government and Indigenous leaders, like the one signed at Mukilteo in 1855 by elders Harriette Dover knew, promised physicians "who shall furnish medicine and advice to the sick" and vaccinations (the only vaccine available was smallpox). But by the turn of the century, according to Tulalip reservation agent Charles Buchanan, Indigenous communities of the Puget Sound had one of the highest infant mortality rates in the United States.[108] (This was long after the terrible ravages of smallpox and other diseases had decimated Indigenous populations across the nation.) The effect on parents and families is hard to imagine. Some scholars estimate that nearly three-fifths of Indigenous children died before the age of five in the opening decades of the twentieth century. Rates of measles, chicken pox, mumps, and smallpox were double or triple national rates.[109] "It was really a stunning, staggering death rate," Dover recalled. Despite the treaties, "there were no doctors, no medications, no care, nothing." Dover had an aunt whose four children all died before they were four years old. Her own mother and father, Ruth and William Shelton, watched three of their six children die. "Most of them I never saw," Dover wrote, "they died when they were babies because of the epidemics that swept the Indian reservations, such as measles and the common cold and pneumonia and tuberculosis."[110]

Having survived to school age, Dover and her sister entered the federal government's system of boarding schools (the Tulalip boarded at a school on the reservation). The schools' proponents imagined these buildings as avenues of civilization and progress. But they also concentrated children within buildings where illness spread quickly. One study estimated that three out of ten boarding school students had trachoma, a painful, contagious bacterial infection of the eye that could lead to blindness.[111] Outbreaks of measles or influenza moved rapidly and were often deadly (the 1918 flu epidemic, among other factors, led to the closure of the Cushman Indian School in Puyallup, Washington, in 1920).[112] Government officials were well aware of the high disease and mortality rates. At the 1915 Congress of Indian Progress the commissioner of Indian Affairs Cato Sells spoke about the Indian Service's duty to protect the "health and constitution of Indian children." A year later he issued a circular for Indian Service employees proclaiming, "We cannot solve the Indian problem without Indians. We cannot educate their children unless they are kept alive."[113]

The incidence of illness and death at the mission and reservation schools was so high that many Indigenous parents hesitated to enroll their children. But Dover's father, William Shelton, believed that getting the education promised by the Treaty of Point Elliott was crucial to his people's survival. (He had to run away from home to join the Tulalip Mission School because his parents were sure he would not survive.) Dover explained her father's attitude by recalling that the chiefs who had signed the treaties wanted the children to read and write: "They said, 'Pay attention to words. We will never be able to catch up. We don't know anything about what the white man thinks or plans unless we know his language—unless we can read it. All of those marks—designs they called them—they make on paper mean something to them.'" So, with the stakes so high, Dover went to school for ten months of every year. She made friends and each fall returned to find some of her friends gone.[114]

When she reflected on her life years later, Dover spoke of how confusing the options available for making sense of her people's suffering, including child loss, felt. Dover remembered how "the Indian women would always get together when there was a child dying." They used to kneel down and say the rosary: "The prayers would go on and on." Father Eugene Casimir Chirouse of the Oblates of Mary Immaculate had established a Catholic mission and school on the reservation in 1857, and so, in the early years of the twentieth century, all the Tulalip students went to a Catholic church service on Sunday. "You would think that Sundays were something nice," Dover recalled, "but I always found Sunday in that school was a big worry for me." The priest in charge "talked to us about being sinners, and how we were going to hell. When I was small, I was afraid to fall asleep because I thought if I went to sleep I would go to hell, and I could see the burning flames—great big flames and nothing but fire."[115]

When she was thirteen or fourteen years old, her teachers allowed Dover to attend Sunday evening chapel as well. But that service, at Agent Buchanan's direction, was Protestant, and "nothing like the Catholic Mass." There students read from the King James Bible rather than the Catholic Bible, inadvertently receiving a primer in the fact that Christians did not agree on what precisely salvation required. Catholic during the week; Protestant on Sunday evenings. Dover worried a great deal that she was a sinner and might go to hell, since she not only attended Protestant services but enjoyed them. Meanwhile, her grandparents maintained and shared her ancestors' (Snohomish, Skay-whah-mish, Puyallup, and Wenatchee on her father's side, and Klallam and Samish on her mother's) beliefs and traditions. "I wonder," Dover mused, looking back, "whether anybody ever had that many religions?"[116]

For generations science had fit well within the lessons of both Catholic and Protestant missionaries who argued that both the Book of Scripture and the Book of Nature proved the white man's God better than Indigenous beliefs. Paley's *Natural Theology* was taught at schools like the Cherokee Female Seminary in the 1850s.[117] Firmly rooted in the mechanical philosophy, this science taught that God alone was spirit, and all of nature a spiritless machine. For generations within American science, nature was supposed to demonstrate the existence and benevolent character of the all-powerful, all-knowing, all-benevolent God of Christianity.

Of course, by the time Dover attended school at the beginning of the twentieth century, the boundaries between science and religion (and scientists and ministers) were changing. And that meant the science classroom could also become a site of confusion. When Dover left the Tulalip reservation to attend Everett High School, her brother wanted her to take the science curriculum, "the hardest course in the school." One day, amid studying "about the very beginning of life as it was explained in general science and zoology," her botany teacher casually remarked that "religion is just a ploy to keep the people subdued, so they don't complain about any hunger or any of the things that happen to them." Dover recalled that her teacher had just been casually walking from one blackboard to another, but Dover never forgot his words: "It seemed as though it hurt in my heart, too. I could feel it in my chest and, of course, in my mind." Things had been confusing enough, and the science teacher's remark hardly helped. "I used to stay awake and just pace the floor," Dover recalled, "tiptoeing along—because I am really torn up about religion. I could almost see the flames of hell for me doubting the Catholic Church."[118]

We don't know which biology or botany textbook Dover's science teacher used in class. But all of the textbooks in circulation in the 1920s would only have added to the confusing mix of explanations offered for suffering, disease, and child mortality. The most popular textbooks of the era embraced White's portrayal of science as the discovery of divinely designed natural laws. Typical were Cornell University–trained biologists David Starr Jordan and Vernon Kellogg's statement in their textbook *Evolution and Animal Life* (1907) that "with the growth of the race has steadily grown our conception of the omnipotence of God," and "we see the hand of the Almighty in nature everywhere; but everywhere he works with law and order."[119] Following White's lead, Jordan and Kellogg defended science education, and biology in particular, as teaching American children the importance of learning and adjusting to natural law in the interest of divinely approved progress. (Both men were also firm proponents of eugenics.)

Other textbooks did not mention God, but all described understanding natural laws as the best means of both material and moral progress. George William Hunter's *A Civic Biology* (the textbook at the center of the Scopes trial) portrayed Darwin as having given "the proofs of the theory on which we to-day base the progress of the world."[120] That progress, Hunter insisted, depended on constant attention to the lessons taught by biological laws. In the section on bacteria, Hunter cited Pasteur's claim that "it is within the power of man to cause all parasitic diseases (diseases mostly caused by bacteria) to disappear from the world" and then wrote that every student had the power to prevent suffering by changing their behavior in the face of better scientific knowledge. "Tuberculosis, typhoid fever, diphtheria, pneumonia, blood poisoning, syphilis, and a score of other germ diseases," Hunter wrote, "ought not to exist.... More than half of the present misery of the world might be prevented and this earth made cleaner and better by the coöperation of the young people now growing up to be our future home makers."[121]

These textbooks contained clear messages about how to explain the suffering caused by infectious disease. They emphasized human agency as key to prevention—and, by implication, ignorance and inaction as the main cause of disease. Hunter included a graph showing that the death rate from tuberculosis, for example, had decreased by half between 1850 and 1906. His young readers learned that the ability to move water and keep it clean via public water systems, careful medical inspection, pasteurization of milk, and better food hygiene were key to explaining the decline of tuberculosis. "It is estimated that bacteria cause annually over 50 per cent of the deaths of the human race," Hunter wrote. "As we will later see, a very large proportion of these diseases might be prevented if people were educated sufficiently to take the proper precautions to prevent their spread. These precautions might save the lives of some 3,000,000 people yearly in Europe and America." Tuberculosis, though still responsible for the greatest number of deaths, was also for Hunter the best example of what might be done: Tuberculosis was "slowly but surely" being overcome by the aid of good laws and sanitary living. Hunter then urged that "the study of biology should be part of the education of every boy and girl, because society is founded upon the principles which biology teaches. Plants and animals are living things, taking what they can from their surroundings; they enter into competition with one another, and those which are the best fitted for life outstrip the others.... Health and strength of body and mind are factors which tell in winning."[122]

What messages did Hunter's biology textbook contain about why the graph of declining tuberculosis rates did not map at all onto what Indigenous parents and children experienced on reservations like the Tulalip? To answer that question one more assumption that pervaded the era's biology textbooks is required—namely, that biologists could group human beings into races and then place those races onto a hierarchical ladder of progress. Here, for example, is Hunter's description of the "varieties of man": "At the present time there exist upon the earth five races or varieties of man, each very different from the other in instincts, social customs, and, to an extent, in structure. These are the Ethiopian or negro type, originating in Africa; the Malay or brown race, from the islands of the Pacific; the American Indian; the Mongolian or yellow race, including the natives of China, Japan, and the Eskimos; and finally, the highest type of all, the Caucasians, represented by the civilized white inhabitants of Europe and America."[123]

As James notes, belief in "progress via a struggle for existence" and long-standing perceptions of Indigenous Americans as "the vanishing race" created ambivalence, if not outright opposition, to doing anything about high mortality on the reservations.[124] Clearly, the differences between mortality rates between white and Indigenous communities from tuberculosis were widely known. Dover spoke of how, as late as the 1970s, whites asked her "why the Indians are always dying of tuberculosis."[125] Historian David S. Jones has described how every reply given for much of the twentieth century (whether constitutional predisposition, heredity, or living conditions) reflected assumptions of white superiority.[126] Each explanation meant poor health could be blamed on either heredity or the persistence of "Indian ways," rather than a recognition of how those ways had been forcibly changed.

Meanwhile, and with tragic irony, Bureau officials developed paternalistic policies that resulted in high rates of poverty and social disruption. Reservations thus exacerbated problems recognized elsewhere as increasing mortality rates from tuberculosis among children and adults. The role of poverty in hindering prevention and treatment of tuberculosis, for example, had been discussed for some time. In New York City, an Upper West Side neighborhood in 1890 had a death rate of 49 per 100,000 from the disease, while a crowded tenement section's death rate was 776.[127] Based on these numbers Dr. Arthur Guerard, bacteriologist in the NYC Department of Health, had argued in 1901 for tenement reform, including the provision of more light, room, and air: "Tuberculosis, though an evil much to be dreaded," Guerard wrote, "is not an inevitable decree of fate, not an unavoidable dispensation of Providence, as it has commonly been thought to be; but, like any other ills from which mankind unhappily suffers, the remedy for it exists to a great extent in ourselves."[128] But what precisely existed within the realm of one's own power, in a society that categorized people and their children onto a hierarchy of both capacity and concern, and distributed resources—including physicians and sanitary infrastructure—accordingly?

Later in life, Dover's answer to those who asked why so many Indians died of tuberculosis was firm: poor living conditions on the reservations and poor food and treatment at the boarding schools.[129] Her father's friend, Tulalip reservation agent Charles Buchanan, agreed. Indeed, Buchanan drew a direct connection between the American tradition of citing nature's laws as the cause of Indigenous suffering and the excuses provided by legislators for doing nothing, if not worse. He wrote in a 1915 appeal to Washingtonians to honor the Treaty of Point Elliott that "it is neither a full nor a direct answer to this question to state that it all comes about by the operation of great natural laws, such as the survival of the fittest, etc. It has come about by the operation of laws which the white man himself has made for the white man's benefit."[130] Buchanan was trying to point out that the "struggle for existence" being touted as the means of progress wasn't actually being played on a level field when one group claimed to have science, the law, and even God on their side.

The testimony of Indigenous physicians like Susan La Flesche Picotte—as well as reservation agents' constant, unsuccessful requests for more funds and support—support Dover's answer to the question of why so many Tulalip, including her beloved sister Ruth, died from tuberculosis. Picotte had often heard her father's, Joseph La Flesche's, defense of his own efforts toward Omaha assimilation: "It is either civilization or extermination."[131] But when tuberculosis hit the Omaha reservation, white civilization (including its science and medicine) had very little to show for itself. The government provided no assistance to help Picotte fight the disease. "The spread of Tuberculosis among my people is something terrible," she wrote in 1907 to the commissioner of Indian Affairs: "So many, many of the young children are marked with it in some form. The physical degeneration in 20 years among my people is terrible. I have talked with them and done all I could to prevent infection and contagion." The answer to her repeated requests for a hospital was, again and again, *no*.[132]

When Picotte finally succeeded in building a hospital on the Omaha reservation, she did so entirely via private funding, including from the Presbyterian Board of Home Missions. Having spent her life as a doctor traveling miles and miles in biting weather to visit sick children, Picotte described how her "greatest desire in having the hospital built was to save the little children." Well aware that the scientific medicine in which she was trained could prevent but not cure tuberculosis, she emphasized that the hospital must be a place for education, not just treatment. She organized classes on sanitation, discouraged the use of communal drinking cups, and encouraged everyone to screen their doors and windows against flies. Then she came up with a plan for taking these teachings to the farthest reaches of the reservation. In 1914 she wrote to Cato Sells asking to borrow the department's tuberculosis exhibit so she could reach "every family this summer." She also requested that all Indian children in government schools be examined every month for tuberculosis. She told the story of a girl coming to her far too late to help, by which time the poor girl had infected her mother and grandmother as well. Both had died. "There is no telling," Picotte warned, "how many of these infected in that large school . . . could have been prevented by proper examination." And then she added, "It is so terribly hard to see the people undergoing hardships from a civilization new to them." But the exhibit and the monthly examinations never arrived. It isn't clear whether Picotte even received a reply.[133]

Eventually, a new school of anthropologists influenced by Columbia University anthropologist Franz Boas called the racial ladders on which the Bureau of Indian Affairs built its paternalistic policies into question. Critics of the Bureau like John Collier even began proposing that Indigenous Americans might in fact have something to teach white Americans about what it meant to be "civilized," including communalism and mutual aid.[134] Here, too, the world war was key, having inspired even some Protestant missionaries to question the Bureau of Indian Affair's driving assumption that white civilization was the hallmark of progress.[135] By the 1920s some observers of "Indian affairs" were calling reservation missions and church services a violation of the separation of church and state. (William Jennings Bryan was at the center of this fight as well, for he countered with resolutions praising the federal government's long-standing support for Christian missions on reservations.) Clearly, despite their agreement that suffering could be ameliorated and "a better world made," Americans were coming to profoundly different conclusions regarding the best foundation for saving children, what precisely they were to be saved from, and who got to say.

As Dover paced the floor of her high school dorm room, worried she was going to hell, she wrestled with various alternative explanations for the suffering of those who lived on the Tulalip reservation. One day, she finally spoke to her father about how worried and mixed up she felt. At the time her mother was Catholic, but William Shelton refused to set foot in the Catholic church. He thought the services beautiful, but "it didn't reach him," Dover recalled, "especially when the priest would tell the Indians, if you don't come to church, if you don't follow this, you are going to hell."[136] Shelton questioned other things as well. He started the Tulalip Improvement Club because, as he argued to Buchanan, how could the agency, which had strict rules against traditional gatherings, object if the people were improving themselves? And he recruited lawyers to fight back when the agency doctors, their services promised by the Treaty of Point Elliott, just took care of their own families and refused to treat tribal members.

Now, as Harriette told her father of her fears, Shelton listened for some time in silence. Then he told her what his uncles had taught him. "Don't worry about all that," he said. "When you stand on this earth, you stand on grass, or the earth that is the creation of a great creator, a great mysterious creator. You are not lost.... You don't have to worry about any of it; it has always been going on—change and time—time is nothing to a great mystery or a great mysterious creator." Her father's words, Dover wrote, made her "feel as though I had come home." After this, when she went to community college and learned more in anthropology classes about the origin of human beings and how the first people to resemble human beings had come from East Africa, "things like that didn't bother me because of what my father said."[137]

Much later in life, Dover recalled the priests and ministers of the reservation school with a complex mix of amused and indignant sympathy: "The poor misguided people. They were bound and determined to save all of our miserable souls."[138] Eventually, Dover chose and fought for a very different vision of salvation for her people. When a group of white women in a reading group in Everett told Dover that it was not right to compare her ancestors' stories and histories to the Bible, since the Bible was the inspired word of God, Dover replied firmly, "I can compare it. I do compare it."[139] Having decided her ancestor's knowledge and ways could indeed be compared to the white man's beliefs, she dedicated her life to activism on Duwamish, Snohomish, Snoqualmie, Skagit, Suiattle, Samish, and Stillaguamish terms. In doing so Dover fought against the idea that the white man's science, religion, medicine, and histories were so superior to Indigenous ways and knowledge. She joined legal fights against the US government for breaking treaty promises, worked for the revival of the Lushootseed language (Dover had been whipped at school for speaking Lushootseed), and attested to the "remarkable" healing practices of the few remaining "Indian doctors."

Dover's successful resurrection of the once outlawed First Salmon Ceremony in 1979 represented an extraordinary rebellion against the mechanistic approach to nature that, for generations, many Americans touted as the best means of progress and civilization. In contrast to mechanistic sciences, the First Salmon Ceremony, in which a single salmon is captured and honored when the fish return to spawn, assumes that salmon have moral and spiritual agency. Salmon are, in other words, much more than just matter. They are much more than just machines. And they are not here solely for human benefit and control. In working for Tulalip autonomy to decide what, of European ways and knowledge, was worth adopting and what must be set aside, Dover was also choosing between different explanations of suffering and alternative routes toward a better future for Tulalip children. Given how claims about the superiority of European science, medicine, and religion had been tied to visions of whose children "shall inherit the earth," these were not disconnected fights.

Conclusion

In some ways this history might seem too distant to have much bearing on the present. Infant and child mortality rates in the United States are now in single-digit percentages. As we saw in the introduction, graphs of that extraordinary drop over the course of the twentieth century are often cited as evidence that numbers in the United States and elsewhere can be lowered still further.

Even by the time of the Scopes trial, some Americans had witnessed dramatic hints that the promises of mechanistic, experimental science might actually, finally be coming to pass. Some stood by as physicians administered diphtheria antitoxin and snatched a beloved child from "the strangling angel." Some observed a course of insulin injections replace a crucial missing part of a child's "machine," thus halting slow death by starvation. Still, in 1925 mortality rates were extremely high relative to what would be achieved within a generation. They also differed enormously when categorized according to racialized groups. In 1925 the infant mortality rate (number of deaths per 1,000 live births) in Richmond, Virginia, was 67.4 for white infants and 131.7 for Black infants. In other words, the difference in mortality rates between white and Black infants was the same as it had been for decades.[1]

Over the course of the twentieth century, continued improvements in nutrition and housing, infrastructure for delivering clean water and milk and removing waste, and an increasing number of medical advances lowered infant and child mortality rates still further. Vaccines dealt the final blow to a number of childhood infectious diseases already on the decline due to rising standards of living and public health measures. A vaccine for diphtheria became widely available, for example, in the 1930s, further lowering mortality rates from that dreaded disease. Vaccinations for whooping cough (pertussis) and tetanus were combined with that for diphtheria into the DTP vaccine in the 1940s. Meanwhile, clear progress in treating, as opposed to preventing, infectious maladies was achieved with the antibiotic revolution of the 1940s. Still later, effective fluid and electrolyte therapy, the development of pediatric hematology, and since the 1970s, the development of intensive care techniques to correct congenital deformities and save babies born prematurely decreased infant mortality even further.[2]

Parents in the United States do not have to worry about most of the diseases that killed children between 1690 and 1925. None of the diseases that appear in these pages make the list of the primary causes of infant and child mortality in the United States in the present.[3] As pediatrician and historian Perri Klass notes, by the 1980s "there was no such thing as 'routine' or 'unavoidable' infant and child mortality; short of very rare and terrible diseases, almost every death was supposed to be preventable—and prevented." Even child leukemia and "mismade" hearts succumbed to the wonders of modern medicine: "Pediatric oncology," notes Klass, "offered a 90 percent cure rate even on the scariest ward; pediatric cardiology could save all but the most severe congenital heart defects."[4] Today, infant mortality rates in particular are used as indicators of population health because they correlate so strongly with structural factors like economic development, living and environmental conditions, and social well-being. When infant deaths due to diseases like cholera, typhoid, polio, diarrhea, or tuberculosis increase, the failures of nation-states, especially the failure to ensure access to vaccines and clean water, rather than a lack of scientific knowledge, are blamed.[5]

In retrospect, all of this progress in moving infant and child mortality rates certainly vindicates the hopes of those who believed that science would lead to progress, whether one saw that possibility as part of God's design or not. And as a result of this progress, today's campaigns to lower infant and child mortality need not rely on a particular vision of God, the fossil record, or evolutionary progress to argue that further progress can be made. They can rely, quite simply and solely, on the evidence of precedence. That is the only evidence required by the Bill & Melinda Gates Foundation, for example, to argue that efforts to lower mortality rates still further are warranted. And yet the past can be a complicated, ambiguous, even dangerous place to go if one is looking for a precedent for one's hopes. The past is a mirror that shows humanity at its best and at its worst. The history of American science, religion, and medicine shows that the study of natural laws as the means of both understanding and preventing suffering could become the means of *both* unprecedented triumph *and* extraordinary tragedy. We have seen, for example, that the same emphasis on natural law that (at long last) led to great triumphs in the laboratory and eventually at the bedside of children was also used to establish boundaries around whose children and whose suffering mattered. In other words, calls for action to ameliorate suffering via science often drew boundaries around whose suffering counted *by appealing to science*. As a result, science became the latest means of declaring some children more worthy of saving than others. This ambiguous quality of influential beliefs, creeds, and ideas was, of course, nothing new. Americans' fights over whether Christ's silence on slavery should take precedence over his call to love one's neighbor clearly demonstrated that scripture could be harnessed to very different visions of the world. This history shows that the same must be said of how Americans drew upon the Book of Nature, whether they read that book as in harmony with the Book of Scripture or not.

The triumphs and tragedies in this past remind us that all visions of progress, whether we categorize them as scientific, religious, or a mix of both, are formed within societies in which judgments of worth, what constitutes progress, and who gets to say, drive what can be imagined as solutions. Some of the legacies of this ambiguous past are, of course, still with us. Although overall infant and child mortality rates are now in single digits in the United States, the disparities within infant and child mortality rates continue. In 2018 the Centers for Disease Control reported infant mortality rates as follows (using then-current US census categories): non-Hispanic Black, 10.8; Native Hawaiian or other Pacific Islander, 9.4; American Indian or Alaska Native, 8.2; Hispanic, 4.9; non-Hispanic white, 4.6; and Asian, 3.6. How these numbers, and the suffering they represent, are explained continues to determine what can be imagined (or not imagined) as solutions.

Tracing the history of ideas about God, nature, and history at the bedside of American children also shows that attributing triumphs or tragedies to either "science" or "religion" fails to capture the complexity of the past. Ultimately, imagining one could decrease infant and child mortality was not the province of science alone or religion alone but rather the legacy of complex interactions between concepts of God (including radical adaptations of and rebellions against traditional views) and ideas about both nature and history. The same goes for the tragedies that we now know arose from dominant assumptions about what constitutes progress and who gets to say. While we might be inclined to pick and choose from this past to highlight the virtues of our own vision of progress, doing so requires that we turn the past into a much more simplistic place than historical evidence suggests it was.

Much has changed since the 1920s, of course, including the landscape of American religion. Amid the continued prevalence of Christianity, diversity both within Christianity and beyond its boundaries has increased. According to the Public Religion Research Institute, in 2020 about 70 percent of the US population self-identifies as Christian (of which 45 percent are Protestant and 22 percent are Catholic). About 23 percent of Americans describe themselves as "religiously unaffiliated" (of which 3 percent identify as agnostic and 3 percent as atheist) while about 4.5 percent identify as Jewish, Muslim, Buddhist, Hindu, or other Indigenous or "world religions."[6] These names and numbers obscure, of course, extraordinary diversity within these traditions. Even within dominant Christian and secular traditions, history shows us that Americans have never agreed on what progress should look like, how it should be achieved, who should benefit most, or how earthly troubles and triumphs relate to the prospect of a hereafter.

Americans' disagreements regarding the best foundation of progress are aggravated by the fact that we remember different parts of this history. Defenders of biomedical developments like the COVID-19 vaccine cite triumphs like germ theory, vaccinations, and public health interventions in lowering child mortality since the nineteenth century. White conservative evangelicals who distrust vaccines justify their skepticism (whether of the Centers for Disease Control, the World Health Organization, the Bill & Melinda Gates Foundation, or all three) by citing philanthropists' and scientists' role in the American eugenics movement of the 1920s. Communities of color remember the ways in which scientific, technical fixes have been emphasized in place of addressing structural health inequities. Attending to which parts of this past Americans remember and why will help us better understand why we hold such different stances on issues that concern science, from evolution to genetic medicine and from vaccines to climate change.

Today, as in the past, Americans do seem to share at least one fundamental value: child loss should be prevented. They continue to differ on the best means of doing so and how the value of preventing child death should be balanced against other values. Is radical socioeconomic reform or philanthropist-funded medical technology most effective, or some complex combination of the two? Is the US child mortality rate from firearms (as of 2020 the top cause of death from age one to eighteen and thirty-six times higher than that of any other industrialized nation) worth the "right to bear arms"? Is noninvasive prenatal genetic testing a triumph of modern science's ability to ameliorate suffering or the means of a new eugenic dystopia? Should future generations of children be included in calculations of the harm present-day policies might have on mortality rates? (In recent years pediatricians have warned that climate change could reverse the reductions in global child mortality made over the past twenty-five years and that such reversals will affect children along historical patterns of disparity.)[7] And at what point after conception does a child whose death should be prevented exist?

As we navigate today's debates over what it means to save children and make a better world, history reminds us that simplistic explanations of stances on all of these questions are not helpful. Making sense of Americans' different visions of progress requires attention to beliefs regarding many things, including the shape of history, human nature, divine governance, the possibilities of human knowledge, the limits of human agency, and the ethical and scientific capacity of one human being to judge the worth of another. Beliefs regarding all of these things are closely related to assumptions regarding the origin of and best response to suffering. They also influence what can be imagined about the past, present, and future in complicated ways. Sometimes these beliefs and assumptions have been explicit; more often they have been, and are, highly implicit. But attending to how underlying beliefs and assumptions about both suffering and progress continue to influence stances on science, faith, and medicine will help us create more accurate maps of today's debates and in doing so, perhaps, improve our ability to build consensus regarding how to ameliorate suffering in the future.

Notes

Introduction

1. "Our Story," Bill & Melinda Gates Foundation (website).

2. For data on the history and present state of child mortality around the world, see "Child and Infant Mortality," Our World in Data (website), Global Change Data Lab.

3. John Hedley Brooke, *Science and Religion: Some Historical Perspectives* (Cambridge: Cambridge University Press, 1991), 5.

4. For the causes of infant mortality in 1926, see United States Bureau of the Census, *Mortality Statistics 1926: Twenty-Seventh Annual Report* (Washington, DC: Government Printing Office, 1929). The means by which infant and child mortality rates (the two categories were not distinguished until the last quarter of the nineteenth century) were lowered are the subject of much debate. Different causation claims entail different lessons, of course, for what should be done to lower mortality rates still further. For classic accounts, see Richard Meckel, *Save the Babies: American Public Health Reform and the Prevention of Infant Mortality, 1850–1929* (Ann Arbor: University of Michigan Press, 1998); Samuel H. Preston and Michael R. Haines, *Fatal Years: Child Mortality in Late Nineteenth-Century* (Princeton, NJ: Princeton University Press, 1991). For a recent account, see Perri Klass, *A Good Time to Be Born: How Science and Public Health Gave Children a Future* (New York: W. W. Norton, 2020).

5. See C. J. Jang and H. C. Lee, "A Review of Racial Disparities in Infant Mortality in the U.S.," *Children* 9, no. 2 (February 14, 2022): 257.

Chapter 1

1. David Levin, *Cotton Mather: The Young Life of the Lord's Remembrancer, 1663–1703* (Cambridge, MA: Harvard University Press, 1978), 137.

2. Cotton Mather, *Right Thoughts in Sad Hours, Representing the Comforts and Duties of Good Men, under All Their Afflictions; and Particularly, That One, the Untimely Death of Children: In a Sermon Delivered . . . Under a Fresh Experience of That Calamity* (London: James Astwood, 1689), 49.

3. For scholarship that undermines Aries's thesis, see Hannah Newton, *The Sick Child in Early Modern England, 1580–1720* (Oxford: Oxford University Press, 2012); Antonia Fraser, *The Weaker Vessel: Woman's Lot in Seventeenth-Century England* (New York: Alfred A. Knopf, 1984), 80–81.

4. Mather, *Right Thoughts*, 18, 48.

5. Mather, 44.

6. Mather, 23, 34, 27, 18, 20. Wormwood signified the transformation of something pleasant into something bitter.

7. For a useful survey of theodicies, see Michael Tooley, "The Problem of Evil," *Stanford Encyclopedia of Philosophy*, Fall 2015 edition, edited by E. N. Zalta.

8. Quoted from Mather's *The Angel of Bethesda* in Lucas Hardy, "'The Practice of Conveying and Suffering the Small-pox': Inoculation as a Means of Spiritual Conversion in Cotton Mather's *Angel of Bethesda*," *Studies in Eighteenth-Century Culture* 44, no. 1 (2015): 67 (Mather's italics).

9. Peter Gregg Slater, *Children in the New England Mind: In Death and in Life* (Hamden, CT: Archon Books, 1977), 20.

10. See Peter Gregg Slater, "From the Cradle to the Coffin: Parental Bereavement and the Shadow of Infant Damnation in Puritan Society," *Psychohistory Review* 6, no. 2–3 (1977): 4–24.

11. Mather, *Right Thoughts*, 56.

12. *The Autobiography of Mrs. Alice Thornton, of East Newton Co. York* (Durham, UK: Andrews, 1875), 126.

13. Philippa Koch, *The Course of God's Providence: Religion, Health, and the Body in Early America* (New York: New York University Press, 2021), 24.

14. Mather, *Right Thoughts*, 48, 53–54, 9, 15.

15. Abijah Perkins Marvin, *The Life and Times of Cotton Mather; or, A Boston Minister of Two Centuries Ago, 1663–1728* (Boston: Congregational Sunday-School and Publishing Society, 1892), 285.

16. Levin, *Cotton Mather*, 138.

17. Mather, *Right Thoughts*, 20–21.

18. Levin, *Cotton Mather*, 306.

19. See, for example, Cotton Mather, *An Epistle to the Christian Indians: Giving Them a Short Account, of What the English Desire Them to Know and to Do, in Order to Their Happiness* (Boston: Bartholomew Green, 1706).

20. E. Merton Coulter, "When John Wesley Preached in Georgia," *Georgia Historical Quarterly* 9, no. 4 (1925): 332–35.

21. Rodney P. Carlisle and J. Geoffrey Golson, eds., *Native America from Prehistory to First Contact* (Santa Barbara, CA: ABC-CLIO, 2007), 184.

22. Alice Lee Marriott and Carol K. Rachlin, eds., *American Indian Mythology* (New York: Signet, 1972), 174–5.

23. Julius H. Rubin, *Tears of Repentance: Christian Indian Identity and Community in Colonial Southern New England* (Lincoln: University of Nebraska Press, 2018), 24, 61.

24. *Diary of Cotton Mather, 1709–1724* (Boston: Massachusetts Historical Society, 1912), 282, 342.

25. *Diary of Cotton Mather, 1709–1724*, 363, 446.

26. See Margaret Humphreys Warner, "Vindicating the Medical Role: Cotton Mather's Concept of the 'Nishmath-Chajim' and the Spiritualization of Medicine," *Journal of the History of Medicine and Allied Sciences* 36, no. 3 (1981): 278–95.

27. Mather, *Right Thoughts*, 10.

28. Edward Taylor, "Upon Wedlock, and Death of Children," in *The New Anthology of*

American Poetry: Traditions and Revolutions, Beginnings to 1900, ed. Steven Gould Axelrod (New Brunswick, NJ: Rutgers University Press, 2003), 91–92.

29. *Diary of Cotton Mather, 1709—1724*, 261.

30. See Paul E. Kopperman and Jeanne Abrams, "Cotton Mather's Medicine, with Particular Reference to Measles," *Journal of Medical Biography* 27, no. 1 (2019): 30–37.

31. Koch, *Course of God's Providence*, 5.

32. For a useful survey, see Ronald Numbers and Darrel W. Amundsen, eds., *Caring and Curing: Health and Medicine in the Western Religious Traditions* (New York: Macmillan, 1986).

33. Basil the Great, "Question 55," in *St. Basil: Ascetical Works*, trans. by Sister M. Monica Wagner (Washington, DC: Catholic University of America Press, 1962), 330–31.

34. Carter Lindberg, "The Lutheran Tradition," in Numbers and Amundsen, *Caring and Curing*, 178.

35. On Wesley, see Harold Y. Vanderpool, "The Wesleyan-Methodist Tradition," in Numbers and Amundsen, *Caring and Curing*, 317–53; Philippa Koch, "Experience and the Soul in Eighteenth-Century Medicine," *Church History* 85, no. 3 (2016): 552–86.

36. For the medieval Christian, see Roy Porter, *The Greatest Benefit to Mankind: A Medical History of Humanity* (New York: W. W. Norton, 1999), 129.

37. Mather is quoted in O. T. Beall and R. H. Shryock, *Cotton Mather: First Significant Figure in American Medicine* (Baltimore: Johns Hopkins University Press, 1954), 215.

38. Mather, *Right Thoughts*, 42.

39. Quoted from Mather's *Wholesome Words* (1703) in Maxine Van de Wetering, "A Reconsideration of the Inoculation Controversy," *New England Quarterly* 58, no. 1 (1985): 60.

40. See Peter Harrison, *The Fall of Man and the Foundations of Science* (Cambridge: Cambridge University Press, 2007); Joanna Picciotto, "Reforming the Garden: The Experimentalist Eden and 'Paradise Lost,'" *English Literary History* 72, no. 1 (Spring 2005): 23–78; Michael Hunter, ed., *Science and the Shape of Orthodoxy: Intellectual Change in Late Seventeenth-Century Britain* (Woodbridge, UK: Boydell, 1995). David Noble has traced Christian visions of technology as a means of recovering Eden back to the medieval period (and forward to the US Space Program) in *The Religion of Technology: The Divinity of Man and the Spirit of Invention* (New York: Knopf, 1997). On the spread of Bacon's ideas to New England, see Zachary McLeod Hutchins, "Building Bensalem at Massachusetts Bay: Francis Bacon and the Wisdom of Eden in Early Modern New England," *New England Quarterly* 83, no. 4 (2010): 577–606.

41. Cotton Mather, *Magnalia Christi Americana: or, The Ecclesiastical History of New-England*, book 4 (London: Thomas Parkhurst, 1702), 132.

42. For herbal remedies, see Cotton Mather, *The Christian Philosopher: A Collection of the Best Discoveries in Nature, with Religious Improvements* (Charlestown, MA: Middlesex Bookstore, 1815; originally published 1721), 145, 147.

43. John Ray, *The Wisdom of God Manifested in the Works of the Creation* (London: Samuel Smith, 1691), 114–15; John Ray, *The Wisdom of God Manifested in the Works of Creation*, 4th ed. (London: Samuel Smith, 1704), 192, 245.

44. For Sydenham, see *The Works of Thomas Sydenham, M.D.*, trans. R. G. Latham from the Latin ed. of William Alexander Greenhill (London: Sydenham Society, 1848), 173.

45. See the classic account, Richard S. Westfall, *Science and Religion in Seventeenth Century England* (Ann Arbor: University of Michigan Press, 1973).

46. On the mechanization of nature as a means of maintaining God's sovereignty, see Gary B. Deason, "Reformation Theology and the Mechanistic Conception of Nature," in *God and Nature: Historical Essays on the Encounter between Christianity and Science*, ed. David C. Lindberg and Ronald L. Numbers (Berkeley: University of California Press, 1986), 167–91.

47. See Steven Shapin, "Descartes the Doctor: Rationalism and Its Therapies," *British Journal for the History of Science* 33, no. 2 (2000): 131–54; Brooke, *Science and Religion*, chap. 4.

48. Brooke, *Science and Religion*, 118.

49. For the problems with the Newtonian version of the design argument and Ray's work as answer, see Neil C. Gillespie, "Natural History, Natural Theology, and Social Order: John Ray and the 'Newtonian Ideology,'" *Journal of the History of Biology* 20, no. 1 (1987): 27.

50. On the dilemma posed by the mechanical philosophy in the wake of the English Civil War and John Ray's solution, see Westfall, *Science and Religion*; Gillespie, "Natural History."

51. Ray, *Wisdom of God* (1691), 102, 105, 11.

52. Ray, 11–12.

53. Ray, 167.

54. Genesis 1:21 (KJV); Ray, 167.

55. On Mather and natural theology, see Jeffrey Jeske, "Cotton Mather: Physico-Theologian," *Journal of the History of Ideas* 47, no. 4 (1986): 583–94. On Mather's use of Ray's work, see Winton U. Solberg, "Science and Religion in Early America: Cotton Mather's 'Christian Philosopher,'" *Church History* 56, no. 1 (1987): 73–92. The numbers are from R. Tindol, "Getting the Pox of All Their Houses: Cotton Mather and the Rhetoric of Puritan Science," *Early American Literature* 46, no. 1 (2011): 1–23.

56. Mather, *Christian Philosopher*, 11.

57. Mather, 234, 289, 274, 238–39. William Cockburn (1669–1739) was a Scottish physician who wrote on the "animal economy," and James Keill (1673–1719) was a Scottish physician who emphasized mathematical methods in physiology. Dr. Sloane was Sir Hans Sloane (1660–1753), an Anglo-Irish physician who served as secretary and editor of the Royal Society's *Philosophical Transactions*.

58. On natural theology and anatomy, see Brooke, *Science and Religion*. On dissection and the mechanical philosophy, see Porter, *Greatest Benefit*. On Harvey, see Richard A. Hunter and Ida Macalpine, "William Harvey and Robert Boyle," *Notes and Records of the Royal Society of London* 13, no. 2 (1958): 115–27.

59. Mather, *Magnalia Christi Americana*, book 1, 7.

60. Mather, *Christian Philosopher*, 288.

61. See "Origin of Disease and Medicine (Cherokee)" in Carlisle and Golson, *Native America*, 184.

62. Vanderpool, "Wesleyan-Methodist Tradition," 324.

63. Mather, *Christian Philosopher*, 291, 308.

64. Mather's prayer is quoted in Kenneth Silverman, *The Life and Times of Cotton Mather* (New York: Harper & Row, 1984), 338.

65. For Mather's description of Onesimus, see *Diary of Cotton Mather, 1681–1708* (Boston: Massachusetts Historical Society, 1911), 579. On Mather and Onesimus's relationship, see Kathryn S. Koo, "Strangers in the House of God: Cotton Mather, Onesimus, and an Experiment in Christian Slaveholding," *Proceedings of the American Antiquarian Society* 117, no. 1 (2007): 143–75. The letter is transcribed in Margot Minardi, "The Boston Inoculation Controversy of 1721–1722: An Incident in the History of Race," *William and Mary Quarterly* 61, no. 1 (2004): 47–76.

66. See Peter Manseau, *One Nation, Under Gods: A New American History* (New York: Little, Brown and Company, 2015).

67. *Diary of Cotton Mather, 1709–1724*, 626.

68. Boylston's numbers are cited in Amalie M. Kass, "Boston's Historic Smallpox Epidemic," *Massachusetts Historical Review* 14 (2010): 34.

69. Edmund Massey, *A Sermon against the Dangerous and Sinful Practice of Inoculation, Preach'd at St. Andrew's Holborn, On Sunday, July the 8th, 1722* (London: William Meadows, 1722), 29, 23–24.

70. On the role of racism in the debate over inoculation, see Minardi, "Boston Inoculation Controversy."

71. On Sewall, see Jill Lepore, *Book of Ages: The Life and Opinions of Jane Franklin* (New York: Alfred A. Knopf, 2013), 17.

72. For Douglass's words, see William Douglass, "Inoculation of the Small Pox as Practised in Boston, Consider'd in a letter to A—— S—— M. D. & F. R. S. in London," (Boston: J. Franklin, 1722), 7.

73. For analysis of the controversy between Mather and the *New England Courant*, see K. Silverman, *Life and Times*, 353–4; Beall and Shyrock, *Cotton Mather*. For the story of the grenade, see *Selected Letters of Cotton Mather*, ed. Kenneth Silverman (Baton Rouge: Louisiana State Press, 1971), 340.

74. *Diary of Cotton Mather, 1681–1708*, 163–64.

75. Job 2:10 (KJV).

76. *Diary of Cotton Mather, 1681–1708*, 164.

77. Claude-Nicolas Le Cat, "A Monstrous Human Foetus, Having Neither Head, Heart, Lungs, Stomach, Spleen, Pancreas, Liver, nor Kidnies," trans. Michael Underwood, *Philosophical Transactions (1683–1775)* 57 (1767): 20.

78. Brooke, *Science and Religion*, 143.

79. See David J. Silverman, *Red Brethren: The Brothertown and Stockbridge Indians and the Problem of Race in Early America* (Cornell, NY: Cornell University Press, 2010), 140–43.

80. Mather, *Magnalia Christi Americana*, book 1, 7 (Mather's italics).

81. David S. Jones, *Rationalizing Epidemics: Meanings and Uses of American Indian Mortality since 1600* (Cambridge, MA: Harvard University Press, 2004), 48–49.

82. Leslie M. Scott, "Indian Diseases as Aids to Pacific Northwest Settlement," *Oregon Historical Quarterly* 29, no. 2 (1928): 146, 148.

Chapter 2

1. *The Autobiography of Benjamin Franklin: Published Verbatim from the Original Manuscript by His Grandson, William Temple Franklin*, ed. Jared Sparks. (London: Henry G. Bohn, 1850), 92.

2. On Franky's death, see From Benjamin Franklin to Jane Mecom, 13 January 1772, in *The Papers of Benjamin Franklin*, vol. 19, *January 1, 1750 through December 31, 1772*, ed. William B. Willcox (New Haven, CT: Yale University Press, 1975), 28–29.

3. Benjamin Franklin, "The Death of Infants," *Pennsylvania Gazette*, June 20, 1734.

4. Benjamin Franklin and William Temple Franklin, *Memoirs of the Life and Writings of Benjamin Franklin* (London: H. Colburn, 1818), 131.

5. *Autobiography of Benjamin Franklin*, 53.

6. See James Turner, *Without God, Without Creed: The Origins of Unbelief in America* (Baltimore: Johns Hopkins University Press), 71.

7. Turner, 44.

8. On Voltaire's criticism of La Mettrie, see Kathleen Anne Wellman, *La Mettrie: Medicine, Philosophy, and Enlightenment* (Durham, NC: Duke University Press, 1992), 275.

9. *Autobiography of Benjamin Franklin*, 53.

10. Roy Porter, *The Creation of the Modern World: The Untold Story of the British Enlightenment* (New York: W. W. Norton, 2000), 22.

11. Quoted in John M. Werner, "David Hume and America," *Journal of the History of Ideas* 33, no. 3 (1972): 448.

12. For Franklin's stance on medicine, see Stanley Finger, *Dr. Franklin's Medicine* (Philadelphia: University of Pennsylvania Press, 2006), 11. For the letter to his brother, see *The Complete Works of Benjamin Franklin* (New York: G. P. Putnam & Sons, 1888), 264–65.

13. *Benjamin Franklin's Autobiographical Writings*, selected and edited by Carl von Doren (New York: Viking, 1945), 69.

14. "From Benjamin Franklin to Joseph Banks, 9 September 1782," Founders Online (website), National Archives.

15. *Autobiography of Benjamin Franklin*, 85.

16. For Franklin's appreciation for Mather, see Jared Sparks, *Benjamin Franklin, "Doer of Good": A Biography* (Edinburgh: William Nimmo, 1865), 21; I. Bernard Cohen, *Franklin and Newton: An Inquiry into Speculative Newtonian Experimental Science and Franklin's Work in Electricity as an Example Thereof* (Philadelphia: American Philosophical Society, 1956), 207.

17. Turner, *Without God, Without Creed*, 87.

18. Benjamin Franklin, *Silence Dogood, the Busy-Body, and Early Writings: Boston and London, 1722–1726, Philadelphia, 1726–1757, London 1757–1775* (New York: Library of America, 2002), 296.

19. On natural theology as a "big tent" in which theological debate could often be avoided, see Brooke, *Science and Religion*, 211–13.

20. James H. Moorhead, "Between Progress and Apocalypse: A Reassessment of Millennialism in American Religious Thought, 1800–1880," *Journal of American History* 71, no. 3 (December 1984): 526.

21. For mortality rates due to smallpox "taken in the common way" versus inoculation,

see "Preface to Dr. Heberden's Pamphlet on Inoculation, 16 February 1759," in *Papers of Benjamin Franklin*, vol. 8, *April 1, 1758, through December 31, 1759*, ed. Leonard W. Labaree (1965), 281–86. Also see Theodore Diller, *Franklin's Contribution to Medicine: Being a Collection of Letters Written by Benjamin Franklin Bearing on the Science and Art of Medicine and Exhibiting His Social and Professional Intercourse with Various Physicians of Europe and America* (Brooklyn, NY: A. T. Huntington, 1912), 76.

22. Lucia McMahon, "'So Truly Afflicting and Distressing to Me His Sorrowing Mother': Expressions of Maternal Grief in Eighteenth-Century Philadelphia," *Journal of the Early Republic* 32, no. 1 (2012): 57.

23. See David Noble, *A World without Women: The Christian Clerical Culture of Western Science* (New York: Alfred K. Knopf, 1992).

24. See Londa Schiebinger, "The Anatomy of Difference: Race and Sex in Eighteenth-Century Science," *Eighteenth-Century Studies* 23, no. 4 (1990): 387–405.

25. Brian Steele, "Thomas Jefferson's Gender Frontier," *Journal of American History* 95, no. 1 (2008): 33.

26. Lepore, *Book of Ages*, xi.

27. Benjamin Franklin and Jane Mecom, *The Letters of Benjamin Franklin and Jane Mecom*, ed. Carl Van Doren (Princeton, NJ: Princeton University Press, 2015), 171.

28. "To Benjamin Franklin from Jane Mecom, [23 October 1767]," Founders Online (website), National Archives.

29. Londa Schiebinger, "Feminist History of Colonial Science," *Hypatia* 19, no. 1 (2004): 234. Also see Schiebinger, *The Mind Has No Sex?: Women in the Origins of Modern Science* (Cambridge, MA: Harvard University Press, 1989); Susan Scott Parrish, "Women's Nature: Curiosity, Pastoral and the New Science in British America," *Early American Literature* 37, no. 2 (2002): 195–245 (the conduct book is quoted on 201 of the latter).

30. On Wheatley, see Henry Louise Gates Jr., *The Trials of Phillis Wheatley: America's First Black Poet and Her Encounters with the Founding Fathers* (New York: Basic Civitas, 2010).

31. *Memoir and Poems of Phillis Wheatley, a Native African and a Slave* (Boston: Geo. W. Light, 1834), 14.

32. *Poems of Phillis Wheatley*, 85.

33. *Poems of Phillis Wheatley*, 86.

34. *Poems of Phillis Wheatley*, 86, 65.

35. *Poems of Phillis Wheatley*, 48.

36. *Poems of Phillis Wheatley*, 59.

37. *Poems of Phillis Wheatley*, 85.

38. *Poems of Phillis Wheatley*, 42.

39. Gates, *Trials of Phillis Wheatley*, 71.

40. On Barbour, see Larry R. Morrison, "The Religious Defense of American Slavery before 1830," *Journal of Religious Thought* 37, no. 2 (1980): 16–29.

41. Matthew 7:12; 22:39 (KJV).

42. Jeffrey Bilbro, "Who Are Lost and How They're Found: Redemption and Theodicy in Wheatley, Newton, and Cowper," *Early American Literature* 47, no. 3 (2012): 565.

43. A. R. Philps, *Parental Obligation: A Sermon* (London: John Snow, 1847), 11.

44. *Poems of Phillis Wheatley*, 75.

45. See Nicholas Hudson, "From 'Nation' to 'Race': The Origin of Racial Classification in Eighteenth-Century Thought," *Eighteenth-Century Studies* 29, no. 3 (1996): 247–64. On Christoph Meiners, see Gustav Jahoda, "Towards Scientific Racism," in *Race and Racialization: Essential Readings*, ed. Tania Das Gupta et al., 24–30 (Toronto: Canadian Scholars' Press, 2018).

46. See James Delbourgo, "The Newtonian Slave Body: Racial Enlightenment in the Atlantic World," *Atlantic Studies* 9, no. 2 (2012): 182–207.

47. Benjamin Waterhouse, *An Essay Concerning Tussis Convulsiva, or, Whooping Cough. With Observations on the Diseases of Children* (Boston: Munroe & Francis, 1822), 17, 20.

48. Waterhouse, 19, 111 (Waterhouse's italics).

49. Elizabeth Wayles Eppes to Jefferson, October 13, 1784, in Thomas Jefferson, *The Papers of Thomas Jefferson*, vol. 7, *2 March 1784 to 25 February 1785*, ed. Julian P. Boyd (Princeton, NJ: Princeton University Press, 1953), 441.

50. James Currie to Jefferson, November 20, 1784, in *Papers of Thomas Jefferson*, 7:538–39.

51. Currie's treatment is noted in Waterhouse, *Essay Concerning Tussis Convulsiva*, 138.

52. Quoted in Fawn Brodie, *Thomas Jefferson: An Intimate History* (New York: W. W. Norton, 1974), 169.

53. Brodie, 169, 173.

54. David Hume, *Dialogues Concerning Natural Religion* (London: William Blackwood & Sons, 1907), 159.

55. *The Life and Selected Writings of Thomas Jefferson*, ed. Adrienne Koch and William Peden (New York: Modern Library, 1944), 645.

56. Thomas Balguy, *Discourses on Various Subjects*, vol. 2, (Cambridge: J. Smith, 1822), 200 (Balguy's italics).

57. Quoted in Maxine Van De Wetering, "Reconsideration"; originally from Cotton Mather, *Wholesome Words* (Boston: B. Green & J. Allen, 1703).

58. Quoted in Hunter, *Science and the Shape of Orthodoxy*, 14–16.

59. On Boyle, see Michael Hunter, "First Steps in Institutionalization," in *Science and Shape of Orthodoxy*, 14–16. For Boyle's citation of Sydenham, see Robert Boyle, *A Free Enquiry into the Vulgarly Received Notion of Nature*, ed. Edward B. Davis and Michael Hunter (Cambridge: Cambridge University Press, 1996), 78.

60. Brooke, *Science and Religion*, 143.

61. *Papers of Thomas Jefferson*, vol. 43, *11 March to 30 June 1804*, ed. James P. McClure (2017), 458–9.

62. Matthew 10:29–31 (KJV).

63. See Brooke, *Science and Religion*, 194. For Jefferson and Rush's correspondence regarding yellow fever, see Thomas A. Appel, *Feverish Bodies, Enlightened Minds: Science and the Yellow Fever Controversy in the Early American Republic* (Palo Alto, CA: Stanford University Press, 2016), chap. 4.

64. On Rush, see John Harvey Powell, *Bring Out Your Dead: The Great Plague of Yellow Fever in Philadelphia in 1793* (Philadelphia: University of Pennsylvania Press, 1949).

65. From Thomas Jefferson to Benjamin Rush, 21 April 1803, in *Papers of Thomas*

Jefferson, vol. 40, *4 March to 10 July 1803*, ed. Barbara B. Oberg (2013), 251–53; From Thomas Jefferson to Moses Robinson, 23 March 1801, in *Papers of Thomas Jefferson*, vol. 33, *17 February to 30 April 1801*, ed. Barbara B. Oberg (2006), 423–24.

66. Turner, *Without God, Without Creed*, 87.

67. Peter S. Onuf, "'We Shall All Be Americans': Thomas Jefferson and the Indians," *Indiana Magazine of History* 95, no. 2 (1999): 134, 141. For Jefferson's views of Linnaeus and other taxonomists, see Thomas Jefferson to John Manners, 22 February 1814, in *Papers of Thomas Jefferson*, Retirement Series, vol. 7, *28 November 1813 to 30 September 1814*, ed. J. Jefferson Looney (2010), 207–11.

68. See Christian B. Keller, "Philanthropy Betrayed: Thomas Jefferson, the Louisiana Purchase, and the Origins of Federal Indian Removal Policy," *Proceedings of the American Philosophical Society* 144, no. 1 (2000): 66.

69. Silverman, *Red Brethren*, 140–43.

70. George Catlin, *Adventures of the Ojibbeway and Ioway Indians in England, France, and Belgium*, vol. 2, 3rd ed. (London: George Catlin, 1852), 41.

71. Benjamin Waterhouse, *A Prospect of Exterminating the Small Pox, Part II* (Cambridge, MA: University Press, 1802), 38.

72. See Andrew C. Isenberg, "An Empire of Remedy: Vaccination, Natives, and Narratives in the North American West," *Pacific Historical Review* 86, no. 1 (2017): 84–113.

73. J. E. Gessler and S. L. Kotar, *Smallpox: A History* (Jefferson, NC: McFarland, 2013), 62–63.

74. Thomas Jefferson, *Notes on the State of Virginia* (London: John Stockdale, 1787), 234.

75. Peter S. Onuf, "'To Declare Them a Free and Independent People': Race, Slavery, and National Identity in Jefferson's Thought," *Journal of the Early Republic* 18, no. 1 (1998): 33.

76. To Thomas Jefferson from Benjamin Banneker, 19 August 1791, in *Papers of Thomas Jefferson*, vol. 22, *6 August 1791 to 31 December 1791*, ed. Charles T. Cullen (1986), 49–54.

77. Ian Frederick Finseth, *Shades of Green: Visions of Nature in the Literature of American Slavery, 1770–1860* (Athens: University of Georgia Press, 2009), 46–47.

78. Waterhouse, *Essay Concerning Tussis Convulsiva*, 23.

79. "To Thomas Jefferson from Benjamin Waterhouse, 8 July 1822," Founders Online (website), National Archives.

80. Waterhouse, *Essay Concerning Tussis Convulsiva*, 74.

81. Benjamin Waterhouse, *The Rise, Progress, and Present State of Medicine. A Discourse, Delivered at Concord, July 6th, 1791. Before the Middlesex Medical Association* (Boston: Thomas & John Fleet, 1792), 25.

82. "Observations Concerning the Increase of Mankind, 1751," in *Papers of Benjamin Franklin*, vol. 4, *July 1, 1750, through June 30, 1753*, ed. Leonard W. Labaree (1961), 225–34.

83. Zohreh Bayatrizi, *Life Sentences: The Modern Ordering of Mortality* (Toronto: University of Toronto Press, 2008), 66–72.

84. Thomas Malthus, *An Essay on the Principle of Population; or, A View of Its Past and Present Effects on Human Happiness, with an Inquiry into Our Prospects Respecting the Future*

Removal or Mitigation of the Evils Which It Occasions, 2nd ed. (London: J. Johnson, 1803), 3, 347.

85. Malthus, *Principle of Population*, 484, 491, 502–3.

86. Thomas Malthus, *First Essay on Population 1798*, with notes by James Bonar (London: MacMillan, 1926), 363.

87. William Paley, *Natural Theology: or, Evidences of the Existence and Attributes of the Deity, Collected from the Appearances of Nature* (London: R. Faulder, 1802), 540.

88. See William Smith, "William Paley's Theological Utilitarianism in America," *William and Mary Quarterly* 11, no. 3 (1954): 417; Wendell Glick, "Bishop Paley in America," *New England Quarterly* 27, no. 3 (September 1954): 347.

89. Waterhouse, *Essay Concerning Tussis Convulsiva*, 130.

90. Andrew Combe, *Treatise on the Physiological and Moral Management of Infancy* (Philadelphia: Carey & Hart, 1840), 24–25, 289.

91. Combe, 49, 18 (Combe's italics).

92. Combe, 49.

93. Paley, *Natural Theology*, 541.

94. Combe, *Treatise*, 27 (Combe's italics).

95. Elizabeth Cady Stanton, *Eighty Years and More* (New York: European Publishing, 1898), 114–15.

96. John S. Haller, *Outcasts from Evolution: Scientific Attitudes of Racial Inferiority, 1859–1900*, 2nd ed. (Carbondale: Southern Illinois University Press, 1996), 77.

97. See Dennis Hodgson, "Malthus' Essay on Population and the American Debate over Slavery," *Comparative Studies in Society and History* 51, no. 4 (2009): 742–70.

98. On Chickering, see Marshall Hall, *The Two-Fold Slavery of the United States* (London: Adam Scott, 1854), 19.

Chapter 3

1. The letters notifying friends and family of Waldo's death may be found in *The Letters of Ralph Waldo Emerson*, vol. 3, ed. Ralph L. Rusk (New York: Columbia University Press, 1939), 6–10; *The Correspondence of Thomas Carlyle and Ralph Waldo Emerson, 1834–1872*, ed. Charles Eliot Norton (Boston: Ticknor, 1886), 390.

2. United States Census Office, *Statistics of the United States, Including Mortality, Property, etc. in 1860* (Washington, DC: Government Printing Office, 1866), 240; Caspar Morris, *Lectures on Scarlet Fever* (Philadelphia: Lindsay & Blakiston, 1851), 1.

3. Morris, *Lectures on Scarlet Fever*, 2, 20.

4. Delores Bird Carpenter, introduction to *Life of Lidian Jackson Emerson*, by Ellen Tucker Emerson (East Lansing: Michigan State University Press, 1992); T. E. C., "Lidian Emerson, Wife of Ralph Waldo Emerson, on the Death of Her Five-Year-Old Son (1842)," *Pediatrics* 83, no. 2. (1989): 192.

5. *Letters of Ralph Waldo Emerson*, 11.

6. Howard R. Murphy, "The Ethical Revolt against Christian Orthodoxy in Early Victorian England," *American Historical Review* 60, no. 4 (1955): 801.

7. John Marsh, *The Beloved Physician* (Hartford, CT: Goodwin, 1825), 10.

8. Paul Bixler, "Horace Mann—Mustard Seed," *American Scholar* 7, no. 1 (1938): 27.

9. See J. D. Bowers, *Joseph Priestley and English Unitarianism in America* (University Park: Pennsylvania State University Press, 2010), 18.

10. *Memoir of William Ellery Channing*, vol. 1, ed. William Henry Channing (London: John Chapman, 1848), 282, 319.

11. *The Works of William Ellery Channing*, vol. 3, ed. Joseph Barker (London: Chapman, 1844), 204–5.

12. William Henry Channing, *The Life of William Ellery Channing* (Boston: American Unitarian Association, 1896), 323–24.

13. Andrew Preston Peabody, *Christian Consolations: Sermons Designed to Furnish Comfort and Strength to the Afflicted* (Boston: W. M. Crosby & H. P. Nichols, 1847), 9.

14. David M. Robinson, "Fields of Investigation: Emerson and Natural History," *American Literature and Science*, ed. Robert J. Scholnick (Lexington: University Press of Kentucky, 2014), 96; Robert Edward James, "Emerson, Unitarianism, and the 1833–1834 Lectures on Science" (PhD diss., University of California, 1995), 31.

15. Karen Kalinevitch, "Turning from the Orthodox: Emerson's Gospel Lectures," *Studies in the American Renaissance* (1986): 100.

16. See Brooke, *Science and Religion*, chap. 7.

17. For an example of someone who found an immanent God in the uniformity of nature's laws, see Aubrey Moore, described in Brooke, 314.

18. Kalinevitch, "Turning from the Orthodox," 71, 99.

19. Paul F. Boller, *American Transcendentalism, 1830–1860: An Intellectual Inquiry* (New York: G. P. Putnam & Sons, 1974), 147.

20. Thoreau to Emerson, March 11, 1842, in Raymond R. Borst, *The Thoreau Log: A Documentary Life of Henry David Thoreau, 1817–1862* (Boston: G. K. Hall, 1992), 77.

21. *The Journals and Miscellaneous Notebooks of Ralph Waldo Emerson*, vol. 2, *1838–1842*, ed. A. W. Plumstead and Harrison Hayford (Cambridge, MA: Belknap, 1969), 132.

22. *Letters of Ralph Waldo Emerson*, 9.

23. *The Selected Letters of Mary Moody Emerson*, ed. Nancy Craig Simmons (Athens: University of Georgia Press, 1993), 442.

24. Phyllis Cole, *Mary Moody Emerson and the Origins of Transcendentalism: A Family History* (Oxford: Oxford University Press, 1998), 263.

25. *Letters of Ralph Waldo Emerson*, 9.

26. Jefferson, *Notes on Virginia*, 83; on the discovery of extinction, see Mark V. Barrow, *Nature's Ghosts: Confronting Extinction from the Age of Jefferson to the Age of Ecology* (Chicago: University of Chicago Press, 2009).

27. Alfred Tennyson, *In Memoriam* (London: Edward Moxon, 1850), 78–80.

28. On Robert Chambers and the story of *Vestiges*, see James A. Secord, *Victorian Sensation: The Extraordinary Publication, Reception, and Secret Authorship of* Vestiges of the Natural History of Creation (Chicago: University of Chicago Press, 2003).

29. Genesis 1:24 (KJV).

30. Robert Chambers, *Vestiges of the Natural History of Creation and Other Evolutionary Writings*, ed. James A. Secord (Chicago: University of Chicago Press, 1994), 156.

31. Chambers, 363–65.

32. Chambers, 368–69.

33. Edward Hitchcock, *A Wreath for the Tomb* (Amherst, MA: J. S. & C. Adams, 1842), 72.
34. Chambers, *Vestiges*, 384.
35. Chambers, 373.
36. *Letters of Ralph Waldo Emerson*, 283.
37. Quoted in Finseth, *Shades of Green*, 327.
38. See John B. Wilson, "Darwin and the Transcendentalists," *Journal of the History of Ideas* 26, no. 2 (1965): 286–90.
39. *The Journals and Miscellaneous Notebooks of Ralph Waldo Emerson*, vol. 8, *1841–1843*, ed. William H. Gilman and J. E. Parsons (Cambridge, MA: Harvard University Press, 1970), 36.
40. *The Later Lectures of Ralph Waldo Emerson, 1843–1871* (Athens: University of Georgia Press, 2010), 123.
41. Quoted in Finseth, *Shades of Green*, 194.
42. Oliver Wendell Holmes, *Ralph Waldo Emerson, John Lothrop Motley: Two Memoirs* (Boston: Houghton, Mifflin, 1892), 313.
43. Oliver Wendell Holmes, *Medical Essays 1842–1882* (Boston: Houghton, Mifflin, 1891), 203.
44. Holmes, *Medical Essays*, 365–66.
45. Holmes, 180–81.
46. Theodore Parker, *False and True Theology: A Sermon, Delivered at the Music Hall, Boston, on Sunday, February 14, 1858* (Boston: B. Marsh, 1859), 6–7. "Old Red Sandstone" was a reference to a book by the Scottish geologist Hugh Miller and sedimentary rock formations (of the same name) in the lands surrounding the North Atlantic.
47. *The Works of Theodore Parker: The World of Matter and the Spirit of Man* (Boston: American Unitarian Association, 1907), 373.
48. Theodore Parker, *Sermons of Theism, Atheism, and the Popular Theology* (Boston: Ticknor & Fields, 1861), 271.
49. Asa Gray, review of *Explanations: A Sequel to "Vestiges of the Natural History of Creation,"* *North American Review* 62, no. 131 (1846): 401, 500, 504. Additional reviewers are examined in Ryan C. MacPherson, "Natural and Theological Science at Princeton, 1845–1859: Vestiges of Creation Meets the Scientific Sovereignty of God," *Princeton University Library Chronicle* 65, no. 2 (2004): 184–236.
50. Robert Chambers, *Explanations: A Sequel to "Vestiges of the Natural History of Creation,"* 2nd ed. (London: John Churchill, 1846), 183–84.
51. Peabody, *Christian Consolations* (1847), 165–67.
52. Andrew Preston Peabody, *Christian Consolations: Sermons Designed to Furnish Comfort and Strength to the Afflicted*, 4th ed. (Boston: Crosby, Nichols, 1858), 287–88.
53. Peabody, *Christian Consolations* (1858), 291.
54. Peabody, 292.
55. Peabody, *Christian Consolations* (1847), 166.
56. Peabody, 170, 101.
57. Nancy Schrom Dye and Daniel Blake Smith, "Mother Love and Infant Death, 1750–1920," *Journal of American History* 73, no. 2 (1996): 343.

58. Walter Clark Palmer, *Life and Letters of Leonidas L. Hamline, DD, Late One of the Bishops of the Methodist Episcopal Church* (New York: Carlton & Porter, 1866), 42–43.

59. Thomas M. Morgan, "The Education and Medical Practice of Dr. James McCune Smith (1813–1865), First Black American to Hold a Medical Degree," *Journal of the National Medical Association* 95, no. 7 (2003): 603–14.

60. For Smith's religious beliefs, see Craig D. Townsend, *Faith in Their Own Color: Black Episcopalians in Antebellum New York City* (New York: Columbia University Press, 2005).

61. *The Works of James McCune Smith: Black Intellectual and Abolitionist*, ed. John Stauffer (Oxford: Oxford University Press, 2006), 49.

62. *Works of James McCune Smith*, 259.

63. American Anti-Slavery Society, *Fifth Annual Report of the Executive Committee of the American Anti-Slavery Society* (New York: William S. Dorr, 1838), 28. On Smith, see Anna Mae Duane, *Educated for Freedom: The Incredible Story of Two Fugitive Schoolboys Who Grew Up to Change a Nation* (New York: New York University Press, 2020).

64. *Works of James McCune Smith*, 308 (Smith's italics).

65. *Works of James McCune Smith*, 72.

66. All quotes are from James McCune Smith, "Colored Orphan's Asylum: Physicians Report," *Colored American* January 26, 1839. On Smith's response to American race theories, see the final chapter of Bruce Dain, *A Hideous Monster of the Mind: American Race Theory in the Early Republic* (Cambridge, MA: Harvard University Press, 2002).

67. *Works of James McCune Smith*, 53.

68. Acts 17:26 (KJV).

69. *Works of James McCune Smith*, 58, 73 (Smith's italics).

70. *Works of James McCune Smith*, 44, 46, 47, 252, 47.

71. *Works of James McCune Smith*, 78.

72. Townsend, *Faith*, 81.

73. *Works of James McCune Smith*, 314, 217.

74. Richard H. Steckel, "A Dreadful Childhood: The Excess Mortality of American Slaves," *Social Science History* 10, no. 4 (1986): 427. Also see John Jackson and Nadine M. Weidman, *Race, Racism, and Science: Social Impact and Interaction* (New Brunswick, NJ: Rutgers University Press, 2006), chap. 1; Harriet A. Washington, *Medical Apartheid: The Dark History of Medical Experimentation on Black Americans from Colonial Times to the Present* (New York: Doubleday, 2006).

75. For Affleck's comments, see Thomas Affleck, "On the Hygiene of Cotton Plantations and the Management of Negro Slaves," in *Steward of the Land: Selected Writings of Nineteenth-Century Horticulturist Thomas Affleck*, ed. Lake Douglas (Baton Rouge: Louisiana State University Press, 2014), 200–207; Nancy Krieger, "Shades of Difference: Theoretical Underpinnings of the Medical Controversy on Black/White Differences in the United States, 1830–1870," *International Journal of Health Services* 17, no. 2 (1987): 259–78.

76. *Works of James McCune Smith*, 308.

77. Townsend, *Faith*, 82.

78. *Autobiography, Correspondence, etc., of Lyman Beecher, D.D.*, vol. 1, ed. Charles Beecher (New York: Harper & Brothers, 1864), 178.

79. *Autobiography of Lyman Beecher*, 429–30.

80. Debby Applegate, *The Most Famous Man in America: The Biography of Henry Ward Beecher* (New York: Doubleday, 2007), 116.

81. Joan D. Hedrick, *Harriet Beecher Stowe: A Life* (Oxford: Oxford University Press, 1995), 153; Catharine Beecher, *Common Sense Applied to Religion; or, The Bible and the People* (New York: Harper & Brothers, 1857), 333. Also see Catharine Beecher, *The Evils Suffered by American Women and American Children: The Causes and the Remedy* (New York: Harper & Brothers, 1846). For the critics, see "Miss Beecher's *Common Sense Applied to Religion*," *Church Review* 10 (1857–1858): 421–44.

82. Tiya Miles, "'Circular Reasoning': Recentering Cherokee Women in the Antiremoval Campaigns," *American Quarterly* 61, no. 2 (2009): 221, 223.

83. *Autobiography of Lyman Beecher*, 391.

84. Hedrick, *Harriet Beecher Stowe*, 279, 283–84.

85. Hedrick, 200.

86. Hedrick, 193.

87. Hedrick, 148. On changing explanations of cholera, see Charles E. Rosenberg, *The Cholera Years: The United States in 1832, 1849, and 1866* (Chicago: University of Chicago Press, 1987).

88. Charles Edward Stowe, *Life of Harriet Beecher Stowe, Compiled from Her Letters and Journals* (Boston: Houghton, Mifflin, 1889), 124.

89. Hedrick, *Harriet Beecher Stowe*, 282.

90. C. Stowe, *Life of Harriet Beecher Stowe*, 198.

91. On attempts to define the different spheres of God's providence, see Charles D. Cashdollar, "The Social Implications of the Doctrine of Divine Providence: A Nineteenth-Century Debate in American Theology," *Harvard Theological Review* 71, no. 3/4 (1978): 265–84.

92. C. Stowe, *Life of Harriet Beecher Stowe*, 198.

93. Harriet Beecher Stowe, *Uncle Tom's Cabin: or, Negro Life in the Slave States of America* (London: Richard Bentley, 1852), 90.

94. C. Stowe, *Life of Harriet Beecher Stowe*, 161.

95. Sarah Lewis, "Circular of the Anti-Slavery Convention of American Women," in *Proceedings of the Third Anti-Slavery Convention of American Women, Held in Philadelphia* (Philadelphia: Merrihew & Thompson, 1839), 25–28.

96. Frances Ellen Watkins, *Poems on Miscellaneous Subjects* (Boston: J. B. Yerrinton, 1854), 6–8.

97. Hedrick, *Harriet Beecher Stowe*, 229.

98. Tom McGehee, "Josiah Clark Nott, M.D.," *Magnolia Messenger* (Spring 2009): 1–5.

99. Josiah C. Nott and Geo. R. Gliddon, *Types of Mankind; or, Ethnological Researches Based upon the Ancient Monuments, Paintings, Sculptures, and Crania of Races, and upon Their Natural, Geographical, Philological, and Biblical History* (London: Trubner, 1854), 398.

100. Samuel A. Cartwright, "The Diseases and Physical Peculiarities of the Negro Race," *Southern Medical Reports* 2 (1851): 423. See Katherine Bankole, "The Human/Subhuman Issue and Slave Medicine in Louisiana," *Race, Gender & Class* 5, no. 3 (1998): 3–11.

101. Martin Pernick, *A Calculus of Suffering: Pain, Professionalism, and Anesthesia in Nineteenth Century America* (New York: Columbia University Press, 1985), 48, 78–79, 155–56.

102. United States Supreme Court, *The Dred Scott Decision: Opinion of Chief Justice Roger B. Taney, in the Dred Scott Case*, 2nd ed. (New York: Van Evrie, Horton, 1863), iii, 19, 45.

103. Josiah C. Nott, "The Mulatto a Hybrid—Probable Extermination of the Two Races if Whites and Blacks Are Allowed to Marry," *American Journal of the Medical Sciences* 5 (1843): 254 (Nott's italics).

104. *Works of James McCune Smith*, 270–71.

105. See Britt Rusert, *Fugitive Science: Empiricism and Freedom in Early African American Culture* (New York: New York University Press, 2017), 52.

Chapter 4

1. Frederick Douglass, *Correspondence*, vol. 2, *1853–1865*, ed. John R. Kaufman-McKivigan, series 3 of *The Frederick Douglass Papers* (New Haven, CT: Yale University Press, 2018), 287–88.

2. David W. Blight, *Frederick Douglass: Prophet of Freedom* (New York: Simon & Schuster, 2020), 319.

3. William D. Carrigan, "In Defense of the Social Order: Racial Thought among Southern White Presbyterians in the Nineteenth Century," *American Nineteenth Century History* 1, no. 2 (2000): 36.

4. Robert Kenny, "From the Curse of Ham to the Curse of Nature: The Influence of Natural Selection on the Debate on Human Unity before the Publication of *The Descent of Man*," *British Journal for the History of Science* 40, no. 3 (2007): 367, 370–71.

5. Blight, *Frederick Douglass*, 115.

6. Quoted in Blight, 115.

7. Adrian Desmond and James Moore, *Darwin's Sacred Cause: How a Hatred of Slavery Shaped Darwin's Views on Human Evolution* (Boston: Houghton Mifflin Harcourt, 2014), 58.

8. "What to the Slave is the Fourth of July?" in Frederick Douglass, *The Essential Douglass: Selected Writings and Speeches*, ed. Nicholas Buccola (Indianapolis, IN: Hackett, 2016), 65.

9. *Life and Times of Frederick Douglass* (Hartford, CT: Park Publishing, 1881), 475.

10. Blight, *Frederick Douglass*, 514–15.

11. See D. H. Dilbeck, *Frederick Douglass: America's Prophet* (Chapel Hill: University of North Carolina Press, 2018), chap. 10.

12. *Life and Times of Frederick Douglass*, new revised ed. (Boston: De Wolfe, Fiske, 1892), 299.

13. James A. Secord's introduction in Chambers, *Vestiges*, xxi.

14. Bill Jenkins, "Phrenology, Heredity and Progress in George Combe's 'Constitution of Man,'" *British Journal for the History of Science* 48, no. 3 (2015): 455–73.

15. Frederic May Holland, *Frederick Douglass: The Colored Orator* (New York: Funk & Wagnalls, 1895), 42.

16. Desmond and Moore, *Darwin's Sacred Cause*, 36, 95, 164–65.

17. Blight, *Frederick Douglass*, 569–70. On debates regarding Douglass's changing autobiographical strategy, see Dilbeck, *Frederick Douglass*.

18. Holland, *Frederick Douglass*, 336–37.

19. Holland, 337.

20. "It Moves, or the Philosophy of Reform," in F. Douglass, *Essential Douglass*, 291.

21. Blight, *Frederick Douglass*, 286.

22. Charles Darwin, *Journal of Researches into the Natural History and Geology of the Countries Visited during the Voyage of the H.M.S. Beagle round the World*, 2nd ed. (London: John Murray, 1845), 500.

23. Charles Darwin, *On the Origin of Species by Means of Natural Selection, or the Preservation of Favoured Races in the Struggle for Life* (London: John Murray, 1859), 3.

24. Darwin, 5.

25. Alfred Russel Wallace, *My Life: A Record of Events and Opinions* (London: Chapman & Hall, 1905), 362.

26. Quoted in James R. Moore, *The Post-Darwinian Controversies: A Study of the Protestant Struggle to Come to Terms with Darwin in Great Britain and America, 1870–1900* (Cambridge: Cambridge University Press, 1981), 196.

27. Darwin, *Origin of Species*, 244, 471.

28. Charles Darwin, *The Descent of Man, and Selection in Relation to Sex*, 1874, 2nd ed., ed. James Moore and Adrian Desmond (London: Penguin Books, 2004), 168.

29. *The Autobiography of Charles Darwin 1809–1882*, ed. and with appendix and notes by Nora Barlow (London: Collins, 1958), 91.

30. Darwin, *Origin of Species*, 471.

31. Darwin, *Journal of Researches*, 435.

32. Darwin, *Origin of Species*, 490.

33. Romans 8:22 (KJV).

34. Asa Gray, *Darwiniana: Essays and Reviews Pertaining to Darwinism* (New York: Appleton, 1888), 378.

35. Asa Gray, *Natural Selection Not Inconsistent with Natural Theology. A Free Examination of Darwin's Treatise on the Origin of Species, and of Its American Reviewers* (Boston: Ticknor & Fields, 1861), 52.

36. Shira Wolosky, "Emily Dickinson's War Poetry: The Problem of Theodicy," *Massachusetts Review* 25, no. 1 (1984): 36.

37. Asa Gray to Charles Darwin, July 7, 1863, Darwin Correspondence Project (website), "Letter no. 4234."

38. Asa Gray to Charles Darwin, November 24, 1862, Darwin Correspondence Project (website), "Letter no. 3823."

39. Holland, *Frederick Douglass*, 339.

40. David Strauss, *The Old Faith and the New: A Confession*, American edition. (New York: Henry Holt, 1873), 205.

41. Brooke, *Science and Religion*, 271.

42. Frederick Douglass, *The Claims of the Negro, Ethnologically Considered* (Rochester, NY: Lee, Mann, 1854), 8.

43. For the place of Darwin's work in debates over race, see Desmond and Moore, *Darwin's Sacred Cause*; John C. Greene, "Darwin as a Social Evolutionist," *Journal of the History of Biology* 10, no. 1 (1977): 1–27. For abolitionists' hopes that Darwin's work would serve the "sacred cause," see Randall Fuller, *The Book That Changed America: How Darwin's Theory of Evolution Ignited a Nation* (New York: Penguin, 2018).

44. Darwin, *Descent of Man*, 203–4.

45. See Matthew Day, "Godless Savages and Superstitious Dogs: Charles Darwin, Imperial Ethnography, and the Problem of Human Uniqueness," *Journal of the History of Ideas* 69, no. 1 (2008): 49–70.

46. Pernick, *Calculus of Suffering*.

47. Quoted in Krieger, "Shades of Difference," 275.

48. Haller, *Outcasts from Evolution*, 85.

49. Lyman Abbott, ed., *Henry Ward Beecher. A Sketch of His Career: With Analyses of His Power as a Preacher, Lecturer, Orator, and Journalist, and Incidents and Reminiscences of His Life* (New York: Funk & Wagnalls, 1883), 362.

50. Applegate, *Most Famous Man*, 269.

51. Abbott, *Henry Ward Beecher*, 362–63.

52. Applegate, *Most Famous Man*, 167, 191.

53. Moore, *Post-Darwinian Controversies*, 219, 239.

54. Applegate, *Most Famous Man*, 269–72.

55. Applegate, 171, 215, 470.

56. Henry Calderwood, "Animal Ethics as Described by Herbert Spencer," *Philosophical Review* 1, no. 3 (1892): 241–52.

57. Moore, *Post-Darwinian Controversies*, 92.

58. Applegate, *Most Famous Man*, 355.

59. Henry Ward Beecher, *Evolution and Religion* (New York: Fords, Howard & Hulbert, 1886), 90, 213.

60. Beecher, 75, 291.

61. Romans 8:22 (KJV).

62. Beecher, *Evolution and Religion*, 76, 233.

63. Beecher, 118.

64. Beecher, 33–34, 118, 124, 247.

65. Beecher, 45–46.

66. Beecher, 49.

67. Jones, *Rationalizing Epidemics*, 138.

68. Darwin to Kingsley, February 6, 1862, Darwin Correspondence Project (website), "Letter no. 3439." The point about moralists is from Darwin, *Descent of Man*, 213.

69. Darwin to William Graham, July 3, 1881, Darwin Correspondence Project (website), "Letter no. 13230."

70. Desmond and Moore, *Darwin's Sacred Cause*, 146–47.

71. Kingsley's stance is summarized in Desmond and Moore, 318.

72. See George M. Fredrickson, *The Black Image in the White Mind: The Debate on Afro-American Character and Destiny, 1817–1914* (Middletown, CT: Wesleyan University Press, 1987).

73. John S. Haller, "The Physician versus the Negro: Medical and Anthropological Concepts of Race in the Late Nineteenth Century," *Bulletin of the History of Medicine* 44, no. 2 (1970): 155.

74. Albion Winegar Tourgée, *An Appeal to Caesar* (New York: Fords, Howard and Hulbert, 1884), 127–28.

75. *The Works of Robert G. Ingersoll*, 12 vols. (New York: Dresden, 1915), 12:423.

76. Ingersoll to Horace Traubel in 1893, quoted in Harold K. Bush, "'Nature Shrieking' and Parasitic Wasps: Mark Twain, Theodicy, and the War of Nature," *Mark Twain Annual* 17, no. 1 (2019): 116–17.

77. *Works of Robert G. Ingersoll*, 11:278–79.

78. Robert Ingersoll, *The Gods, and Other Lectures* (Peoria, IL: C. P. Farrell, 1877), 70–71.

79. Ingersoll, 71.

80. Herman E. Kittredge, *Ingersoll: A Biographical Appreciation* (New York: Dresden, 1911), 377–79.

81. Kittredge, 115–16.

82. Ingersoll, *Gods*, 72–73.

83. Ingersoll, 72–73.

84. Ingersoll, 44–45.

85. Ingersoll, 86.

86. Frank M. Turner, "The Victorian Crisis of Faith and the Faith That Was Lost," in *Victorian Faith in Crisis: Essays on Continuity and Change in Nineteenth-Century Religious Belief*, ed. Richard J. Helmstadter and Bernard V. Lightman (London: Palgrave Macmillan, 1990), 16, 17, 19.

87. *Letters of Travel from Caspar Morris, M. D., 1871–1872, to His Family* (Philadelphia: Times Printing House, 1896), 66.

88. Michael Ruse, *The Darwinian Revolution: Science Red in Tooth and Claw* (Chicago: University of Chicago Press, 1979), 73.

89. *Works of Robert G. Ingersoll*, 8:609.

90. Elizabeth Cady Stanton et al., *The Woman's Bible* (New York: European Publishing, 1895), 27.

91. Bonnie S. Anderson, *The Rabbi's Atheist Daughter: Ernestine Rose, International Feminist and Pioneer* (Oxford: Oxford University Press, 2017), 55.

92. Carla Bittel, "Science, Suffrage, and Experimentation: Mary Putnam Jacobi and the Controversy over Vivisection in Late Nineteenth-Century America," *Bulletin of the History of Medicine* 79, no 4 (2005): 677.

93. John Weiss, *Life and Correspondence of Theodore Parker*, vol. 2 (New York: D. Appleton, 1864), 423.

94. "It Moves, or the Philosophy of Reform," in F. Douglass, *Essential Douglass*, 289.

95. See Jean Max Charles, "The Slave Revolt That Changed the World and the Conspiracy against It: The Haitian Revolution and the Birth of Scientific Racism," *Journal of Black Studies* 51, no. 4 (2020): 275–94.

96. Harold E. B. Speight, "From Magic, to Science, to Faith," *Christian Century* 43, no. 5 (1926): 141.

Chapter 5

1. Ronald Numbers, introduction to *Galileo Goes to Jail: And Other Myths about Science and Religion*, ed. Ronald Numbers (Cambridge, MA: Harvard University Press, 2010), 3; Donald Fleming, *John William Draper and the Religion of Science* (Philadelphia: University of Pennsylvania Press, 1950), 31. On Draper, see Fleming's book; and James C. Ungureanu, "A Yankee at Oxford: John William Draper at the British Association for the Advancement of Science at Oxford, 30 June 1860," *Notes and Records of the Royal Society of London* 70, no. 2 (2016): 135–50.

2. John William Draper, *History of the Conflict between Religion and Science* (New York: D. Appleton, 1874), xv.

3. M. Alper Yalcinkaya, "Science as an Ally of Religion: A Muslim Appropriation of 'The Conflict Thesis,'" *British Journal for the History of Science* 44, no. 2 (2011): 161–81.

4. Draper, *History of Conflict*, 323.

5. *Autobiography of Andrew Dickson White*, vol. 2 (New York: Century, 1905), 514–22.

6. Edward Beecher and C. K. Tuckerman, "Lyman Beecher and Infant Damnation," *North American Review* 150, no. 401 (1890): 529–31.

7. *Autobiography of Andrew Dickson White*, 533–4, 520.

8. Andrew Dickson White, *A History of the Warfare of Science with Theology in Christendom*, 2 vols. (New York: D. Appleton and Company, 1896), 1:42.

9. *Autobiography of Andrew Dickson White*, 559.

10. See David C. Lindberg, "Medieval Science and Its Religious Context," *Osiris* 10, no. 1 (1995): 60–79.

11. Andrew Dickson White, *The Warfare of Science* (New York: D. Appleton, 1876), 94.

12. *Autobiography of Andrew Dickson White*, 194, 453.

13. White, *History of Warfare*, 2:395.

14. White, 1:23–24.

15. White, *Warfare of Science*, 122.

16. White, *History of Warfare*, 2:291–92.

17. White, 1:65–66.

18. See Glenn C. Altschuler, "From Religion to Ethics: Andrew D. White and the Dilemma of a Christian Rationalist," *Church History* 47, no. 3 (1978): 308–24. Also see James Ungureanu, "From Divine Oracles to the Higher Criticism: Andrew D. White and the Warfare of Science with Theology in Christendom," *Zygon* 56, no. 1 (2021): 209–33; Richard Schaefer, "Andrew Dickson White and the History of a Religious Future," *Zygon* 50, no. 1 (2015): 7–27.

19. Glenn C. Altschuler, *Better Than Second Best: Love and Work in the Life of Helen Magill* (Champaign: University of Illinois Press, 1990), 90–91.

20. *Autobiography of Andrew Dickson White*, 535.

21. White, *Warfare of Science*, 116, 71, 107.

22. David C. Lindberg and Ronald L. Numbers, "Beyond War and Peace: A Reappraisal of the Encounter between Christianity and Science," *Church History* 55, no. 3 (1986): 340. Also see Ronald Numbers, "Science and Religion," *Osiris* 1, no. 1 (1985): 59–80; Moore, *Post-Darwinian Controversies*.

23. White, *History of Warfare*, 2:63.

24. *Autobiography of Andrew Dickson White*, 54.
25. White, *History of Warfare*, 2:107.
26. White, 2:95.
27. J. S. Lippincott, "The Critics of Evolution," *American Naturalist* 14, no. 5 (1880): 414.
28. John Burroughs, "The Decadence of Theology," *North American Review* 156, no. 438 (1893): 581.
29. White, *History of Warfare*, 2:95, citing Roger S. Tracy, "Village Sanitary Associations," in *A Treatise on Hygiene and Public Health*, vol. 2, ed. Albert H. Buck (New York: William Wood, 1879), 573–5. The numbers are taken from Tracy.
30. "Science and Theology," *Church Quarterly* 46, no. 91 (1898): 131–32.
31. "Texas' Sin of Omission—Her Sanitary Needs," *Texas Medical Journal* 14 (1898–1899): 397.
32. *Autobiography of Andrew Dickson White*, 2:200–201.
33. *Autobiography of Andrew Carnegie* (Boston: Houghton Mifflin, 1920), 23.
34. James Joseph Walsh, *Makers of Modern Medicine* (New York: Fordham University Press, 1915), 317–18.
35. James Joseph Walsh, *Religion and Health* (Boston: Little, Brown, 1920), 62.
36. Paul Carus, *The History of the Devil and the Idea of Evil from the Earliest Times to the Present Day* (Chicago: Open Court, 1900), 369.
37. David L. McMahan, "Modernity and the Early Discourse of Scientific Buddhism," *Journal of the American Academy of Religion* 72, no. 4 (2004): 897–933.
38. J. M. Wheeler, "Buddhist and Christian Missions," *Freethinker*, December 5, 1879, 770.
39. Paul Carus, *Monism and Meliorism: A Philosophical Essay on Causality and Ethics* (New York: F. W. Christern, 1885), 75, 82.
40. John Walter Cross, *George Eliot's Life: As Related in Her Letters and Journals*, vol. 3 (Edinburgh: William Blackwood & Sons, 1880), 429.
41. Junius Henri Browne, "The Philosophy of Meliorism," *Forum* 22 (January 1897): 629.
42. Elizabeth Stuart Phelps, "The Gates Ajar—Twenty-Five Years After," *North American Review* 156, no. 438 (1893): 572, 575.
43. Altschuler, "From Religion to Ethics," 320.
44. Jeanne Campbell Reesman, "Mark Twain vs. God: The Story of a Relationship," *Mark Twain Journal* 52, no. 2 (2014): 114.
45. *Mark Twain's Letters*, vol. 5, ed. Lin Salamo and Harriet Elinor Smith (Berkeley: University of California Press, 1997), 99, 97, 101.
46. The story "Little Bessie Would Assist Providence" can be found in Albert Bigelow Paine, *Mark Twain: A Biography*, vol. 4 (New York: Harper & Brothers, 1912).
47. Mark Twain, *What Is Man? and Other Irreverent Essays* (New York: Prometheus, 2009), 121.
48. *Works of Robert G. Ingersoll*, 9:482.
49. Canadian Social Hygiene Council, "Diphtheria: A Popular Health Article," *Public Health Journal* 18 (1927): 574.

50. George Bernard Shaw, *Back to Methuselah: A Metabiological Pentateuch* (New York: Brentano's, 1921), l–li.

51. Edward Headlam Greenhow, *On Diphtheria* (London: J. W. Parker & Son, 1860), 212.

52. J. Lewis Smith. *A Treatise on the Diseases of Infancy and Childhood* (Philadelphia: Lea Brothers, 1890), 356.

53. John Ballard Blake, *Public Health in the Town of Boston, 1630–1822* (Cambridge, MA: Harvard University Press, 1959), 50–51.

54. T. E. Mitchell. "Diphtheria," *Southern Practitioner* 19, no. 3 (1897): 111.

55. Ira Jefferson Bush, "Diphtheria," *Medical Brief* 24 (1896): 1017.

56. George Buchanan, "Tracheotomy in Diphtheria and Croup," *British Medical Journal* 1, no. 1006 (1880): 555.

57. "The Rabbeth Memorial," *Maryland Medical Journal* 12 (November 29, 1884): 87.

58. Abraham Jacobi, *A Treatise on Diphtheria* (New York: William Wood, 1880), 223–24.

59. Charles D. Meigs, *Obstetrics: The Science and the Art* (Philadelphia: Lea & Blanchard, 1849), 616.

60. Carla Bittel, *Mary Putnam Jacobi and the Politics of Medicine in Nineteenth-Century America* (Chapel Hill: University of North Carolina Press, 2012), 96.

61. Bittel, "Science, Suffrage, and Experimentation," 670.

62. Eugene Link, "Abraham and Mary Jacobi, Humanitarian Physicians," *Journal of the History of Medicine and Allied Sciences* 4, no. 4 (1949): 382–92. The quotes above are from Mary Jacobi's *The Value of Life*, as quoted in Link.

63. James Joseph Walsh, "Biographical Notes on Joseph O'Dwyer, M.D.—A.D. 1841–1898. The Inventor of Intubation," *Records of the American Catholic Historical Society of Philadelphia* 14, no. 4 (1903): 419.

64. Howard Atwood Kelly, *A Cyclopedia of American Medical Biography, Comprising the Lives of Eminent Deceased Physicians and Surgeons from 1610 to 1910*, vol. 2 (Philadelphia: W. B. Saunders, 1912), 226.

65. Quoted in Perri Klass, *Good Time*, 144.

66. See Nancy Tomes, *The Gospel of Germs: Men, Women, and the Microbe in American Life* (Cambridge, MA: Harvard University Press, 1999).

67. Clifton F. Hodge. "The Vivisection Question," *Popular Science Monthly* 49 (1896): 615.

68. Hodge, 618.

69. Craig Buettinger. "Women and Antivivisection in Late Nineteenth-Century America," *Journal of Social History* 30, no. 4 (1997): 858.

70. Albert Leffingwell, "Does Vivisection Pay?" *Scribner's Monthly* 20, no. 3 (July 1880): 395 (Leffingwell's italics).

71. Hodge, "Vivisection Question," 620, 624.

72. Arthur Newsholme, *Fifty Years in Public Health: A Personal Narrative with Comments* (London: George Allen & Unwin, 1935), 190.

73. W. P. Northrup, "Antitoxin Treatment of Diphtheria a Pronounced Success," *Forum* 22 (September 1896): 59–60, 64.

74. William Cheatham, "The Present Status of the Serum Treatment of Diphtheria," *Southern Practitioner* 21 (1899): 216.

75. "Diphtheria," 573–4.

76. Jacob R. Johns, "Thoughts on Antitoxin," *North American Practitioner* 9, no. 1 (January 1897): 8. A misprint in one journal replaced "logic" with "magic."

77. George Wilkinson, "Welch on Diphtheria," *Omaha Clinic* 8 (1895): 221.

78. Evelynn Hammonds, *Childhood's Deadly Scourge: The Campaign to Control Diphtheria in New York City, 1880–1930* (Baltimore: Johns Hopkins University Press, 2002), 6–7.

79. William Williams Keen. *Animal Experimentation and Medical Progress* (Boston: Houghton Mifflin, 1914), 231, 233, 285.

80. F. E. McCann, "Put This before Your Clientele," *American Journal of Clinical Medicine* 23 (1916): 444–45.

81. "Relied on Prayer to Cure Their Ills: The Five Hansens Almost Dead of Diphtheria Before the Physicians Arrived," *New York Times*, September 24, 1910, 6.

82. C. J. B. C., "A New Cure for the Plague," *New Unity* 5, no. 23 (1897): 473–74.

83. Shaw, *Back to Methuselah*, l–li.

84. W. E. B. Du Bois, *The Souls of Black Folk: Essays and Sketches* (Chicago: A. C. McClurg, 1903), 211.

85. Shannon Mariotti, "On the Passing of the First-Born Son: Emerson's 'Focal Distancing,' Du Bois' 'Second Sight,' and Disruptive Particularity," *Political Theory* 37, no. 3 (2009): 351–74.

86. W. E. B. Du Bois, "I Bury My Wife," *Chicago Globe*, July 15, 1950.

87. See Meckel, *Save the Babies*.

88. Du Bois, "I Bury My Wife."

89. H. M. Folkes, "The Negro as a Health Problem," in *Transactions of the Section on Diseases of Children of the American Medical Association at the 61st Annual Session, Held at St. Louis, Mo., 1910* (American Medical Association Press, Chicago, 1910), 63–65.

90. Helen R. Deese, "Caroline Healey Dall and the American Social Science Movement," in *Toward a Female Genealogy of Transcendentalism*, ed. Jana L. Argersinger and Phyllis Cole (Athens: University of Georgia Press, 2014), 303–24.

91. Sharon D. Jones-Eversley and Lorraine T. Dean, "After 121 Years, It's Time to Recognize WEB Du Bois as a Founding Father of Social Epidemiology," *Journal of Negro Education* 87, no. 3 (2018): 230–45.

92. Du Bois, *Souls of Black Folk*, chap. 11.

93. Edward J. Blum, "'There Won't Be Any Rich People in Heaven': The Black Christ, White Hypocrisy, and the Gospel According to W. E. B. Du Bois," *Journal of African American History* 90, no. 4 (2005): 368–86.

94. Edward J. Blum, "The Spiritual Scholar: W. E. B. Du Bois," *Journal of Blacks in Higher Education* 57 (Autumn 2007): 73.

95. Patrick S. Allen, "'We Must Attack the System': The Print Practice of Black 'Doctresses'" *American Quarterly* 74, no. 4 (2018): 99.

96. Gregory Michael Dorr, "Assuring America's Place in the Sun: Ivey Foreman Lewis and the Teaching of Eugenics at the University of Virginia, 1915–1953," *Journal of Southern History* 66, no. 2 (2000): 269, 273, 292.

97. Alison M. Parker, *Unceasing Militant: The Life of Mary Church Terrell* (Chapel Hill: University of North Carolina Press, 2020), 46.

98. Mary Church Terrell, *A Colored Woman in a White World* (Washington, DC: Ransdell, 1940), 119, 136.

99. Terrell, 172.

100. Mary Church Terrell, "Harriet Beecher Stowe," 1911, *Mary Church Terrell Papers: Speeches and Writings, 1911–1953*, Library of Congress (website), manuscript/mixed material.

101. Parker, *Unceasing Militant*, 52.

102. Terrell, *Colored Woman*, 163.

103. Terrell, 118.

104. Terrell, 401.

105. Terrell, 119.

106. H. G. Wells, introduction to Terrell, *Colored Woman*, 1–13.

107. "The Sacrifice of the Innocents," prepared under the advice and suggestions of Dr. Henry Koplik, *Cosmopolitan* 47 (1909): 423.

108. "Sacrifice of the Innocents," 434.

109. Dye and Smith, "Mother Love," 353. On the role that experience with child loss played in women's commitment to child welfare policies, see Molly Ladd-Taylor, *Mother-Work: Women, Child Welfare, and the State, 1890–1930* (Champaign: University of Illinois Press, 1994).

110. Paul S. Boyer, Janet Wilson James, and Edward T. James, *Notable American Women, 1607–1950: A Biographical Dictionary* (Cambridge, MA: Belknap, 1971), 42–43.

111. Laura Arksey, "Dutiful Daughter to Independent Woman: The Diaries of Reba Hurn, 1907–1908," *Pacific Northwest Quarterly* 95, no. 4 (2004): 182–93.

112. Horatio N. Parker, review of "Disease in Milk: The Remedy, Pasteurization," *American Journal of Public Health* 4, no. 2 (1914): 151–52.

113. Susannah J. Link and William A. Link, ed., *The Gilded Age and Progressive Era: A Documentary Reader* (Hoboken, NJ: Wiley, 2012), 169–71.

114. Emil G. Hirsch, "The Philosophy of the Reform Movement of American Judaism," in *Yearbook of the Central Conference of American Rabbis* (Cincinnati, OH: Bloch, 1892), 95, 100.

115. Lina Gutherz Straus, *Disease in Milk: The Remedy Pasteurization; The Life Work of Nathan Straus*, 2nd ed. (New York: E. P. Dutton, 1917), 365.

116. Straus, 296.

117. Pernick, *Calculus of Suffering*, 80.

118. Meckel, *Save the Babies*, 136.

119. Gary Dorrien, *Social Ethics in the Making: Interpreting an American Tradition* (Hoboken, NJ: Wiley, 2011), 88.

120. Thomas Hagerty, *Why Physicians Should Be Socialists* (Terre Haute, IN: Standard, 1902), 15. On Hagerty, see Robert E. Doherty, "Thomas J. Hagerty, the Church, and Socialism," *Labor History* 3, no. 1 (1962): 39–56.

121. Hagerty, *Why Physicians*, 12.

122. Hagerty, 16.

123. Thomas Hagerty, *Economic Discontent and Its Remedy* (Terre Haute, IN: Standard, 1902).

124. Margaret Sanger, *The Case for Birth Control: A Supplementary Brief and Statement of Facts* (May 1917), 6.

125. *Margaret Sanger: An Autobiography* (New York: W. W. Norton, 1938), 91–92.

126. *Margaret Sanger: An Autobiography*, 23.

127. Sanger, *Case for Birth Control*, 5.

128. *Margaret Sanger: An Autobiography*, 85.

129. *Margaret Sanger: An Autobiography*, 81.

130. Sanger, *Case for Birth Control*, 11.

131. Ralph Bevan, "God's Call to Birth Control Eugenics," *Birth Control Review* 8, no. 9 (September 1924): 250–52.

132. *The Selected Letters of Florence Kelley, 1869–1931* (Champaign: University of Illinois Press, 2009), 140–41.

133. Kathryn Kish Sklar, *Florence Kelley and the Nation's Work: The Rise of Women's Political Culture, 1830–1900* (New Haven, CT: Yale University Press, 1995), 60.

134. Quoted in Dye and Smith, "Mother Love and Infant Death," 347; Kelley's words are from her April 1915 article "Children in the Cities," published in the *National Municipal Review*.

135. Louis L. Athey, "Florence Kelley and the Quest for Negro Equality," *Journal of Negro History* 56, no. 4 (1971): 252–4, 257.

136. Athey, 255.

137. On the AMA's opposition to the Sheppard-Towner Act, see Howard Pearson, "The History of Pediatrics in America," in *Oski's Pediatrics: Principles and Practice*, ed. Julia A. Macmillan et al., 4th ed. (Philadelphia: Lippincott, Williams & Wilkins, 2006). Pusey's stance can be found in Eugene Link, *The Social Ideas of American Physicians (1776–1976): Studies of the Humanitarian Tradition in Medicine* (Selinsgrove, PA: Susquehanna University Press, 1992), 129.

138. United States Congress, House, Committee on Interstate and Foreign Commerce, *Child Welfare Extension Service: Hearing(s) before the Committee on Interstate and Foreign Commerce on H. R. 14070, to Provide a Child Welfare Extension Service and for Other Purposes*, 70th Cong., 2nd sess., January 24 and 25, 1929, 248–90.

Chapter 6

1. See Martin Pernick, *The Black Stork: Eugenics and the Death of "Defective" Babies in American Medicine and Motion Pictures since 1915* (Oxford: Oxford University Press, 1996).

2. "Was the Doctor Right? Some Independent Opinions," *Independent*, January 3, 1916, 27.

3. Pernick, *Black Stork*, 90, 107.

4. Helen Keller, "Physicians' Juries for Defective Babies," *New Republic*, December 18, 1915, 173–74.

5. Pernick, *Black Stork*, 86, 98.

6. Malone Duggan, "Sane at Last," *Texas Medical Journal*, 31, no. 7 (1915): 240–42.

7. Duggan, 240–42.

8. "Grief Kills Mother of Bollinger Baby," *Chicago Examiner*, July 28, 1917.

9. Beecher, *Evolution and Religion*, 215–17.

10. Darwin, *Descent of Man*, 159. See Diane Paul, "Darwin, Social Darwinism and Eugenics," in *The Cambridge Companion to Darwin*, ed. Jonathan Hodge and Gregory Radick (Cambridge: Cambridge University Press, 2006), 214–39.

11. Darwin, *Descent of Man*, 617.

12. William Jennings Bryan, *The Menace of Darwinism* (New York: Fleming H. Revell, 1921), 39.

13. Holmes, *Medical Essays*, 200.

14. James Young Simpson, *The Spiritual Interpretation of Nature*, 2nd ed. (London: Hodder & Stoughton, 1912), 230–31.

15. David Starr Jordan, *War and the Breed: The Relation of War to the Downfall of Nations* (Boston: Beacon, 1915), 12.

16. The number of college courses on eugenics is from Hamilton Cravens, *The Triumph of Evolution: American Scientists and the Heredity-Environment Controversy, 1900–1941* (Philadelphia: University of Pennsylvania Press, 1978), 53.

17. Francis Galton, *Memories of My Life* (London, Methuen, 1908), 323.

18. Ethel M. Elderton and Karl Pearson, "Further Evidence of Natural Selection in Man," *Biometrika* 10, no. 4 (1915): 506.

19. Paul Popenoe and Roswell Hill Johnson, *Applied Eugenics* (New York: Macmillan, 1920), 122, 412.

20. Popenoe and Johnson, 414.

21. Albert Wiggam, *The New Decalogue of Science* (Indianapolis, IN: Bobbs-Merrill, 1923), 60–62.

22. Wiggam, 20–23.

23. Wiggam, 17–18.

24. Wiggam, 100–101. For more on Wiggam and the use of "God talk" in eugenics writings, see Christine Rosen, *Preaching Eugenics: Religious Leaders and the American Eugenics Movement* (Oxford: Oxford University Press, 2004).

25. Marshall Dawson, *Nineteenth Century Evolution and After* (New York: Macmillan, 1924), 139.

26. "Infant Mortality and Natural Selection," *American Journal of Public Health* 6, no. 2 (June 1916): 182.

27. Wiggam, *New Decalogue of Science*, 67.

28. Edward M. East, *Mankind at the Crossroads* (New York: Charles Scribner's & Sons, 1923), 111.

29. Edward Alsworth Ross, *Changing America: Studies in Contemporary Society* (New York: Century, 1912), 46–48.

30. W. Grant Hague, *The Eugenic Mother and Baby: A Complete Home Guide* (New York: Hague, 1913), 6.

31. William L. Holt, "Eugenics in the United States," *American Journal of Public Health* 4, no. 2 (1914): 152.

32. Hague, *Eugenic Mother and Baby*, 6, 55.

33. See Diane Paul, *Controlling Human Heredity: 1865 to the Present* (New York: Humanity Books, 1995).

34. John Langdon-Davies, *The New Age of Faith* (New York: Viking, 1925), 109.

35. See Diane Paul, "Eugenics and the Left," *Journal of the History of Ideas* 45, no. 4 (1984): 567–90.

36. James Joseph Walsh, "Our Latest Panacea—Eugenics," *Common Cause* 3, no. 2 (1913): 81.

37. G. K. Chesterton, *Eugenics and Other Evils* (London: Cassell & Company, 1922), 50.

38. Franz Boas, "Eugenics," *Scientific Monthly* 3, no 5 (1916): 477.

39. Charles Davenport, "Eugenics," in *Contributions of Science to Religion*, ed. Shailer Mathews (New York: D. Appleton, 1924), 295, 297. For Sanger's efforts to show eugenics and birth control were compatible, see Ellen Chesler, *Woman of Valor: Margaret Sanger and the Birth Control Movement in America* (New York: Simon & Schuster, 2007), 216.

40. Margaret Sanger, *The Pivot of Civilization* (New York: Brentado's, 1922), 181–82.

41. On Du Bois's support for birth control, see W. E. B. Du Bois, "Opinion," *Crisis* 24, no. 144 (October 1922): 247–53; "Black Folk and Birth Control," *Birth Control Review* 16, no. 6 (June 1932): 166–67. For a useful discussion of Sanger's campaign for birth control in Black communities, see Dorothy E. Roberts, *Killing the Black Body: Race, Reproduction, and the Meaning of Liberty* (New York: Vintage, 1999), 72–81.

42. Gregor Michael Dorr and Angela Logan, "Quality, Not Mere Quantity Counts: Black Eugenics and the NAACP Baby Contests," in *A Century of Eugenics in America: From the Indiana Experiment to the Human Genome Era*, ed. Paul Lombardo (Bloomington: Indiana University Press, 2011), 68–94.

43. Julia E. Liss, "Diasporic Identities: The Science and Politics of Race in the Work of Franz Boas and W. E. B. Du Bois, 1894–1919," *Cultural Anthropology* 13, no. 2 (1998): 128, 130, 141.

44. Billy Sunday, "Feeding the Five Thousand," *Pittsburgh Press*, February 12, 1914.

45. The phrase is from William Bell Riley, "The Faith of the Fundamentalists," *Current History* 26 (June 1927): 434–36.

46. See Paul Keith Conkin, *When All the Gods Trembled: Darwinism, Scopes, and American Intellectuals* (Lanham, MD: Rowman & Littlefield, 2001); and the classic works by George Marsden, including *Fundamentalism and American Culture* (Oxford: Oxford University Press, 2006).

47. Michael Kazin, *A Godly Hero: The Life of William Jennings Bryan* (New York: Alfred A. Knopf, 2006), 125.

48. Edward Larson, *Summer for the Gods: The Scopes Trial and America's Continuing Debate over Science and Religion* (New York: Basic Books, 1997), 97.

49. "Billy Sunday on Social Religion," *Literary Digest*, March 6, 1915, 482; William T. Ellis, *"Billy" Sunday: The Man and His Message, with His Own Words That Have Won Thousands for Christ* (Philadelphia: John C. Winston, 1917), 332.

50. William Jennings Bryan, *In His Image* (New York: Fleming H. Revell, 1922), 13, 170.

51. Wiggam, *New Decalogue of Science*, 105.

52. Wiggam, 20.

53. Shailer Mathews, *New Faith for Old: An Autobiography* (New York: MacMillan, 1936), 122.

54. Quoted in Larson, *Summer for the Gods*, 39.

55. Ozora S. Davis, "A Quarter-Century of American Preaching," *Journal of Religion* 6, no. 2 (1926): 142.

56. Clarence Walworth Alvord, "Musings of an Inebriated Historian," *American Mercury* 5, no. 20 (1925): 441.

57. Lynn Thorndike, "The Historical Background of Modern Science," *Scientific Monthly* 16, no. 5 (1923): 492.

58. Bryan, *In His Image*, 124.

59. Bryan, 107.

60. Beecher, *Evolution and Religion*, 190–92.

61. Henry George to E. R. Taylor, January 21, 1881, quoted in Robert C. Bannister, "'The Survival of the Fittest Is Our Doctrine': History or Histrionics?" *Journal of the History of Ideas* 31, no. 3 (1970): 388.

62. *The World's Most Famous Court Trial: Tennessee Evolution Case* (Cincinnati: National Book Company, 1925), 334.

63. Bryan, *In His Image*, 135.

64. Bryan, 117 (Bryan's italics).

65. See Peter J. Bowler, *The Eclipse of Darwinism: Anti-Darwinian Evolution Theories in the Decades around 1900* (Baltimore: Johns Hopkins University Press, 1992).

66. Bryan, *In His Image*, 108–9.

67. William J. Robinson, "The Billy Sunday Disgrace," *Medical Critic and Guide*, May 1917, 167–68.

68. See Edward B. Davis, "Science and Religious Fundamentalism in the 1920s: Religious Pamphlets by Leading Scientists of the Scopes Era Provide Insight into Public Debates about Science and Religion," *American Scientist*, 93, no. 3 (2005): 253–60.

69. Hugh Weir, "The Dawn Man," *McClure's* 55, no. 1 (March 1923): 19.

70. "Scientific Notes and News," *Science* 61, no. 1590 (1925): 629.

71. E. Davenport, "What Science Has Done for Agriculture," in Mathews, *Contributions of Science*, 322.

72. John M. Dodson, "Recent Contributions of Medicine to Human Welfare," in Mathews, *Contributions of Science*, 280, 283.

73. H. A. Delano, "Kind Words for the Doctor: A Minister Excusing Disbelieving Physicians," *Illinois Medical Journal* 40 (September 1921): 247–48.

74. Homer D. Lindgren, ed., *Modern Speeches* (New York: F. S. Crofts, 1926), 149.

75. See Thea Cooper and Arthur Ainsberg, *Breakthrough: Elizabeth Hughes, the Discovery of Insulin, and the Making of a Medical Miracle* (New York: St. Martin's, 2010).

76. John M. Dodson, "Recent Contributions of Medicine to Human Welfare," in Mathews, *Contributions of Science*, 280, 283.

77. Ellsworth Faris, "Social Evolution," in Mathews, *Contributions of Science*, 233.

78. James Young Simpson, "Religious Beliefs in the Light of Modern Science," *Journal of Religion* 5, no. 3 (1925): 323.

79. Davenport, "Eugenics," 297.

80. Maurizio Esposito, "More than the Parts: W. E. Ritter, the Scripps Marine Association, and the Organismal Conception of Life," *Historical Studies in the Natural Sciences* 45, no. 2 (2015): 286. On both Ritter and Davenport, see Philip J. Pauly, *Biologists and the Promise of American Life: From Meriwether Lewis to Alfred Kinsey* (Princeton, NJ: Princeton University Press, 2000).

81. Edwin Grant Conklin, *The Human Direction of Evolution* (New York: Scribner's Sons, 1922), 237–40, 247.

82. For historical scholarship on the Scopes Trial, see Edward Larson's classic work, *Summer for the Gods*; Michael Lienesch, *In the Beginning: Fundamentalism, the Scopes Trial, and the Making of the Antievolution Movement* (Chapel Hill: University of North Carolina Press, 2007); Adam R. Shapiro, *Trying Biology: The Scopes Trial, Textbooks, and the Antievolution Movement in American Schools* (Chicago: University of Chicago Press, 2013); Adam Laats, *Fundamentalism and Education in the Scopes Era: God, Darwin, and the Roots of America's Culture Wars* (New York: Palgrave Macmillan, 2010).

83. The Bryan and Darrow quotes are from Larson, *Summer for the Gods*, 128, 103.

84. Larson, 136.

85. Clarence Darrow and Will Durant, *Debate: Is Man a Machine?* (New York: League for Public Discussion, 1927), 49.

86. *Most Famous Court Trial*, 174–75, 325.

87. Dominic Erdozain, *The Soul of Doubt: The Religious Roots of Unbelief from Luther to Marx* (Oxford University Press, 2015), 174.

88. Larson, *Summer for the Gods*, 22.

89. *Most Famous Court Trial*, 251.

90. *Most Famous Court Trial*, 252.

91. Edward B. Davis, "Altruism and the Administration of the Universe: Kirtley Fletcher Mather on Science and Values," *Zygon* 46, no. 3 (2011): 525.

92. Kirtley Fletcher Mather, *Science in Search of God* (New York: H. Holt, 1928), 107, 116.

93. John H. Dietrich, *The Fathers of Evolution and Other Addresses* (Minneapolis: First Unitarian Society, 1927), 249–56, 272.

94. Speight, "From Magic," 141.

95. Jeffrey Moran, "Reading Race into the Scopes Trial: African American Elites, Science, and Fundamentalism," *Journal of American History* 90, no. 3 (2003): 901.

96. Jeffrey Moran, "The Scopes Trial and Southern Fundamentalism in Black and White: Race, Region, and Religion," *Journal of Southern History* 70, no. 1 (2004): 101, 118.

97. Quoted in Pernick, *Black Stork*, 96.

98. Clarence Darrow, "The Eugenics Cult," *American Mercury* 8, no. 30 (1926): 129, 133.

99. Michael S. Lief and H. Mitchell Caldwell, *The Devil's Advocates: Greatest Closing Arguments in Criminal Law* (New York: Simon & Schuster, 2006), 286.

100. Darrow, "Eugenics Cult," 137, 135.

101. Harriette Shelton Dover, *Tulalip, from My Heart: An Autobiographical Account of a Reservation Community* (Seattle: University of Washington Press, 2013), 131.

102. Dover, 132.

103. Dover, 133.

104. Sheila M. Rothman, *Living in the Shadow of Death: Tuberculosis and the Social Experience of Illness in American History* (New York: Basic Books, 1994), 131.

105. Elizabeth James, "'Hardly a Family Is Free from the Disease': Tuberculosis, Health Care, and Assimilation Policy on the Nez Perce Reservation, 1908–1942," *Oregon Historical Quarterly* 112, no. 2 (2011): 146–47.

106. United States Congress, Senate, *Contagious and Infectious Diseases Among the Indians*, January 27, 1913, 62nd Cong., 3rd sess., 1913, S. Doc 1038 40. Cited in James, "Hardly a Family," 144.

107. Clifford E. Trafzer, "Coughing Blood: Tuberculosis Deaths and Data on the Yakama Indian Reservation, 1911–64," *Canadian Bulletin of Medical History* 15, no. 2 (1998): 251–76.

108. Dover, *Tulalip*, 132.

109. See David H. DeJong, *"If You Knew the Conditions": A Chronicle of the Indian Medical Service and American Indian Health Care, 1908–1955* (Lanham, MD: Lexington Books, 2008), 26–29; Diane T. Putney, "Fighting the Scourge: American Indian Morbidity and Federal Policy, 1897–1928," (PhD diss., Marquette University, 1980), 178–9, 258.

110. Dover, *Tulalip*, 45–46, 180–81.

111. DeJong, *If You Knew*, 258.

112. See Charles Roberts, "The Cushman Indian Trades School and World War I," *American Indian Quarterly* 11, no. 3 (1987): 221–39.

113. David H. DeJong, "'Unless They Are Kept Alive': Federal Indian Schools and Student Health, 1878–1918," *American Indian Quarterly* 31, no. 2 (2007): 256.

114. Dover, *Tulalip*, 79, 144.

115. Dover, 113.

116. Dover, 88, 113–15.

117. Devon Abbott, "'Commendable Progress': Acculturation at the Cherokee Female Seminary," *American Indian Quarterly* 11, no. 3 (1987): 187–201.

118. Dover, *Tulalip*, 180, 189–91.

119. David Starr Jordan and Vernon Kellogg, *Evolution and Animal Life* (New York: D. Appleton, 1907), 468.

120. George William Hunter, *A Civic Biology Presented in Problems* (New York: American Book Company, 1914), 405.

121. Hunter, 152.

122. Hunter, 152–53.

123. Hunter, 196.

124. James, "Hardly a Family," 164.

125. Dover, *Tulalip*, 123.

126. See Jones, *Rationalizing Epidemics*.

127. Rothman, *Living in Shadow*, 184.

128. Arthur Guerard, "The Relation of Tuberculosis to the Tenement House Problem," *Medical News*, February 16, 1901, 15. Reprinted in pamphlet form.

129. Dover, *Tulalip*, 123.

130. Charles M. Buchanan, "Rights of the Puget Sound Indians to Game and Fish," *Washington Historical Quarterly* 6, no. 2 (1915): 111.

131. Joe Starita, *A Warrior of the People: How Susan La Flesche Overcame Racial and Gender Inequality to Become America's First Indian Doctor* (New York: St. Martin's, 2016), 37.

132. Starita, 233.

133. Starita, 267–69.

134. See Lawrence C. Kelly, *The Assault on Assimilation: John Collier and the Origins of Indian Policy Reform* (Albuquerque: University of New Mexico Press, 1983), 292.

135. James B. LaGrand, "The Changing 'Jesus Road': Protestants Reappraise American Indian Missions in the 1920s and 1930s," *Western Historical Quarterly* 27, no. 4 (1996): 479–504.

136. Dover, *Tulalip*, 64.

137. Dover, 166, 189–92.

138. Dover, 278.

139. Dover, 278–79.

Conclusion

1. S. J. Crumbine, "A Statistical Report of Infant Mortality for 1926," *American Journal of Public Health* 17, no. 9 (1927): 922–27.

2. On these developments, see Meckel, *Save the Babies*; Klass, *Good Time*.

3. In 2016 the primary causes of mortality between the ages of one and nineteen were car accidents (20 percent of a total of 20,360 deaths), firearms (15 percent, more than half of which were ruled homicides, 35 percent suicides, and 4 percent accidental discharge), cancer (9 percent), suffocation (7 percent), and drowning (5 percent). See Rebecca M. Cunningham, Maureen A. Walton, and Patrick M. Carter, "The Major Causes of Death in Children and Adolescents in the United States," *New England Journal of Medicine* 379, no. 25 (2018): 2468–2475. In 2022 firearms moved ahead of car accidents. See Jason E. Goldstick, Rebecca M. Cunningham, and Patrick M. Carter. "Current Causes of Death in Children and Adolescents in the United States." *New England Journal of Medicine* 386, no. 20 (2022): 1955–56.

4. Klass, *Good Time*, 7.

5. See Daniel D. Reidpath and Pascale Allotey, "Infant Mortality Rate as an Indicator of Population Health," *Journal of Epidemiology & Community Health* 57, no. 5 (2003): 344–46.

6. PRRI Staff, "The American Religious Landscape in 2020," PRRI (website), Public Religion Research Institute.

7. See Rebecca Pass Philipsborn and Kevin Chan, "Climate Change and Global Child Health," *Pediatrics* 141, no. 6 (2018): e20173774; Charmian M. Bennett and Sharon Friel, "Impacts of Climate Change on Inequities in Child Health," *Children* 1, no. 3 (2014): 461–73.

Bibliography

Abbott, Devon. "'Commendable Progress': Acculturation at the Cherokee Female Seminary." *American Indian Quarterly* 11, no. 3 (1987): 187–201.

Abbott, Lyman, ed. *Henry Ward Beecher. A Sketch of His Career: With Analyses of His Power as a Preacher, Lecturer, Orator, and Journalist, and Incidents and Reminiscences of His Life*. New York: Funk & Wagnalls, 1883.

Affleck, Thomas. "On the Hygiene of Cotton Plantations and the Management of Negro Slaves." In *Steward of the Land: Selected Writings of Nineteenth-Century Horticulturist Thomas Affleck*, edited by Lake Douglas, 200–207. Baton Rouge: Louisiana State University Press, 2014.

Allen, Patrick S. "'We Must Attack the System': The Print Practice of Black 'Doctresses.'" *American Quarterly* 74, no. 4 (2018): 161–86.

Altschuler, Glenn C. *Better Than Second Best: Love and Work in the Life of Helen Magill*. Champaign: University of Illinois Press, 1990.

———. "From Religion to Ethics: Andrew D. White and the Dilemma of a Christian Rationalist." *Church History* 47, no. 3 (1978): 308–24.

Alvord, Clarence Walworth. "Musings of an Inebriated Historian." *American Mercury* 5, no. 20 (1925): 434–44.

American Anti-Slavery Society. *Fifth Annual Report of the Executive Committee of the American Anti-Slavery Society*. New York: William S. Dorr, 1838.

Anderson, Bonnie S. *The Rabbi's Atheist Daughter: Ernestine Rose, International Feminist and Pioneer*. Oxford: Oxford University Press, 2017.

Appel, Thomas A. *Feverish Bodies, Enlightened Minds: Science and the Yellow Fever Controversy in the Early American Republic*. Palo Alto, CA: Stanford University Press, 2016.

Applegate, Debby. *The Most Famous Man in America: The Biography of Henry Ward Beecher*. New York: Doubleday, 2007.

Arksey, Laura. "Dutiful Daughter to Independent Woman: The Diaries of Reba Hurn, 1907–1908." *Pacific Northwest Quarterly* 95, no. 4 (2004): 182–93.

Athey, Louis L. "Florence Kelley and the Quest for Negro Equality." *Journal of Negro History* 56, no. 4 (1971): 249–61.

Balguy, Thomas. *Discourses on Various Subjects*. Vol. 2. Cambridge: J. Smith, 1822.

Bankole, Katherine. "The Human/Subhuman Issue and Slave Medicine in Louisiana." *Race, Gender & Class* 5, no. 3 (1998): 3–11.

Bannister, Robert C. "'The Survival of the Fittest Is Our Doctrine': History or Histrionics?" *Journal of the History of Ideas* 31, no. 3 (1970): 377–98.

Barrow, Mark V. *Nature's Ghosts: Confronting Extinction from the Age of Jefferson to the Age of Ecology*. Chicago: University of Chicago Press, 2009.

Basil the Great. "Question 55." In *St. Basil: Ascetical Works*, translated by Sister M. Monica Wagner, 330–37. Washington, DC: Catholic University of America Press, 1962.

Bayatrizi, Zohreh. *Life Sentences: The Modern Ordering of Mortality*. Toronto: University of Toronto Press, 2008.

Beall, O. T., and R. H. Shyrock. *Cotton Mather: First Significant Figure in American Medicine*. Baltimore: Johns Hopkins University Press, 1954.

Beecher, Catharine. *Common Sense Applied to Religion; or, The Bible and the People*. Montreal: Harper & Brothers, 1857.

———. *The Evils Suffered by American Women and American Children: The Causes and the Remedy*. New York: Harper & Brothers, 1846.

Beecher, Edward, and C. K. Tuckerman. "Lyman Beecher and Infant Damnation." *North American Review* 150, no. 401 (1890): 529–31.

Beecher, Henry Ward. *Evolution and Religion*. New York: Fords, Howard & Hulbert, 1886.

Beecher, Lyman. *Autobiography, Correspondence, etc., of Lyman Beecher, D.D.* Vol. 1. Edited by Charles Beecher. New York: Harper & Brothers, 1864.

Bennett, Charmian M., and Sharon Friel. "Impacts of Climate Change on Inequities in Child Health." *Children* 1, no. 3 (2014): 461–73.

Bevan, Ralph. "God's Call to Birth Control Eugenics." *Birth Control Review* 8, no. 9 (September 1924): 250–52.

Bilbro, Jeffrey. "Who Are Lost and How They're Found: Redemption and Theodicy in Wheatley, Newton, and Cowper." *Early American Literature* 47, no. 3 (2012): 561–89.

Bittel, Carla. *Mary Putnam Jacobi and the Politics of Medicine in Nineteenth-Century America*. Chapel Hill: University of North Carolina Press, 2012.

———. "Science, Suffrage, and Experimentation: Mary Putnam Jacobi and the Controversy over Vivisection in Late Nineteenth-Century America." *Bulletin of the History of Medicine* 79, no. 4 (2005): 664–94.

Bixler, Paul. "Horace Mann—Mustard Seed." *American Scholar* 7, no. 1 (1938): 24–38.

Blake, John Ballard. *Public Health in the Town of Boston, 1630–1822*. Cambridge, MA: Harvard University Press, 1959.

Blight, David W. *Frederick Douglass: Prophet of Freedom*. New York: Simon & Schuster, 2020.

Blum, Edward J. "The Spiritual Scholar: W. E. B. Du Bois." *Journal of Blacks in Higher Education* 57 (Autumn 2007): 73–79.

———. "'There Won't Be Any Rich People in Heaven': The Black Christ, White Hypocrisy, and the Gospel According to W. E. B. Du Bois." *Journal of African American History* 90, no. 4 (2005): 368–86.

Boas, Franz. "Eugenics." *Scientific Monthly* 3, no. 5 (1916): 471–78.

Boller, Paul F. *American Transcendentalism, 1830–1860: An Intellectual Inquiry*. New York: G. P. Putnam & Sons, 1974.

Borst, Raymond R. *The Thoreau Log: A Documentary Life of Henry David Thoreau, 1817–1862*. Boston: G. K. Hall, 1992.

Bowers, J. D. *Joseph Priestley and English Unitarianism in America*. University Park: Pennsylvania State University Press, 2010.
Bowler, Peter J., *The Eclipse of Darwinism: Anti-Darwinian Evolution Theories in the Decades around 1900*. Baltimore: Johns Hopkins University Press, 1992.
Boyer, Paul S., Janet Wilson James, and Edward T. James. *Notable American Women, 1607–1950: A Biographical Dictionary*. Cambridge, MA: Belknap, 1971.
Boyle, Robert. *A Free Enquiry into the Vulgarly Received Notion of Nature*. Edited by Edward B. Davis and Michael Hunter. Cambridge: Cambridge University Press, 1996.
Brodie, Fawn. *Thomas Jefferson: An Intimate History*. New York: W. W. Norton, 1974.
Brooke, John Hedley. *Science and Religion: Some Historical Perspectives*. Cambridge: Cambridge University Press, 1991.
Browne, Junius Henri. "The Philosophy of Meliorism." *Forum* 22 (January 1897): 624–32.
Bryan, William Jennings. *In His Image*. New York: Fleming H. Revell, 1922.
———. *The Menace of Darwinism*. New York: Fleming H. Revell, 1921.
Buchanan, Charles M. "Rights of the Puget Sound Indians to Game and Fish." *Washington Historical Quarterly* 6, no. 2 (1915): 109–18.
Buchanan, George. "Tracheotomy in Diphtheria and Croup." *British Medical Journal* 1, no. 1006 (1880): 554–55.
Buettinger, Craig. "Women and Antivivisection in Late Nineteenth-Century America." *Journal of Social History* 30, no. 4 (1997): 857–72.
Burroughs, John. "The Decadence of Theology." *North American Review* 156, no. 438 (1893): 576–85.
Bush, Harold K. "'Nature Shrieking' and Parasitic Wasps: Mark Twain, Theodicy, and the War of Nature." *Mark Twain Annual* 17, no. 1 (2019): 112–28.
Bush, Ira Jefferson. "Diphtheria." *Medical Brief* 24 (1896): 1016–17.
Calderwood, Henry. "Animal Ethics as Described by Herbert Spencer." *Philosophical Review* 1, no. 3 (1892): 241–52.
Canadian Social Hygiene Council. "Diphtheria: A Popular Health Article." *Public Health Journal* 18 (1927): 572–76.
Carlisle, Rodney P., and J. Geoffrey Golson, eds. *Native America from Prehistory to First Contact*. Santa Barbara, CA: ABC-CLIO, 2007.
Carnegie, Andrew. *Autobiography of Andrew Carnegie*. Boston: Houghton Mifflin, 1920.
Carpenter, Delores Bird. Introduction to *Life of Lidian Jackson Emerson*, by Ellen Tucker Emerson, xii-lvi. East Lansing: Michigan State University Press, 1992.
Carrigan, William D. "In Defense of the Social Order: Racial Thought among Southern White Presbyterians in the Nineteenth Century." *American Nineteenth Century History* 1, no. 2 (2000): 31–52.
Cartwright, Samuel A. "The Diseases and Physical Peculiarities of the Negro Race." *Southern Medical Reports* 2 (1851): 421–29.
Carus, Paul. *The History of the Devil and the Idea of Evil from the Earliest Times to the Present Day*. Chicago: Open Court, 1900.
———. *Monism and Meliorism: A Philosophical Essay on Causality and Ethics*. New York: F. W. Christern, 1885.
Cashdollar, Charles D. "The Social Implications of the Doctrine of Divine Providence: A

Nineteenth-Century Debate in American Theology." *Harvard Theological Review* 71, no. 3/4 (1978): 265–84.

Catlin, George. *Adventures of the Ojibbeway and Ioway Indians in England, France, and Belgium.* Vol. 2. 3rd ed. London: George Catlin, 1852.

Chambers, Robert. *Explanations: A Sequel to "Vestiges of the Natural History of Creation."* 2nd ed. London: John Churchill, 1846.

———. *Vestiges of the Natural History of Creation and Other Evolutionary Writings.* Edited with a new introduction by James A. Secord. Chicago: University of Chicago Press, 1994.

Channing, William Ellery. *Memoir of William Ellery Channing.* Vol. 1. Edited by William Henry Channing. London: John Chapman, 1848.

———. *The Works of William Ellery Channing.* Vol. 3. Edited by Joseph Barker. London: Chapman, 1844.

Channing, William Henry. *The Life of William Ellery Channing.* Boston: American Unitarian Association, 1896.

Charles, Jean Max. "The Slave Revolt That Changed the World and the Conspiracy against It: The Haitian Revolution and the Birth of Scientific Racism." *Journal of Black Studies* 51, no. 4 (2020): 275–94.

Cheatham, William. "The Present Status of the Serum Treatment of Diphtheria." *Southern Practitioner* 21(1899): 216–18.

Chesler, Ellen. *Woman of Valor: Margaret Sanger and the Birth Control Movement in America.* New York: Simon & Schuster, 2007.

Chesterton, G. K. *Eugenics and Other Evils.* London: Cassell & Company, 1922.

C. J. B. C. "A New Cure for the Plague." *New Unity* 5, no. 23 (1897): 473–74.

Cohen, I. Bernard. *Franklin and Newton: An Inquiry into Speculative Newtonian Experimental Science and Franklin's Work in Electricity as an Example Thereof.* Philadelphia: American Philosophical Society, 1956.

Cole, Phyllis. *Mary Moody Emerson and the Origins of Transcendentalism: A Family History.* Oxford: Oxford University Press, 1998.

Combe, Andrew. *Treatise on the Physiological and Moral Management of Infancy.* Philadelphia: Carey & Hart, 1840.

Conkin, Paul Keith. *When All the Gods Trembled: Darwinism, Scopes, and American Intellectuals.* Lanham, MD: Rowman & Littlefield, 2001.

Conklin, Edwin Grant. *The Human Direction of Evolution.* New York: Scribner's Sons, 1922.

Cooper, Thea, and Arthur Ainsberg. *Breakthrough: Elizabeth Hughes, the Discovery of Insulin, and the Making of a Medical Miracle.* New York: St. Martin's, 2010.

Coulter, E. Merton. "When John Wesley Preached in Georgia." *Georgia Historical Quarterly* 9, no. 4 (1925): 317–51.

Cravens, Hamilton. *The Triumph of Evolution: American Scientists and the Heredity-Environment Controversy, 1900–1941.* Philadelphia: University of Pennsylvania Press, 1978.

Cross, John Walter. *George Eliot's Life: As Related in Her Letters and Journals.* Vol. 3. Edinburgh: William Blackwood & Sons, 1880.

Crumbine, S. J. "A Statistical Report of Infant Mortality for 1926." *American Journal of Public Health* 17, no. 9 (1927): 922–27.

Cunningham, Rebecca M., Maureen A. Walton, and Patrick M. Carter. "The Major Causes of Death in Children and Adolescents in the United States." *New England Journal of Medicine* 379, no. 25 (2018): 2468–75.
Dain, Bruce. *A Hideous Monster of the Mind: American Race Theory in the Early Republic.* Cambridge, MA: Harvard University Press, 2002.
Darrow, Clarence. "The Eugenics Cult." *American Mercury* 8, no. 30 (1926): 129–37.
Darrow, Clarence, and Will Durant. *Debate: Is Man a Machine?* New York: League for Public Discussion, 1927.
Darwin, Charles. *The Autobiography of Charles Darwin 1809–1882.* Edited and with appendix and notes by Nora Barlow. London: Collins, 1958.
———. *The Descent of Man, and Selection in Relation to Sex.* 2nd ed. 1874. Edited by James Moore and Adrian Desmond. London: Penguin, 2004.
———. *Journal of Researches into the Natural History and Geology of the Countries Visited during the Voyage of the H.M.S. Beagle round the World.* 2nd ed. London: John Murray, 1845.
———. *On the Origin of Species by Means of Natural Selection, or the Preservation of Favoured Races in the Struggle for Life.* London: John Murray, 1859.
Davenport, Charles. "Eugenics." In Mathews, *Contributions of Science*, 285–301.
Davenport, E. "What Science Has Done for Agriculture." In Mathews, *Contributions of Science*, 302–22.
Davis, Edward B. "Altruism and the Administration of the Universe: Kirtley Fletcher Mather on Science and Values." *Zygon* 46, no. 3 (2011): 517–35.
———. "Science and Religious Fundamentalism in the 1920s: Religious Pamphlets by Leading Scientists of the Scopes Era Provide Insight into Public Debates about Science and Religion." *American Scientist* 93, no. 3 (2005): 253–60.
Davis, Ozora S. "A Quarter-Century of American Preaching." *Journal of Religion* 6, no. 2 (1926): 135–53.
Dawson, Marshall. *Nineteenth Century Evolution and After.* New York: Macmillan, 1924.
Day, Matthew. "Godless Savages and Superstitious Dogs: Charles Darwin, Imperial Ethnography, and the Problem of Human Uniqueness." *Journal of the History of Ideas* 69, no. 1 (2008): 49–70.
Deason, Gary B., "Reformation Theology and the Mechanistic Conception of Nature." In *God and Nature: Historical Essays on the Encounter between Christianity and Science*, edited by David C. Lindberg and Ronald L. Numbers, 167–91. Berkeley: University of California Press, 1986.
Deese, Helen R. "Caroline Healey Dall and the American Social Science Movement." In *Toward a Female Genealogy of Transcendentalism*, edited by Jana L. Argersinger and Phyllis Cole, 303–24. Athens: University of Georgia Press, 2014.
DeJong, David H. *"If You Knew the Conditions": A Chronicle of the Indian Medical Service and American Indian Health Care, 1908–1955.* Lanham, MD: Lexington Books, 2008.
———. "'Unless They Are Kept Alive': Federal Indian Schools and Student Health, 1878–1918." *American Indian Quarterly* 31, no. 2 (2007): 256–82.
Delano, H. A. "Kind Words for the Doctor: A Minister Excusing Disbelieving Physicians." *Illinois Medical Journal* 40 (September 1921): 247–48.

Delbourgo, James. "The Newtonian Slave Body: Racial Enlightenment in the Atlantic World." *Atlantic Studies* 9, no. 2 (2012): 182–207.

Desmond, Adrian, and James Moore. *Darwin's Sacred Cause: How a Hatred of Slavery Shaped Darwin's Views on Human Evolution*. Boston: Houghton Mifflin Harcourt, 2014.

Dietrich, John H. *The Fathers of Evolution and Other Addresses*. Minneapolis: First Unitarian Society, 1927.

Dilbeck, D. H. *Frederick Douglass: America's Prophet*. Chapel Hill: University of North Carolina Press, 2018.

Diller, Theodore. *Franklin's Contribution to Medicine: Being a Collection of Letters Written by Benjamin Franklin Bearing on the Science and Art of Medicine and Exhibiting His Social and Professional Intercourse with Various Physicians of Europe and America*. Brooklyn, NY: A. T. Huntington, 1912.

Dodson, John M. "Recent Contributions of Medicine to Human Welfare." In Mathews, *Contributions of Science*, 269–84.

Doherty, Robert E. "Thomas J. Hagerty, the Church, and Socialism." *Labor History* 3, no. 1 (1962): 39–56.

Dorr, Gregory Michael. "Assuring America's Place in the Sun: Ivey Foreman Lewis and the Teaching of Eugenics at the University of Virginia, 1915–1953." *Journal of Southern History* 66, no. 2 (2000): 257–96.

Dorr, Gregory Michael, and Angela Logan. "Quality, Not Mere Quantity Counts: Black Eugenics and the NAACP Baby Contests." In *A Century of Eugenics in America: From the Indiana Experiment to the Human Genome Era*, edited by Paul Lombardo, 68–94. Bloomington: Indiana University Press, 2011.

Dorrien, Gary. *Social Ethics in the Making: Interpreting an American Tradition*. Hoboken, NJ: Wiley, 2011.

Douglass, Frederick. *The Claims of the Negro, Ethnologically Considered*. Rochester, NY: Lee, Mann, 1854.

———. *Correspondence*. Vol. 2, *1853–1865*. Edited by John R. Kaufman-McKivigan. Series 3 of *The Frederick Douglass Papers*. New Haven, CT: Yale University Press, 2018.

———. *The Essential Douglass: Selected Writings and Speeches*. Edited by Nicholas Buccola. Indianapolis, IN: Hackett, 2016.

———. *Life and Times of Frederick Douglass*. Hartford, CT: Park Publishing, 1881.

———. *Life and Times of Frederick Douglass*. New revised edition. Boston: De Wolfe, Fiske, 1892.

Douglass, William. "Inoculation of the Small Pox as Practised in Boston, Consider'd in a Letter to A—— S—— M. D. & F. R. S., in London." Boston: Printed and sold by J. Franklin, at his printing-house in Queen-Street, over against Mr. Sheaf's school, 1722.

Dover, Harriette Shelton. *Tulalip, from My Heart: An Autobiographical Account of a Reservation Community*. Seattle: University of Washington Press, 2013.

Draper, John William. *History of the Conflict between Religion and Science*. New York: D. Appleton, 1874.

Duane, Anna Mae. *Educated for Freedom: The Incredible Story of Two Fugitive Schoolboys Who Grew Up to Change a Nation*. New York: New York University Press, 2020.

Du Bois, W. E. B. "Black Folk and Birth Control." *Birth Control Review* 16, no. 6 (June 1932): 166–67.

———. "I Bury My Wife." *Chicago Globe*, July 15, 1950.

———. "Opinion." *Crisis* 24, no. 144 (October 1922): 247–53.

———. *The Souls of Black Folk: Essays and Sketches*. Chicago: A. C. McClurg, 1903.

Duggan, Malone. "Sane at Last." *Texas Medical Journal* 31, no. 7 (1915): 240-242.

Dye, Nancy Schrom, and Daniel Blake Smith. "Mother Love and Infant Death, 1750–1920." *Journal of American History* 73, no. 2 (1996): 329–53.

East, Edward M. *Mankind at the Crossroads*. New York: Charles Scribner's & Sons, 1923.

Elderton, Ethel M., and Karl Pearson. "Further Evidence of Natural Selection in Man." *Biometrika* 10, no. 4 (1915): 488–506.

Ellis, William T. *"Billy" Sunday: The Man and His Message, with His Own Words That Have Won Thousands for Christ*. Philadelphia: John C. Winston, 1917.

Emerson, Mary Moody. *The Selected Letters of Mary Moody Emerson*. Edited by Nancy Craig Simmons. Athens: University of Georgia Press, 1993.

Emerson, Ralph Waldo. *The Journals and Miscellaneous Notebooks of Ralph Waldo Emerson*. Vol. 2, *1838–1842*. Edited by A. W. Plumstead and Harrison Hayford. Cambridge, MA: Belknap, 1969.

———. *The Journals and Miscellaneous Notebooks of Ralph Waldo Emerson*. Vol. 8, *1841–1843*. Edited by William H. Gilman and J. E. Parsons. Cambridge, MA: Harvard University Press, 1970.

———. *The Later Lectures of Ralph Waldo Emerson, 1843–1871*. Athens: University of Georgia Press, 2010.

———. *The Letters of Ralph Waldo Emerson*. Vol. 3. Edited by Ralph L. Rusk. New York: Columbia University Press, 1939.

Emerson, Ralph Waldo, and Thomas Carlyle. *The Correspondence of Thomas Carlyle and Ralph Waldo Emerson, 1834–1872*. Edited by Charles Eliot Norton. Boston: Ticknor, 1886.

Erdozain, Dominic. *The Soul of Doubt: The Religious Roots of Unbelief from Luther to Marx*. Oxford: Oxford University Press, 2015.

Esposito, Maurizio. "More than the Parts: W. E. Ritter, the Scripps Marine Association, and the Organismal Conception of Life." *Historical Studies in the Natural Sciences* 45, no. 2 (2015): 273–302.

Faris, Ellsworth. "Social Evolution." In Mathews, *Contributions of Science*, 211–42.

Finger, Stanley. *Dr. Franklin's Medicine*. Philadelphia: University of Pennsylvania Press, 2006.

Finseth, Ian Frederick. *Shades of Green: Visions of Nature in the Literature of American Slavery, 1770–1860*. Athens: University of Georgia Press, 2009.

Fleming, Donald. *John William Draper and the Religion of Science*. Philadelphia: University of Pennsylvania Press, 1950.

Folkes, H. M. "The Negro as a Health Problem." In *Transactions of the Section on Diseases of Children of the American Medical Association at the Sixty-First Annual Session,*

Held at St. Louis, Mo., 1910, 63–68. Chicago: American Medical Association Press, 1910.

Franklin, Benjamin. *The Autobiography of Benjamin Franklin: Published Verbatim from the Original Manuscript by His Grandson, William Temple Franklin*. Edited by Jared Sparks. London: Henry G. Bohn, 1850.

———. *Benjamin Franklin's Autobiographical Writings*. Selected and edited by Carl von Doren. New York: Viking, 1945.

———. *The Complete Works of Benjamin Franklin*. New York: G. P. Putnam & Sons, 1888.

———. "The Death of Infants." *Pennsylvania Gazette*, June 20, 1734.

———. *The Papers of Benjamin Franklin*. 43 vols. New Haven, CT: Yale University Press, 1959–2018.

———. *Silence Dogood, the Busy-Body, and Early Writings: Boston and London, 1722–1726, Philadelphia, 1726–1757, London 1757–1775*. New York: Library of America, 2002.

Franklin, Benjamin, and William Temple Franklin. *Memoirs of the Life and Writings of Benjamin Franklin*. London: H. Colburn, 1818.

Franklin, Benjamin, and Jane Mecom. *The Letters of Benjamin Franklin and Jane Mecom*. Edited by Carl Van Doren. Princeton, NJ: Princeton University Press, 2015.

Fraser, Antonia. *The Weaker Vessel: Woman's Lot in Seventeenth-Century England*. New York: Alfred A. Knopf, 1984.

Fredrickson, George M. *The Black Image in the White Mind: The Debate on Afro-American Character and Destiny, 1817–1914*. Middletown, CT: Wesleyan University Press, 1987.

Fuller, Randall. *The Book That Changed America: How Darwin's Theory of Evolution Ignited a Nation*. New York: Penguin, 2018.

Galton, Francis. *Memories of My Life*. London: Methuen, 1908.

Gates, Henry Louis, Jr. *The Trials of Phillis Wheatley: America's First Black Poet and Her Encounters with the Founding Fathers*. New York: Basic Civitas, 2010.

Gessler, J. E., and S. L. Kotar. *Smallpox: A History*. Jefferson, NC: McFarland, 2013.

Gillespie, Neil C. "Natural History, Natural Theology, and Social Order: John Ray and the 'Newtonian Ideology.'" *Journal of the History of Biology* 20, no. 1 (1987): 1–49.

Glick, Wendell. "Bishop Paley in America." *New England Quarterly* 27, no. 3 (September 1954): 347–54.

Goldstick, Jason E., Rebecca M. Cunningham, and Patrick M. Carter. "Current Causes of Death in Children and Adolescents in the United States." *New England Journal of Medicine* 386, no. 20 (2022): 1955–56.

Gray, Asa. *Darwiniana: Essays and Reviews Pertaining to Darwinism*. New York: D. Appleton, 1888.

———. *Natural Selection Not Inconsistent with Natural Theology. A Free Examination of Darwin's Treatise on the Origin of Species, and of Its American Reviewers*. Boston: Ticknor & Fields, 1861.

———. Review of *Explanations: A Sequel to "Vestiges of the Natural History of Creation."* *North American Review* 62, no. 131 (1846): 465–506.

Greene, John C. "Darwin as a Social Evolutionist." *Journal of the History of Biology* 10, no. 1 (1977): 1–27.

Greenhow, Edward Headlam. *On Diphtheria*. London: J. W. Parker & Son, 1860.
"Grief Kills Mother of Bollinger Baby." *Chicago Examiner*, July 28, 1917.
Guerard, Arthur. "The Relation of Tuberculosis to the Tenement House Problem." *Medical News*, February 16, 1901. Reprinted in pamphlet form.
Hagerty, Thomas. *Economic Discontent and Its Remedy*. Terre Haute, IN: Standard, 1902.
———. *Why Physicians Should Be Socialists*. Terre Haute, IN: Standard, 1902.
Hague, W. Grant. *The Eugenic Mother and Baby: A Complete Home Guide*. New York: Hague, 1913.
Hall, Marshall. *The Two-Fold Slavery of the United States*. London: Adam Scott, 1854.
Haller, John S. *Outcasts from Evolution: Scientific Attitudes of Racial Inferiority, 1859–1900*. 2nd ed. Carbondale: Southern Illinois University Press, 1996.
———. "The Physician versus the Negro: Medical and Anthropological Concepts of Race in the Late Nineteenth Century." *Bulletin of the History of Medicine* 44, no. 2 (1970): 154–67.
Hammonds, Evelynn. *Childhood's Deadly Scourge: The Campaign to Control Diphtheria in New York City, 1880–1930*. Baltimore: Johns Hopkins University Press, 2002.
Hardy, Lucas. "'The Practice of Conveying and Suffering the Small-pox': Inoculation as a Means of Spiritual Conversion in Cotton Mather's *Angel of Bethesda*." *Studies in Eighteenth-Century Culture* 44, no. 1 (2015): 61–79.
Harrison, Peter. *The Fall of Man and the Foundations of Science*. Cambridge: Cambridge University Press, 2007.
Hedrick, Joan D. *Harriet Beecher Stowe: A Life*. Oxford: Oxford University Press, 1995.
Hirsch, Emil G. "The Philosophy of the Reform Movement of American Judaism." In *Yearbook of the Central Conference of American Rabbis*, 90–112. Cincinnati, OH: Bloch, 1892.
Hitchcock, Edward. *A Wreath for the Tomb*. Amherst, MA: J. S. & C. Adams, 1842.
Hodge, Clifton F. "The Vivisection Question." *Popular Science Monthly* 49 (1896): 614–24.
Hodgson, Dennis. "Malthus' Essay on Population and the American Debate over Slavery." *Comparative Studies in Society and History* 51, no. 4 (2009): 742–70.
Holland, Frederic May. *Frederick Douglass: The Colored Orator*. New York: Funk & Wagnalls, 1895.
Holmes, Oliver Wendell. *Medical Essays 1842–1882*. Boston: Houghton, Mifflin, 1891.
———. *Ralph Waldo Emerson, John Lothrop Motley: Two Memoirs*. Boston: Houghton, Mifflin, 1892.
Holt, William L. "Eugenics in the United States." *American Journal of Public Health* 4, no. 2 (1914): 152.
Hudson, Nicholas. "From 'Nation' to 'Race': The Origin of Racial Classification in Eighteenth-Century Thought." *Eighteenth-Century Studies* 29, no. 3 (1996): 247–64.
Hume, David. *Dialogues Concerning Natural Religion*. London: William Blackwood & Sons, 1907.
Hunter, George William. *A Civic Biology Presented in Problems*. New York: American Book Company, 1914.
Hunter, Michael, ed. *Science and the Shape of Orthodoxy: Intellectual Change in Late Seventeenth-Century Britain*. Woodbridge, UK: Boydell, 1995.

Hunter, Richard A., and Ida Macalpine. "William Harvey and Robert Boyle." *Notes and Records of the Royal Society of London* 13, no. 2 (1958): 115–27.

Hutchins, Zachary McLeod. "Building Bensalem at Massachusetts Bay: Francis Bacon and the Wisdom of Eden in Early Modern New England." *New England Quarterly* 83, no. 4 (2010): 577–606.

"Infant Mortality and Natural Selection." *American Journal of Public Health* 6, no. 2 (June 1916): 182.

Ingersoll, Robert. *The Gods, and Other Lectures*. Peoria, IL: C. P. Farrell, 1877.

———. *The Works of Robert G. Ingersoll*. 12 vols. New York: Dresden, 1915.

Isenberg, Andrew C. "An Empire of Remedy: Vaccination, Natives, and Narratives in the North American West." *Pacific Historical Review* 86, no. 1 (2017): 84–113.

Jackson, John P., and Nadine M. Weidman. *Race, Racism, and Science: Social Impact and Interaction*. New Brunswick, NJ: Rutgers University Press, 2006.

Jacobi, Abraham. *A Treatise on Diphtheria*. New York: William Wood, 1880.

Jahoda, Gustav. "Towards Scientific Racism." In *Race and Racialization: Essential Readings*, edited by Tania Das Gupta et al., 24–30. Toronto: Canadian Scholars' Press, 2018.

James, Elizabeth. "'Hardly a Family Is Free from the Disease': Tuberculosis, Health Care, and Assimilation Policy on the Nez Perce Reservation, 1908–1942." *Oregon Historical Quarterly* 112, no. 2 (2011): 142–69.

James, Robert Edward. "Emerson, Unitarianism, and the 1833–1834 Lectures on Science." PhD diss., University of California, 1995.

Jang, C. J., and H. C. Lee. "A Review of Racial Disparities in Infant Mortality in the U.S." *Children* (Basel) 9, no. 2 (February 14, 2022): 257.

Jefferson, Thomas. *The Life and Selected Writings of Thomas Jefferson*. Edited by Adrienne Koch and William Peden. New York: Modern Library, 1944.

———. *Notes on the State of Virginia*. London: John Stockdale, 1787.

———. *The Papers of Thomas Jefferson*. 46 vols. Princeton, NJ: Princeton University Press, 1950–2023.

Jenkins, Bill. "Phrenology, Heredity and Progress in George Combe's 'Constitution of Man.'" *British Journal for the History of Science* 48, no. 3 (2015): 455–73.

Jeske, Jeffrey. "Cotton Mather: Physico-Theologian." *Journal of the History of Ideas* 47, no. 4 (1986): 583–94.

Johns, Jacob R. "Thoughts on Antitoxin." *North American Practitioner* 9, no. 1 (January 1897): 7–9.

Jones, David S. *Rationalizing Epidemics: Meanings and Uses of American Indian Mortality since 1600*. Cambridge, MA: Harvard University Press, 2004.

Jones-Eversley, Sharon D., and Lorraine T. Dean, "After 121 Years, It's Time to Recognize WEB Du Bois as a Founding Father of Social Epidemiology." *Journal of Negro Education* 87, no. 3 (2018): 230–45.

Jordan, David Starr. *War and the Breed: The Relation of War to the Downfall of Nations*. Boston: Beacon, 1915.

Jordan, David Starr, and Vernon Kellogg. *Evolution and Animal Life*. New York: D. Appleton, 1907.

Kalinevitch, Karen. "Turning from the Orthodox: Emerson's Gospel Lectures." *Studies in the American Renaissance* (1986): 69–112.

Kass, Amalie M. "Boston's Historic Smallpox Epidemic." *Massachusetts Historical Review* 14 (2010): 1–51.

Kazin, Michael. *A Godly Hero: The Life of William Jennings Bryan*. New York: Alfred A. Knopf, 2006.

Keen, William Williams. *Animal Experimentation and Medical Progress*. Boston: Houghton Mifflin, 1914.

Keller, Christian B. "Philanthropy Betrayed: Thomas Jefferson, the Louisiana Purchase, and the Origins of Federal Indian Removal Policy." *Proceedings of the American Philosophical Society* 144, no. 1 (2000): 39–66.

Keller, Helen. "Physicians' Juries for Defective Babies." *New Republic*, December 18, 1915, 173–74.

Kelley, Florence. *The Selected Letters of Florence Kelley, 1869–1931*. Champaign: University of Illinois Press, 2009.

Kelly, Howard Atwood. *A Cyclopedia of American Medical Biography, Comprising the Lives of Eminent Deceased Physicians and Surgeons from 1610 to 1910*. Vol. 2. Philadelphia: W. B. Saunders, 1912.

Kelly, Lawrence C. *The Assault on Assimilation: John Collier and the Origins of Indian Policy Reform*. Albuquerque: University of New Mexico Press, 1983.

Kenny, Robert. "From the Curse of Ham to the Curse of Nature: The Influence of Natural Selection on the Debate on Human Unity before the Publication of *The Descent of Man*." *British Journal for the History of Science* 40, no. 3 (2007): 367–88.

Kittredge, Herman E. *Ingersoll: A Biographical Appreciation*. New York: Dresden, 1911.

Klass, Perri. *A Good Time to Be Born: How Science and Public Health Gave Children a Future*. New York: W. W. Norton, 2020.

Koch, Philippa. *The Course of God's Providence: Religion, Health, and the Body in Early America*. New York: New York University Press, 2021.

———. "Experience and the Soul in Eighteenth-Century Medicine." *Church History* 85, no. 3 (2016): 552–86.

Koo, Kathryn S. "Strangers in the House of God: Cotton Mather, Onesimus, and an Experiment in Christian Slaveholding." *Proceedings of the American Antiquarian Society* 117, no. 1 (2007): 143–75.

Kopperman, P. E. and J. Abrams. "Cotton Mather's Medicine, with Particular Reference to Measles." *Journal of Medical Biography* 27, no. 1 (2016): 30–37.

Krieger, Nancy. "Shades of Difference: Theoretical Underpinnings of the Medical Controversy on Black/White Differences in the United States, 1830–1870." *International Journal of Health Services* 17, no. 2 (1987): 259–78.

Laats, Adam. *Fundamentalism and Education in the Scopes Era: God, Darwin, and the Roots of America's Culture Wars*. New York: Palgrave Macmillan, 2010.

Ladd-Taylor, Molly. *Mother-Work: Women, Child Welfare, and the State, 1890–1930*. Champaign: University of Illinois Press, 1994.

LaGrand, James B. "The Changing 'Jesus Road': Protestants Reappraise American Indian

Missions in the 1920s and 1930s." *Western Historical Quarterly* 27, no. 4 (1996): 479–504.

Langdon-Davies, John. *The New Age of Faith*. New York: Viking, 1925.

Larson, Edward. *Summer for the Gods: The Scopes Trial and America's Continuing Debate over Science and Religion*. New York: Basic Books, 1997.

Le Cat, Claude-Nicolas. "A Monstrous Human Foetus, Having Neither Head, Heart, Lungs, Stomach, Spleen, Pancreas, Liver, nor Kidnies." Translated by Michael Underwood. *Philosophical Transactions (1683–1775)* 57 (1767): 1–20.

Leffingwell, Albert. "Does Vivisection Pay?" *Scribner's Monthly* 20, no. 3 (July 1880): 391–99.

Lepore, Jill. *Book of Ages: The Life and Opinions of Jane Franklin*. New York: Alfred A. Knopf, 2013.

Levin, David. *Cotton Mather: The Young Life of the Lord's Remembrancer, 1663–1703*. Cambridge, MA: Harvard University Press, 1978.

Lewis, Sarah. "Circular of the Anti-Slavery Convention of American Women." In *Proceedings of the Third Anti-Slavery Convention of American Women, Held in Philadelphia*, 25–28. Philadelphia: Merrihew & Thompson, 1839.

Lief, Michael S., and H. Mitchell Caldwell. *The Devil's Advocates: Greatest Closing Arguments in Criminal Law*. New York: Simon & Schuster, 2006.

Lienesch, Michael. *In the Beginning: Fundamentalism, the Scopes Trial, and the Making of the Antievolution Movement*. Chapel Hill: University of North Carolina Press, 2007.

Lindberg, Carter. "The Lutheran Tradition." In *Caring and Curing: Health and Medicine in the Western Religious Traditions*, edited by Ronald L. Numbers and Darrel W. Amundsen, 173–203. New York: Macmillan, 1986.

Lindberg, David C. "Medieval Science and Its Religious Context." *Osiris* 10, no. 1 (1995): 60–79.

Lindberg, David C., and Ronald L. Numbers. "Beyond War and Peace: A Reappraisal of the Encounter between Christianity and Science." *Church History* 55, no. 3 (1986): 338–54.

Lindgren, Homer D., ed. *Modern Speeches*. New York: F. S. Crofts, 1926.

Link, Eugene. "Abraham and Mary Jacobi, Humanitarian Physicians." *Journal of the History of Medicine and Allied Sciences* 4, no. 4 (1949): 382–92.

———. *The Social Ideas of American Physicians (1776–1976): Studies of the Humanitarian Tradition in Medicine*. Selinsgrove, PA: Susquehanna University Press, 1992.

Link, Susannah J., and William A. Link, eds. *The Gilded Age and Progressive Era: A Documentary Reader*. Hoboken, NJ: Wiley, 2012.

Lippincott, J. S. "The Critics of Evolution." *American Naturalist* 14, no. 5 (1880): 319–416.

Liss, Julia E. "Diasporic Identities: The Science and Politics of Race in the Work of Franz Boas and W. E. B. Du Bois, 1894–1919." *Cultural Anthropology* 13 no. 2 (1998): 127–66.

MacPherson, Ryan C. "Natural and Theological Science at Princeton, 1845–1859: Vestiges of Creation Meets the Scientific Sovereignty of God." *Princeton University Library Chronicle* 65, no. 2 (2004): 184–236.

Malthus, Thomas. *An Essay on the Principle of Population; or, A View of Its Past and Present*

Effects on Human Happiness, with an Inquiry into Our Prospects Respecting the Future Removal or Mitigation of the Evils Which It Occasions. 2nd ed. London: J. Johnson, 1803.

———. *First Essay on Population 1798*. With notes by James Bonar. London: MacMillan, 1926.

Manseau, Peter. *One Nation, Under Gods: A New American History*. New York: Little, Brown, 2015.

Mariotti, Shannon. "On the Passing of the First-Born Son: Emerson's 'Focal Distancing,' Du Bois' 'Second Sight,' and Disruptive Particularity." *Political Theory* 37, no. 3 (2009): 351–74.

Marriott, Alice Lee, and Carol K. Rachlin, eds. *American Indian Mythology*. New York: Signet, 1972.

Marsden, George. *Fundamentalism and American Culture*. Oxford: Oxford University Press, 2006.

Marsh, John. *The Beloved Physician*. Hartford, CT: Goodwin, 1825.

Marvin, Abijah Perkins. *The Life and Times of Cotton Mather; or, A Boston Minister of Two Centuries Ago, 1663–1728*. Boston: Congregational Sunday-School and Publishing Society, 1892.

Massey, Edmund. *A Sermon against the Dangerous and Sinful Practice of Inoculation, Preach'd at St. Andrew's Holborn, on Sunday, July the 8th, 1722*. London: William Meadows, 1722.

Mather, Cotton. *The Christian Philosopher: A Collection of the Best Discoveries in Nature, with Religious Improvements*. Charlestown, MA: Middlesex Bookstore, 1815. Originally published 1721.

———. *Diary of Cotton Mather, 1681–1708*. Boston: Massachusetts Historical Society, 1911.

———. *Diary of Cotton Mather, 1709–1724*. Boston: Massachusetts Historical Society, 1912.

———. *An Epistle to the Christian Indians: Giving Them a Short Account, of What the English Desire Them to Know and to Do, in Order to Their Happiness*. Boston: Bartholomew Green, 1706.

———. *Magnalia Christi Americana; or, The Ecclesiastical History of New-England*. London: Thomas Parkhurst, 1702.

———. *Right Thoughts in Sad Hours, Representing the Comforts and Duties of Good Men, under All Their Afflictions; and Particularly, That One, the Untimely Death of Children: In a Sermon Delivered . . . Under a Fresh Experience of That Calamity*. London: James Astwood, 1689.

———. *Selected Letters of Cotton Mather*. Edited by Kenneth Silverman. Baton Rouge: Louisiana State Press, 1971.

———. *Wholesome Words*. Boston: B. Green & J. Allen, 1703.

Mather, Kirtley Fletcher. *Science in Search of God*. New York: H. Holt, 1928.

Mathews, Shailer, ed. *Contributions of Science to Religion*. New York: D. Appleton, 1924.

———. *New Faith for Old: An Autobiography*. New York: MacMillan, 1936.

McCann, F. E. "Put This before Your Clientele." *American Journal of Clinical Medicine* 23 (1916): 444–45.

McGehee, Tom. "Josiah Clark Nott, M.D." *Magnolia Messenger*, Spring 2009, 1–5.
McMahan, David L. "Modernity and the Early Discourse of Scientific Buddhism." *Journal of the American Academy of Religion* 72, no. 4 (2004): 897–933.
McMahon, Lucia. "'So Truly Afflicting and Distressing to Me His Sorrowing Mother': Expressions of Maternal Grief in Eighteenth-Century Philadelphia." *Journal of the Early Republic* 32, no. 1 (2012): 27–60.
Meckel, Richard. *Save the Babies: American Public Health Reform and the Prevention of Infant Mortality, 1850–1929*. Ann Arbor: University of Michigan Press, 1998.
Meigs, Charles D., *Obstetrics: The Science and the Art*. Philadelphia: Lea & Blanchard, 1849.
Miles, Tiya. "'Circular Reasoning': Recentering Cherokee Women in the Antiremoval Campaigns." *American Quarterly* 61, no. 2 (2009): 221–43.
Minardi, Margot. "The Boston Inoculation Controversy of 1721–1722: An Incident in the History of Race." *William and Mary Quarterly* 61, no. 1 (2004): 47–76.
"Miss Beecher's *Common Sense Applied to Religion*." *Church Review* 10 (1857–1858): 421–44.
Mitchell, T. E. "Diphtheria." *Southern Practitioner* 19, no. 3 (1897): 109–17.
Moore, James R. *The Post-Darwinian Controversies: A Study of the Protestant Struggle to Come to Terms with Darwin in Great Britain and America, 1870–1900*. Cambridge: Cambridge University Press, 1981.
Moorhead, James H. "Between Progress and Apocalypse: A Reassessment of Millennialism in American Religious Thought, 1800–1880." *Journal of American History* 71, no. 3 (1984): 525–42.
Moran, Jeffrey. "Reading Race into the Scopes Trial: African American Elites, Science, and Fundamentalism." *Journal of American History* 90, no. 3 (2003): 891–911.
———. "The Scopes Trial and Southern Fundamentalism in Black and White: Race, Region, and Religion." *Journal of Southern History* 70, no. 1 (2004): 95–120.
Morgan, Thomas M. "The Education and Medical Practice of Dr. James McCune Smith (1813–1865), First Black American to Hold a Medical Degree." *Journal of the National Medical Association* 95, no. 7 (2003): 603–14.
Morris, Caspar. *Lectures on Scarlet Fever*. Philadelphia: Lindsay & Blakiston, 1851.
———. *Letters of Travel from Caspar Morris, M. D., 1871–1872, to His Family*. Philadelphia: Times Printing House, 1896.
Morrison, Larry R. "The Religious Defense of American Slavery before 1830." *Journal of Religious Thought* 37, no. 2 (1980): 16–29.
Murphy, Howard R. "The Ethical Revolt against Christian Orthodoxy in Early Victorian England." *American Historical Review* 60, no. 4 (1955): 800–817.
Newsholme, Arthur. *Fifty Years in Public Health: A Personal Narrative with Comments*. London: George Allen & Unwin, 1935.
Newton, Hannah. *The Sick Child in Early Modern England, 1580–1720*. Oxford: Oxford University Press, 2012.
Noble, David. *The Religion of Technology: The Divinity of Man and the Spirit of Invention*. New York: Alfred A. Knopf, 1997.
———. *A World without Women: The Christian Clerical Culture of Western Science*. New York: Alfred A. Knopf, 1992.

Northrup, W. P. "Antitoxin Treatment of Diphtheria a Pronounced Success." *Forum* 22 (September 1896): 53–64.

Nott, Josiah C. "The Mulatto a Hybrid—Probable Extermination of the Two Races if Whites and Blacks Are Allowed to Marry." *American Journal of the Medical Sciences* 5 (1843): 252–56.

Nott, Josiah C., and Geo. R. Gliddon. *Types of Mankind; or, Ethnological Researches Based upon the Ancient Monuments, Paintings, Sculptures, and Crania of Races, and upon Their Natural, Geographical, Philological, and Biblical History*. London: Trubner, 1854.

Numbers, Ronald. Introduction to *Galileo Goes to Jail: And Other Myths about Science and Religion*, edited by Ronald Numbers, 1–7. Cambridge, MA: Harvard University Press, 2010.

———. "Science and Religion." *Osiris* 1, no. 1 (1985): 59–80.

Numbers, Ronald, and Darrel W. Amundsen, eds. *Caring and Curing: Health and Medicine in the Western Religious Traditions*. New York: Macmillan, 1986.

Onuf, Peter S. "'To Declare Them a Free and Independent People': Race, Slavery, and National Identity in Jefferson's Thought." *Journal of the Early Republic* 18, no. 1 (1998): 1–46.

———. "'We Shall All Be Americans': Thomas Jefferson and the Indians." *Indiana Magazine of History* 95, no. 2 (1999): 103–41.

Paine, Albert Bigelow. *Mark Twain: A Biography*. Vol. 4. New York: Harper & Brothers, 1912.

Paley, William. *Natural Theology; or, Evidences of the Existence and Attributes of the Deity, Collected from the Appearances of Nature*. London: R. Faulder, 1802.

Palmer, Walter Clark. *Life and Letters of Leonidas L. Hamline, DD, Late One of the Bishops of the Methodist Episcopal Church*. New York: Carlton & Porter, 1866.

Parker, Alison M. *Unceasing Militant: The Life of Mary Church Terrell*. Chapel Hill: University of North Carolina Press, 2020.

Parker, Horatio N. Review of *Disease in Milk: The Remedy, Pasteurization*, by Lina Gutherz Straus. *American Journal of Public Health* 4, no. 2 (1914): 151–52.

Parker, Theodore. *False and True Theology: A Sermon, Delivered at the Music Hall, Boston, on Sunday, February 14, 1858*. Boston: B. Marsh, 1859.

———. *Sermons of Theism, Atheism, and the Popular Theology*. Boston: Ticknor & Fields, 1861.

———. *The Works of Theodore Parker: The World of Matter and the Spirit of Man*. Boston: American Unitarian Association, 1907.

Parrish, Susan Scott. "Women's Nature: Curiosity, Pastoral and the New Science in British America." *Early American Literature* 37, no. 2 (2002): 195–245.

Paul, Diane. *Controlling Human Heredity: 1865 to the Present*. New York: Humanity Books, 1995.

———. "Darwin, Social Darwinism and Eugenics." In *The Cambridge Companion to Darwin*, edited by Jonathan Hodge and Gregory Radick, 214–39. Cambridge: Cambridge University Press, 2006.

———. "Eugenics and the Left." *Journal of the History of Ideas* 45, no. 4 (1984): 567–90.

Pauly, Philip J. *Biologists and the Promise of American Life: From Meriwether Lewis to Alfred Kinsey*. Princeton, NJ: Princeton University Press, 2000.

Peabody, Andrew Preston. *Christian Consolations: Sermons Designed to Furnish Comfort and Strength to the Afflicted*. Boston: W. M. Crosby & H. P. Nichols, 1847.

———. *Christian Consolations: Sermons Designed to Furnish Comfort and Strength to the Afflicted*. 4th ed. Boston: Crosby, Nichols, 1858.

Pearson, Howard. "The History of Pediatrics in America." In *Oski's Pediatrics: Principles and Practice*, edited by Julia A. Macmillan et al., 4th ed., 2–7. Philadelphia: Lippincott, Williams & Wilkins, 2006.

Pernick, Martin. *The Black Stork: Eugenics and the Death of "Defective" Babies in American Medicine and Motion Pictures since 1915*. Oxford: Oxford University Press, 1996.

———. *A Calculus of Suffering: Pain, Professionalism, and Anesthesia in Nineteenth Century America*. New York: Columbia University Press, 1985.

Phelps, Elizabeth Stuart. "The Gates Ajar—Twenty-Five Years After." *North American Review* 156, no. 438 (1893): 567–76.

Philipsborn, Rebecca Pass, and Kevin Chan. "Climate Change and Global Child Health." *Pediatrics* 141, no. 6 (June 2018): e20173774.

Philps, A. R. *Parental Obligation: A Sermon*. London: John Snow, 1847.

Picciotto, Joanna. "Reforming the Garden: The Experimentalist Eden and 'Paradise Lost.'" *English Literary History* 72, no. 1 (Spring 2005): 23–78.

Popenoe, Paul, and Roswell Hill Johnson. *Applied Eugenics*. New York: Macmillan, 1920.

Porter, Roy. *The Creation of the Modern World: The Untold Story of the British Enlightenment*. New York: W. W. Norton, 2000.

———. *The Greatest Benefit to Mankind: A Medical History of Humanity*. New York: W. W. Norton, 1999.

Powell, John Harvey. *Bring Out Your Dead: The Great Plague of Yellow Fever in Philadelphia in 1793*. Philadelphia: University of Pennsylvania Press, 1949.

Preston, Samuel H., and Michael R. Haines. *Fatal Years: Child Mortality in Late Nineteenth-Century*. Princeton, NJ: Princeton University Press, 1991.

Putney, Diane T. "Fighting the Scourge: American Indian Morbidity and Federal Policy, 1897–1928." PhD diss., Marquette University, 1980.

"The Rabbeth Memorial." *Maryland Medical Journal* 12 (November 29, 1884): 87.

Ray, John. *The Wisdom of God Manifested in the Works of the Creation*. London: Samuel Smith, 1691.

———. *The Wisdom of God Manifested in the Works of the Creation*. 4th ed. London: Samuel Smith, 1704.

Reesman, Jeanne Campbell. "Mark Twain vs. God: The Story of a Relationship." *Mark Twain Journal* 52, no. 2 (2014): 112–35.

Reidpath, Daniel D., and Pascale Allotey. "Infant Mortality Rate as an Indicator of Population Health." *Journal of Epidemiology & Community Health* 57, no. 5 (2003): 344–46.

"Relied on Prayer to Cure Their Ills: The Five Hansens Almost Dead of Diphtheria before the Physicians Arrived." *New York Times*, September 24, 1910.

Riley, William Bell. "The Faith of the Fundamentalists." *Current History* 26 (June 1927): 434–36.

Roberts, Charles. "The Cushman Indian Trades School and World War I." *American Indian Quarterly* 11, no. 3 (1987): 221–39.

Roberts, Dorothy E. *Killing the Black Body: Race, Reproduction, and the Meaning of Liberty.* New York: Vintage, 1999.

Robinson, David M. "Fields of Investigation: Emerson and Natural History." In *American Literature and Science*, edited by Robert J. Scholnick, 94–109. Lexington: University Press of Kentucky, 1992.

Robinson, William J. "The Billy Sunday Disgrace." *Medical Critic and Guide*, May 1917.

Rosen, Christine. *Preaching Eugenics: Religious Leaders and the American Eugenics Movement.* Oxford: Oxford University Press, 2004.

Rosenberg, Charles E. *The Cholera Years: The United States in 1832, 1849, and 1866.* Chicago: University of Chicago Press, 1987.

Ross, Edward Alsworth. *Changing America: Studies in Contemporary Society.* New York: Century, 1912.

Rothman, Sheila M. *Living in the Shadow of Death: Tuberculosis and the Social Experience of Illness in American History.* New York: Basic Books, 1994.

Rubin, Julius H. *Tears of Repentance: Christian Indian Identity and Community in Colonial Southern New England.* Lincoln: University of Nebraska Press, 2018.

Ruse, Michael. *The Darwinian Revolution: Science Red in Tooth and Claw.* Chicago: University of Chicago Press, 1979.

Rusert, Britt. *Fugitive Science: Empiricism and Freedom in Early African American Culture.* New York: New York University Press, 2017.

"The Sacrifice of the Innocents." Prepared under the advice and suggestions of Dr. Henry Koplik. *Cosmopolitan* 47 (1909): 423–35.

Sanger, Margaret. *The Case for Birth Control: A Supplementary Brief and Statement of Facts.* May 1917.

———. *Margaret Sanger: An Autobiography.* New York: W. W. Norton, 1938.

———. *The Pivot of Civilization.* New York: Brentado's, 1922.

Schaefer, Richard. "Andrew Dickson White and the History of a Religious Future." *Zygon* 50, no. 1 (2015): 7–27.

Schiebinger, Londa. "The Anatomy of Difference: Race and Sex in Eighteenth-Century Science." *Eighteenth-Century Studies* 23, no. 4 (1990): 387–405.

———. "Feminist History of Colonial Science." *Hypatia* 19, no. 1 (2004): 233–54.

———. *The Mind Has No Sex?: Women in the Origins of Modern Science.* Cambridge, MA: Harvard University Press, 1989.

"Science and Theology." *Church Quarterly* 46, no. 91 (1898): 121–41.

"Scientific Notes and News." *Science* 61, no. 1590 (1925): 626–29.

Scott, Leslie M. "Indian Diseases as Aids to Pacific Northwest Settlement." *Oregon Historical Quarterly* 29, no. 2 (1928): 144–61.

Secord, James A. *Victorian Sensation: The Extraordinary Publication, Reception, and Secret Authorship of Vestiges of the Natural History of Creation.* Chicago: University of Chicago Press, 2003.

Shapin, Steven. "Descartes the Doctor: Rationalism and Its Therapies." *British Journal for the History of Science* 33, no. 2 (2000): 131–54.

Shapiro, Adam R. *Trying Biology: The Scopes Trial, Textbooks, and the Antievolution Movement in American Schools*. Chicago: University of Chicago Press, 2013.

Shaw, George Bernard. *Back to Methuselah: A Metabiological Pentateuch*. New York: Brentano's, 1921.

Silverman, David J. *Red Brethren: The Brothertown and Stockbridge Indians and the Problem of Race in Early America*. Cornell, NY: Cornell University Press, 2010.

Silverman, Kenneth. *The Life and Times of Cotton Mather*. New York: Harper & Row, 1984.

Simpson, James Young. "Religious Beliefs in the Light of Modern Science." *Journal of Religion* 5, no. 3 (1925): 321–24.

———. *The Spiritual Interpretation of Nature*. 2nd ed. London: Hodder & Stoughton, 1912.

Sklar, Kathryn Kish. *Florence Kelley and the Nation's Work: The Rise of Women's Political Culture, 1830–1900*. New Haven, CT: Yale University Press, 1995.

Slater, Peter Gregg. *Children in the New England Mind: In Death and in Life*. Hamden, CT: Archon Books, 1977.

———. "From the Cradle to the Coffin: Parental Bereavement and the Shadow of Infant Damnation in Puritan Society." *Psychohistory Review* 6, no. 2–3 (1977): 4–24.

Smith, James McCune. "Colored Orphan's Asylum: Physicians Report." *Colored American*, January 26, 1839.

———. *The Works of James McCune Smith: Black Intellectual and Abolitionist*. Edited by John Stauffer. Oxford: Oxford University Press, 2006.

Smith, J. Lewis. *A Treatise on the Diseases of Infancy and Childhood*. Philadelphia: Lea Brothers, 1890.

Smith, William. "William Paley's Theological Utilitarianism in America." *William and Mary Quarterly* 11, no. 3 (1954): 402–24.

Solberg, Winton U. "Science and Religion in Early America: Cotton Mather's 'Christian Philosopher.'" *Church History* 56, no. 1 (1987): 73–92.

Sparks, Jared. *Benjamin Franklin, "Doer of Good": A Biography*. Edinburgh: William Nimmo, 1865.

Speight, Harold E. B. "From Magic, to Science, to Faith." *Christian Century* 43, no. 5 (1926): 140–42.

Stanton, Elizabeth Cady. *Eighty Years and More*. New York: European Publishing, 1898.

Stanton, Elizabeth Cady, et al. *The Woman's Bible*. New York: European Publishing, 1895.

Starita, Joe. *A Warrior of the People: How Susan La Flesche Overcame Racial and Gender Inequality to Become America's First Indian Doctor*. New York: St. Martin's, 2016.

Steckel, Richard H. "A Dreadful Childhood: The Excess Mortality of American Slaves." *Social Science History* 10, no. 4 (1986): 427–65.

Steele, Brian. "Thomas Jefferson's Gender Frontier." *Journal of American History* 95, no. 1 (2008): 17–42.

Stowe, Charles Edward. *Life of Harriet Beecher Stowe, Compiled from Her Letters and Journals*. Boston: Houghton, Mifflin, 1889.

Stowe, Harriet Beecher. *Uncle Tom's Cabin: or, Negro Life in the Slave States of America*. London: Richard Bentley, 1852.

Straus, Lina Gutherz. *Disease in Milk: The Remedy, Pasteurization; The Life Work of Nathan Straus*. 2nd ed. New York: E. P. Dutton, 1917.

Strauss, David. *The Old Faith and the New: A Confession.* American edition. New York: Henry Holt, 1873.

Sunday, Billy. "Billy Sunday on Social Religion." *Literary Digest,* March 6, 1915.

———. "Feeding the Five Thousand." *Pittsburgh Press,* February 12, 1914.

Sydenham, Thomas. *The Works of Thomas Sydenham, M.D.* Translated by R. G. Latham from the Latin edition of William Alexander Greenhill. London: Sydenham Society, 1848.

Taylor, Edward. "Upon Wedlock, and Death of Children." In *The New Anthology of American Poetry: Traditions and Revolutions, Beginnings to 1900,* edited by Steven Gould Axelrod, 91–92. New Brunswick, NJ: Rutgers University Press, 2003.

T. E. C. "Lidian Emerson, Wife of Ralph Waldo Emerson, on the Death of Her Five-Year-Old Son (1842)." *Pediatrics* 83, no. 2 (February 1989): 192.

Tennyson, Alfred. *In Memoriam.* London: Edward Moxon, 1850.

Terrell, Mary Church. *A Colored Woman in a White World.* Washington, DC: Ransdell, 1940.

———. "Harriet Beecher Stowe." 1911. *Mary Church Terrell Papers: Speeches and Writings, 1866–1953.* Library of Congress (website). Manuscript/Mixed Material.

"Texas' Sin of Omission—Her Sanitary Needs." *Texas Medical Journal* 14 (1898–1899): 392–401.

Thorndike, Lynn. "The Historical Background of Modern Science." *Scientific Monthly* 16, no. 5 (1923): 488–97.

Thornton, Alice. *The Autobiography of Mrs. Alice Thornton, of East Newton Co. York.* Durham, UK: Andrews, 1875.

Tindol, R., "Getting the Pox of All Their Houses: Cotton Mather and the Rhetoric of Puritan Science." *Early American Literature* 46, no. 1 (2011): 1–23.

Tomes, Nancy. *The Gospel of Germs: Men, Women, and the Microbe in American Life.* Cambridge, MA: Harvard University Press, 1999.

Tooley, Michael. "The Problem of Evil." *Stanford Encyclopedia of Philosophy,* Fall 2015 edition, edited by E. N. Zalta.

Tourgée, Albion Winegar. *An Appeal to Caesar.* New York: Fords, Howard & Hulbert, 1884.

Townsend, Craig D. *Faith in Their Own Color: Black Episcopalians in Antebellum New York City.* New York: Columbia University Press, 2005.

Trafzer, Clifford E. "Coughing Blood: Tuberculosis Deaths and Data on the Yakama Indian Reservation, 1911–64." *Canadian Bulletin of Medical History* 15, no. 2 (1998): 251–76.

Turner, Frank M. "The Victorian Crisis of Faith and the Faith That Was Lost." In *Victorian Faith in Crisis: Essays on Continuity and Change in Nineteenth-Century Religious Belief,* edited by Richard J. Helmstadter and Bernard V. Lightman, 9–38. London: Palgrave Macmillan, 1990.

Turner, James. *Without God, Without Creed: The Origins of Unbelief in America.* Baltimore: Johns Hopkins University Press, 1985.

Twain, Mark. *Mark Twain's Letters.* Vol. 5. Edited by Lin Salamo and Harriet Elinor Smith. Berkeley: University of California Press, 1997.

———. *What Is Man? and Other Irreverent Essays.* New York: Prometheus, 2009.

Ungureanu, James C. "From Divine Oracles to the Higher Criticism: Andrew D. White and the Warfare of Science with Theology in Christendom." *Zygon* 56, no. 1 (2021): 209–33.

———. "A Yankee at Oxford: John William Draper at the British Association for the Advancement of Science at Oxford, 30 June 1860." *Notes and Records of the Royal Society of London* 70, no. 2 (2016): 135–50.

United States Bureau of the Census. *Mortality Statistics 1926: Twenty-Seventh Annual Report.* Washington, DC: Government Printing Office, 1929.

United States Census Office. *Statistics of the United States, Including Mortality, Property, etc. in 1860.* Washington, DC: Government Printing Office, 1866.

United States Congress. House. Committee on Interstate and Foreign Commerce. *Child Welfare Extension Service: Hearing(s) before the Committee on Interstate and Foreign Commerce on H.R. 14070, to Provide a Child Welfare Extension Service and for Other Purposes.* 70th Cong., 2nd sess., January 24 and 25, 1929.

United States Congress. Senate. *Contagious and Infectious Diseases Among the Indians.* January 27, 1913. 62nd Cong., 3rd sess., 1913. S. Doc. 1038 40.

United States Supreme Court. *The Dred Scott Decision: Opinion of Chief Justice B. Taney, in the Dred Scott Case.* 2nd ed. New York: Van Evrie, Horton, 1863.

Vanderpool, Harold Y. "The Wesleyan-Methodist Tradition." In *Caring and Curing: Health and Medicine in the Western Religious Traditions,* edited by Ronald L. Numbers and Darrel W. Amundsen, 317–53. New York: MacMillan, 1986.

Wallace, Alfred Russel. *My Life: A Record of Events and Opinions.* London: Chapman & Hall, 1905.

Walsh, James Joseph. "Biographical Notes on Joseph O'Dwyer, M.D.—A.D. 1841–1898. The Inventor of Intubation." *Records of the American Catholic Historical Society of Philadelphia* 14, no. 4 (1903): 391–422.

———. *Makers of Modern Medicine.* New York: Fordham University Press, 1915.

———. "Our Latest Panacea—Eugenics." *Common Cause* 3, no. 2 (1913): 73–81.

———. *Religion and Health.* Boston: Little, Brown, 1920.

Warner, Margaret Humphreys. "Vindicating the Medical Role: Cotton Mather's Concept of the 'Nishmath-Chajim' and the Spiritualization of Medicine." *Journal of the History of Medicine and Allied Sciences* 36, no 3 (1981): 278–95.

Washington, Harriet A. *Medical Apartheid: The Dark History of Medical Experimentation on Black Americans from Colonial Times to the Present.* New York: Doubleday, 2006.

"Was the Doctor Right? Some Independent Opinions." *Independent,* January 3, 1916.

Waterhouse, Benjamin. *An Essay Concerning Tussis Convulsiva, or, Whooping Cough. With Observations on the Diseases of Children.* Boston: Munroe & Francis, 1822.

———. *A Prospect of Exterminating the Small Pox, Part II.* Cambridge, MA: University Press, 1802.

———. *The Rise, Progress, and Present State of Medicine. A Discourse, Delivered at Concord, July 6th, 1791. Before the Middlesex Medical Association.* Boston: Thomas & John Fleet, 1792.

Watkins, Frances Ellen. *Poems on Miscellaneous Subjects.* Boston: J. B. Yerrinton, 1854.

Weir, Hugh. "The Dawn Man." *McClure's* 55, no. 1 (March 1923): 19–28.

Weiss, John. *Life and Correspondence of Theodore Parker*. Vol. 2. New York: D. Appleton, 1864.
Wellman, Kathleen Anne. *La Mettrie: Medicine, Philosophy, and Enlightenment*. Durham, NC: Duke University Press, 1992.
Werner, John M. "David Hume and America." *Journal of the History of Ideas* 33, no. 3 (1972): 439–56.
Westfall, Richard S. *Science and Religion in Seventeenth Century England*. Ann Arbor: University of Michigan Press, 1973.
Wetering, Maxine Van de. "A Reconsideration of the Inoculation Controversy." *New England Quarterly* 58, no. 1 (1985): 46–67.
Wheatley, Phillis. *Memoir and Poems of Phillis Wheatley, a Native African and a Slave*. Boston: Geo. W. Light, 1834.
Wheeler, J. M. "Buddhist and Christian Missions." *Freethinker*, December 5, 1879, 770.
White, Andrew Dickson. *Autobiography of Andrew Dickson White*. Vol. 2. New York: Century, 1905.
——— . *A History of the Warfare of Science with Theology in Christendom*. 2 vols. New York: D. Appleton, 1896.
——— . *The Warfare of Science*. New York: D. Appleton, 1876.
Wiggam, Albert Edward. *The New Decalogue of Science*. Indianapolis, IN: Bobbs-Merrill, 1923.
Wilkinson, George. "Welch on Diphtheria." *Omaha Clinic* 8 (1895): 221.
Wilson, John B. "Darwin and the Transcendentalists." *Journal of the History of Ideas* 26, no. 2 (1965): 286–90.
Wolosky, Shira. "Emily Dickinson's War Poetry: The Problem of Theodicy." *Massachusetts Review* 25, no. 1 (1984): 22–41.
The World's Most Famous Court Trial: Tennesssee Evolution Case. A complete stenographic report of the famous court test of the Tennessee anti-evolution act, at Dayton, July 10 to 21, 1925, including speeches and arguments of attorneys. Cincinnati: National Book Company, 1925.
Yalcinkaya, M. Alper. "Science as an Ally of Religion: A Muslim Appropriation of 'The Conflict Thesis.'" *British Journal for the History of Science* 44, no. 2 (2011): 161–81.

Index

abolition, 43, 48, 51–52, 61, 88, 92, 94–95, 102, 122, 128; Charles Darwin and, 112–13; Frederick Douglass and, 105–6, 120–22; Harriet Beecher Stowe and, 98–100
abortion, 178, 226
Adams, John, 56
Addams, Jane, 182
Affleck, Thomas, 92–93
agnosticism, 131, 173, 174, 183, 211–12, 226. *See also* Darrow, Clarence; Ingersoll, Robert; Jacobi, Abraham; Jacobi, Mary Putnam
Alvord, Clarence Walworth, 197
American Anti-Slavery Society, 87
American Civil Liberties Union, 205, 208
American Genetic Association, 187
American Journal of Public Health, 189, 190
American Medical Association, 162, 180, 201
American Pediatric Society, 157, 158
American Philosophical Society, 42, 44, 46
American Social Science Association, 163
American Society for the Study and Prevention of Infant Mortality, 190
anatomy, 25, 26–27, 35, 81–82, 143, 154, 182
anesthetics, 83, 102, 154
Angel of Bethesda, The, 17
antibiotics, 223
antitoxin, 138, 150, 157–59, 160, 223; criticism of, 175–76
Applegate, Debby, 124

Aptheker, Herbert, 165
Aries, Philip, 7
Assing, Ottilie, 108–9
atheism, 25, 41–42, 50, 109, 117–18, 131, 134, 142, 173

Bacon, Francis, 18–20, 21, 90, 126
Bacon, Roger, 140, 144, 145, 160, 164, 208
bacteriology, 154–55, 157, 170, 218–19
Balguy, Thomas, 55
Banks, Joseph, 42
Banneker, Benjamin, 62
Barbour, James, 51
Barnet, Malvina, 91
Beecher, Catharine, 95
Beecher, Edward, 95, 139
Beecher, Eunice, 122, 123
Beecher, Henry Ward, 2, 122–29, 142, 184, 198, 209
Beecher, Lyman, 95–95, 124, 129, 139
Beecher, Roxana, 94
Berkman, Alexander, 178
Bernard, Claude, 156
Besant, Annie, 150–51
Bevan, Ralph, 178
biblical criticism. *See* higher criticism
Bilbro, Jeffrey, 51
Bill and Melinda Gates Foundation, 1, 224, 226
Bilsius, Ludovicus, 27
biology textbooks, 217–18
birth control, 166, 176–78, 192
Bittel, Carla, 152
Black Codes, 162

Black Stork, The, 182
Blake, John Ballard, 151
Blight, David, 107
Blum, Edward J., 165
Boas, Franz, 192, 220
Bollinger, Anna, 3, 181, 183–84
Bonifacius: An Essay Upon the Good, 43
Boyle, Robert, 21, 35, 56, 81, 143
Boylston, Zabdiel, 31, 38
Brooke, John Hedley, 3, 23, 27, 35, 44
Browne, Junius Henry, 147
Bruno, Giordano, 143
Bryan, William Jennings, 3, 185, 195–201, 206, 207, 221
Buchanan, Charles, 214, 216, 219
Buchanan, George, 151
Buddhism, 147, 226
Bureau of Indian Affairs, 213, 219, 220
Burroughs, John, 145
Bush, Ira Jefferson, 151
Bushnell, Horace, 69

cancer, 132
Callis, A. B., 210
capitalism, 171, 174–76, 179–80, 198
Carlyle, Thomas, 70
Carnegie, Andrew, 145
Carnegie Institute of Washington, 146
Cartwright, Samuel A., 101
Carus, Paul, 147
Catholicism, 67, 74, 98, 146, 153, 175, 191; on the Tulalip Indian Reservation, 215–16, 217. *See also* Chesterton, G. K.; Dover, Harriette Shelton; Draper, Elizabeth; Hagerty, Thomas J.; Mersenne, Marin; Pasteur, Louis; Turner, Thomas Wyatt; Walsh, James Joseph
Centers for Disease Control, 226
Chambers, Robert, 79–80, 84–85. See also *Vestiges of the Natural History of Creation*
Channing, William Ellery, 74–75
Cherokee, 13, 28, 95

Chesterton, G. K., 191–92
Chickering, Jesse, 69
Chief Little Turtle, 60
child mortality: change over time, 1, 4, 63, 67, 138–39, 145, 157–59, 170, 174, 181, 185–86, 187, 202, 223–24; debates over the best means of lowering, 170–80, 188, 226–27; explanations of differential rates of, 36–37, 58–60, 68–69, 88–93, 101, 116–17, 122, 127, 161–69, 218–20, 225. *See also* medicine; sanitation
Children's Bureau, 171
Chirouse, Eugene Casimir, 215
cholera, 2, 93, 97, 143
cholera infantum. *See* summer diarrhea
Christianity and the Social Crisis, 175
Christian Philosopher, The, 26, 30, 43
Christian Scientists, 159
Chrysostom, John, 16–17, 26
Civic Biology, 207, 217–18
Civil War, 118, 120, 143
Clarke, Joseph, 67
Clemens, Langdon, 149
Clemens, Samuel, 2, 149, 160
Clement of Alexandria, 16
climate change, 226
Cole, Rebecca J., 165–66
Collier, John, 220
Colored American, 88–89
Colored Orphan Asylum, 87, 88, 91
Colored Woman in a White World, A, 167
Columbus, Christopher, 143
Combe, Andrew, 66–68, 108, 134
Combe, George, 108, 124
communism, 165, 174, 175, 180, 198
Condorcet, Nicolas de, 63, 64
congenital malformation, 2, 3, 35, 182
Conkin, Paul, 194
Conklin, Edwin Grant, 201
Constitution of Man, The, 108
Contributions of Science to Religion, 201–2, 203
Copernicus, 29

Currie, James, 53
Cushman Indian School, 215
Cuvier, Georges, 79

Dall, Caroline, 163
Darrow, Clarence, 3, 205, 206, 211–12
Darwin, Annie, 115, 116
Darwin, Charles, 82, 103, 111–17, 120–21, 128, 129; on eugenics, 184–85, 198, 200–201
Darwin, Emma, 116
Davenport, Charles, 183, 192, 201, 203
Davenport, Eugene, 201
deism, 23, 41, 42, 44
Delano, H. A., 202
Descartes, René, 22
Descent of Man, The, 115–17, 120–21, 200
design argument: criticisms of, 41–42, 54–55, 79–80, 112, 114–15, 132, 133–34, 135; eighteenth century versions of, 39, 41, 55, 62; William Paley's version, 65, 75; seventeenth century version of, 22–25, 26–28, 35–36;
Desmond, Adrian, 128
Dharmapāla, Anagarika, 147
diabetes, 203
Dialogues Concerning Natural Religion, 54
Dickinson, Emily, 118
Dietrich, John H., 210
Dilbeck, D. H., 107
diphtheria, 2, 3, 138, 149, 150–54, 157–60, 170
disease: ultimate origin of, 10, 13, 14, 17–18, 28, 35
dissection, 27
Divine Benevolence Asserted; and Vindicated from the Objections of Ancient and Modern Sceptics, 55
Dodson, John M., 201–2, 203
Douglass, Anna, 104, 105
Douglass, Annie, 104, 123
Douglass, Frederick, 2, 104–11, 118–20, 122, 123, 127, 135–36

Douglass, Rosetta, 104, 109
Douglass, William, 32
Dover, Harriette Shelton, 3, 212–17, 219, 221–22
Draper, Elizabeth, 137
Draper, John William, 137–38, 197
Draper, William, 137
drapetomania, 102
Dred Scott decision, 102
Du Bois, Burghardt, 160–61, 164
Du Bois, Nina, 160, 161–62
Du Bois, W. E. B., 2, 160–66, 169, 179, 192–93
Dysaethesia Aethiopis, 101
dysentery, 1, 2, 86, 161, 170, 172, 173

East, Edward M., 189
Eaton, Mary, 148
Efendi, Ahmed Midhat, 138
Einstein, Albert, 173
Eliot, George. *See* Evans, Mary Ann
Emerson, Ellen, 75–76
Emerson, Lydia, 71–72
Emerson, Ralph Waldo, 2, 70–79, 124, 161, 162
Emerson, Waldo, 70, 77–78
Engels, Friedrich, 109, 179
Eppes, Elizabeth Wayles, 53
Equiano, Olaudah, 52
Essay on the Principle of Population, An, 64
eternal damnation. *See* hell
Eugenic Mother and Baby, The, 190
eugenics, 147, 171, 174, 182–93; opposition to, 196, 198–99, 204, 211–12, 226
Eugenics and other Evils, 191
Eugenics Record Office, 171, 183
euthenics, 188–89
Evans, Mary Ann, 147
evil, origin of. *See* Fall of Man; original sin; theodicy
evolution: belief in divinely governed progress and, 117–18, 124–27, 130, 134, 141–42, 149, 158–60, 201–2, 209–10;

Charles Darwin's theory of, 113–17; eugenics and, 183–93; Frederick Douglass and, 119–22, 135–36; historians and, 137–38, 141–42; Robert Ingersoll and, 130, 134; Jewish tradition and, 173; opposition to, 25, 193–94, 197–202, 205–12; pre-Darwinian, 79–85; Elizabeth Cady Stanton and, 134. *See also* natural selection; science
Evrie, J. H. Van, 102
extinction, 79, 112, 113

Fall of Man, 10, 18–19, 20, 29, 34, 43; criticisms of belief in, 73, 75, 80, 84, 119, 125–26, 134, 135, 137, 140, 141–42, 145, 207, 210
Faris, Ellsworth, 203
Feuerbach, Ludwig, 109
Finney, Charles Grandison, 94
Finseth, Ian Frederick, 62
First Great Awakening, 151
First Salmon Ceremony, 222
Folkes, H. M., 162
Forry, Samuel, 93
Fourteenth Amendment, 162
Franklin, Benjamin, 2, 33, 38–45, 63
Franklin, Franky, 38, 45
Franklin, James, 33
Franklin, John, 42
freethinkers, 139, 147, 177. *See also* agnosticism; atheism; Unitarianism
French Revolution, 64
Freud, Sigmund, 173
Fugitive Slave Act, 99
fundamentalist Christianity, 194–95, 207–8

Galileo, 29, 143
Galton, Francis, 184, 187, 190
Gates, Bill, 1
Gates, Henry Louis, Jr., 50
Gates, Melinda, 1
gender, separation of spheres based upon, 45–48, 98, 134–35, 152, 163, 171

genetics. *See* heredity
geology, 83, 84, 124, 142, 193
germ theory, 71, 154–55
Godwin, William, 63, 64
Goldman, Emma, 178
Graham, William, 128
Grant, Madison, 189, 192
Graunt, John, 63
Gray, Asa, 3, 84, 117–18
Grew, Nehemiah, 27
Guerard, Arthur, 219

Hagerty, Thomas J., 175–76
Hague, W. Grant, 190–91
Haiselden, Harry, 3, 181–84, 211
Haiti, 136
Haller, John, 68, 129
Hamline, Leonides, 86
Hammonds, Evelynn, 158
Hansen, Gerhard Armauer, 155
Harper, Frances Ellen Watkins, 100
Harriman, Mary, 171
Harvey, William, 25, 27, 29
Hays, Arthur Garfield, 208
heaven: belief in, 10–11, 12, 15–16, 45, 49, 57, 75, 78, 93, 95, 96, 115, 142, 146, 195, 199; Frederick Douglass on, 109; Ben Franklin on, 40; repudiation of, 131–32, 147, 148, 152, 196–97; *Vestiges* on, 82, 84
hell: belief in, 10–11, 194, 217, 221; repudiation of, 72, 73, 75, 83, 95–96, 125–26, 130, 131, 133, 135, 139, 201, 207–8. *See also* fundamentalist Christianity
Hemmings, Sally, 54, 61
heredity, 184, 185, 186–88, 190–91, 192, 212, 218–19. *See also* eugenics
higher criticism: Jewish, 173; Protestant, 76–77, 80, 83–84, 109, 113, 119
Hinduism, 226
Hirsch, Emil, 173
histories of science, 137, 140–46, 179, 186, 203, 208

INDEX | 285

History of the Conflict between Religion and Science, 137–38
History of the Warfare of Science with Theology in Christendom, 139, 141–47, 203, 208, 210, 217
Hitchcock, Edward, 81, 124
Hodge, Clifton F., 155–57
Holmes, Oliver Wendell, 83, 185
Holt, Luther Emmett, 157, 190
Holt, William L., 190
Hughes, Antoinette, 203
Hughes, Charles Evans, 203
Hughes, Elizabeth, 203
Human Direction of Evolution, The, 204
Hume, David, 41–42, 54–55
Hunter, George W., 207, 217–18
Huxley, Thomas Henry, 131

Indian Removal Act, 59, 95
Industrial Workers of the World, 176, 178
infanticide, 178
influenza, 194
Ingersoll, Robert, 130–34, 150, 160, 177, 194
inoculation, 30–34, 38, 44–45, 143
insulin, 156, 203, 223
Isenberg, Andrew, 60
Islam, 138, 226

Jackson, Andrew, 59
Jacobi, Abraham, 2, 151, 152–54, 172, 173–74, 182
Jacobi, Ernst, 152
Jacobi, Mary Putnam, 2, 134–35, 152–54
Jefferson, Lucy, 53
Jefferson, Martha, 54
Jefferson, Patsy, 54
Jefferson, Thomas, 2, 46–47, 53–62, 69, 74, 79, 101, 103
Jenner, Edward, 60
Job, 8–9, 34
Johnson, James Wheldon, 180
Jones, David S., 36, 218

Jordan, David Starr, 144, 186, 217
Journal of Heredity, 187
Judaism, 172–73, 226

Kazin, Michael, 195
Keen, William Williams, 158–59
Keller, Helen, 182
Kelley, Florence, 144, 178–80
Kellogg, Vernon, 198, 217
Kingsley, Charles, 129
Kirkland, Samuel, 59
Klass, Perri, 224
Klebs, Edwin, 157
Koch, Philippa, 11, 16
Koch, Robert, 154
Kristof, Nicholas, 1
Kropotkin, Peter, 198
Ku Klux Klan, 162

La Flesche, Joseph, 219
Lamarck, Jean-Baptiste, 84
Langdon-Davies, John, 191
Larson, Edward, 195
Le Cat, Claude-Nicolas, 35
Leffingwell, Albert, 156
Leopold and Loeb Trial, 206
Leibniz, Gottfried, 9
Lepore, Jill, 47
leprosy, 209
Lewis, Ivey Foreman, 166
Linnaeus, Carl, 52, 58
Lister, Joseph, 154, 155
Loeffler, Friedrich, 157
Longfellow, Fanny, 86
Luther, Martin, 17
Lyell, Charles, 112

MacDonald, James, 88–89
Magill, Helen, 144
malaria, 1, 123
Malthus, Thomas, 63–69, 74, 113–14, 178
Mann, Horace, 73
Mariotti, Shannon, 161
Marsh, John, 73

Marx, Karl, 109, 176, 179
Massey, Edmund, 31
Mather, Abigail (daughter), 7–8, 34
Mather, Abigail (wife), 8, 12, 30
Mather, Cotton, 2, 7–20, 26–27, 29–37, 143, 182; compared to Ralph Waldo Emerson, 72; compared to Franklin, 38, 40, 43; compared to Jefferson, 54, 55–56, 58
Mather, Increase, 33
Mather, Joseph, 34–35, 182
Mather, Kirtley Fletcher, 209–10
Mathews, Shailer, 196
McCann, F. E., 159
measles, 2, 16, 87, 171
mechanical philosophy, 20, 21–23, 24–25, 28, 30, 38, 39, 46; and anatomy, 27, 71, 154; Indigenous knowledge systems and, 28–29, 216, 222; materialism and, 206
Meckel, Richard, 161, 170
Mecon, Jane, 4
medicine: Bureau of Indian Affairs and, 214, 220, 222; debates over means of progress in, 33, 44, 56, 66–67, 74–75, 81–82, 83–84, 87, 92, 138, 140, 146–47, 150, 174, 203, 206, 209; Indigenous, 28–29; seventeenth-century Christian attitudes toward, 16–20, 28, 32; treatments, 42, 45, 53, 54, 71, 83, 138, 150–60, 181, 182, 208. *See also* antitoxin; eugenics; inoculation; sanitation; science, exclusion based on; science, mechanistic foundations of; science, natural law and progress in; vaccination
Meigs, Charles, 152
Meiners, Christoph, 52
meliorism (coined), 147
Mendenhall, Dorothy Reed, 171
Mersenne, Marin, 22
Metcalf, Maynard, 209
Mettrie, Julien Offray de La, 42
Milbank, Elizabeth, 171

Miles, Tiya, 95
milk stations, 172, 174
Miller, William, 97
miracles, 64–65, 67, 74, 76–77, 80, 84; criticisms of belief in, 107–8, 113, 119, 132, 135, 137, 142, 146, 148–49, 202; defenses of, 194, 196; general and special providence versus, 98, 107
Mitchell, John, 52
monogenism, 88, 89, 103, 120–22
Moody, Mary, 78, 83
Moore, James, 123, 128
Moran, Jeffrey, 210
Morris, Caspar, 70, 133
mumps, 2, 122
Murphy, Howard, 72

National Association for the Advancement of Colored People (NAACP), 179, 180, 192
National Medical Association, 162
National Tuberculosis Association, 214
natural law. *See* science
natural selection: Charles Darwin and, 114–15, 116, 117; eugenics and, 180, 184, 187, 188, 189; explanations of suffering and, 123, 128, 129, 149; opposition to, 200. *See also* evolution
natural theology: American education and, 65, 75, 81, 139, 216; anatomy and, 27–28; defined, 20; deist versions of, 41–42, 44; eugenics and, 183; Asa Gray and, 117–18; Thomas Jefferson and, 55; slavery and, 106, 119; Phillis Wheatley and, 50. *See also* design argument; theodicy
Natural Theology; or, Evidences of the Existence and Attributes of the Deity, Collected from the Appearances of Nature, 65, 75, 216
Neu-mon-ya, 60
New England Courant, The, 33
Newton, Isaac, 21, 23, 41, 65, 73, 137

New York Society for the Promotion of Education Among Colored Children, 88
Northrup, W. P., 157–58
Nott, Josiah, 2, 101–3, 122

Occom, Samson, 36
O'Dwyer, Joseph, 153
Omaha Reservation, 219–20
Onesimus, 14–15, 30–31, 32, 143
On the Origin of Species by Natural Selection, 84, 103, 104, 105, 111, 114–15, 117, 120, 129, 135, 140
opium, 20
original sin, 10, 13, 58, 73, 75. *See also* Fall of Man
Osborn, Henry Fairfield, 201
Owen, Richard, 82

Paley, William, 65, 75, 106, 110, 117
Parker, Theodore, 84, 135, 142
Pasteur, Louis, 146, 154, 170, 196, 217
Paustoobee, 13
Peabody, Andrew Preston, 75, 85–86, 115, 161
Pearson, Karl, 187
pediatrics, 190, 223–24. *See also* Holt, Luther Emmett; Jacobi, Abraham
Pernick, Martin, 174, 183
Phelps, Elizabeth Stuart, 148
Philadelphia Negro, The, 163, 164
philanthropists, 1, 171–74, 226
phrenology, 108, 124
Pickens, William, 210
Picotte, Susan La Flesche, 3, 219–20
Plessy v. Ferguson, 162
Plymouth Church, Brooklyn, 124, 125
Plymouth Platform, 172
polio, 2
polygenism, 90, 101–3, 120–22, 162–63
Popenoe, Paul, 187–88
Porter, Roy, 42
postmillennialism, 19, 43, 58, 122, 133, 150, 178; criticism of, 132–33, 149–50, 194; and eugenics, 191

prayer: belief in, 12, 14, 16, 31, 45, 51, 57, 94, 153, 159, 194; repudiation of, 107–8, 110, 133, 135, 148, 152
premillennialism, 19, 44, 97, 194–95
prenatal genetic testing, 226
Priestley, Joseph, 74
Primitive Physick: or, An Easy and Natural Method for Curing Most Diseases, 17
Proctor, Adelaide A., 176
progress. *See* medicine; postmillenialism; science; *names of individuals*
Protestants: mechanical philosophy and, 21–22; medicine and, 17; miracles and, 67, 98, 148. *See also* fundamentalist Christianity; higher criticism; *names of individuals*
providence, general vs. special, 23, 34, 36–37, 40–41, 57, 74, 82, 86, 98, 107–8, 138, 202. *See also* miracles
public health, 145, 190–91. *See also* eugenics; Kelley, Florence; sanitation; Straus, Nathan
Pusey, William Allen, 180

racism: debates over inoculation and, 33; in courts, 212; genocide and, 37, 127–29; scientific, 61–62, 68–69, 88–93, 100–103, 105, 120–22, 128–29, 136, 161–69, 179, 189–90, 218
Rappleyea, George, 207
Rauschenbusch, Walter, 174–75
Ray, John, 19, 21, 22–26, 36
Recollections of the Development of My Mind and Character, 116
Reconstruction, 162
Right Management of the Sick under the Distemper of the Measles, 16
Rittenhouse, David, 52
Ritter, William, 204
Robinson, William J., 201
Rockefeller, John D., 171
Rose, Ernestine, 134, 171
Ross, Edward Alsworth, 189
Roux, Emile, 157

Royal Society of London, 21–23, 28, 41, 46
Rubin, Julius, 14
Ruse, Michael, 133–34
Rush, Benjamin, 57, 58
Russian Revolution, 194

Salem Witch Trials, 33, 35
Sanger, Margaret, 2, 176–78, 192
sanitation, 138, 144, 145, 154, 155, 171, 187, 214, 217–18, 220
scarlet fever, 2, 3, 70–71, 94
science: exclusion based on, 45–52, 58–62, 67–69, 88–94, 100–103, 119–22, 127–29, 136, 162–63, 166, 169, 179–80, 187–93, 210–12, 218–19, 224–25; mechanistic, foundations of, 20–22, 28, 29, 137; natural law and progress in, 18–19, 55, 56, 58, 63, 64–68, 74, 80–82, 84, 108, 109–11, 126–27, 130, 137–38, 140–42, 147–49, 159–60, 188, 196, 201–3, 209–10. *See also* eugenics; evolution; mechanical philosophy; medicine
Science in Search of God, 209
Scopes, John T., 205, 208, 209
Scopes Trial, 5, 205–12, 217
Scott, Leslie M., 37
Second Great Awakening, 94
Sedgwick, Adam, 79
segregation, 162–63, 164, 166, 168–69, 179–80, 211
Sells, Cato, 215, 220
Semmelweiss, Ignaz, 155
Sergeant, John, Jr., 59
Seventh Day Adventists, 195
Shaw, George Bernard, 150, 160
Shelton, Ruth (daughter), 212–13
Shelton, Ruth (mother), 214
Shelton, William, 214, 215, 221
Sheppard-Towner Act, 180
Simpson, James Young, 102, 185
Slater, Peter Gregg, 11
slavery, 36, 43, 48–52, 61, 120, 139, 163, 225; defenses of, 50–51, 52, 69, 92, 100–101, 104–5, 106, 119; opposition to, 51, 87–88, 90, 94–95, 98–103, 108, 112–13, 122. *See also* abolition
smallpox, 2, 12; differential mortality from, 37, 214; Indigenous explanations of, 14, 60; inoculation, 30–34, 37–38, 44–45; vaccination, 63, 160, 170, 174, 175, 184, 200
Smith, Gerrit, 91, 93
Smith, James McCune, 2, 87–94, 103, 107, 111, 162
Smith, Job Lewis, 151
Snow, John, 97, 155
social determinants of health, 164, 175, 219–20
Social Gospel, 175
socialism, 175–76, 177, 178, 191
Socialist Party of America, 176
sociology, 163–64, 203–4
Souls of Black Folk, The, 160
Speene, Robin, 14
Speight, Harold E. B., 136, 210
Spencer, Herbert, 124–25, 140
Stanton, Elizabeth Cady, 3, 68, 134, 171
Stauffer, John, 91
St. Basil the Great, 17, 51
Stiles, Ezra, 40
Stoddard, Lothrop, 189, 192
Stowe, Calvin, 97
Stowe, Charley, 97
Stowe, Harriet Beecher, 2, 94–100, 167
Straus, Nathan, 2, 171–74, 190
Strauss, David, 109, 119
struggle for existence, 112, 114–15, 116, 117, 122; as a justification of the status quo, 68–69, 92, 127, 181, 218–21; progress and the, 74, 126, 128, 149, 184, 204–5; war and the, 198
suffering, explanations of. *See* theodicy
summer diarrhea. *See* dysentery
Sumner, William Graham, 163–64
Sunday, Billy, 3, 193–94, 195–96
surgery, 181, 182, 185
Sweets, Ossian, 212

Sydenham, Thomas, 20, 56, 81

Taney, Roger B., 102
Taylor, Edward, 16
Tennyson, Alfred, 79, 204
Terrell, Mary Church, 2, 166–69
Terrell, Robert Heberton, 166, 167–68
theodicy: Henry Ward Beecher's, 123, 126–27; Robert Chambers's, 80–81; William Ellery Channing's, 74; Andrew Combe's, 66–67; Edwin Grant Conklin's, 204–5; criticism of, 130–31, 149; Charles Darwin's, 115–17, 128; definition of, 9; Frederick Douglass's, 107–8, 110; Ralph Waldo Emerson's, 77, 82–83; eugenics and, 183, 186, 188–89, 190–91; Benjamin Franklin's, 40; Asa Gray's, 117–18; Indigenous, 12–14; Thomas Jefferson's, 55–56, 57; Thomas Malthus's, 64–65; Cotton Mather's, 9–12, 15; Kirtley Fletcher Mather's, 209–10; natural theology and, 27–28, 35–36, 117, 124; Andrew Preston Peabody's, 85–86; Harriet Beecher Stowe's, 97–98; Billy Sunday's, 197; Phillis Wheatley's, 50; Andrew Dickson White's, 144
theory of special creation, 56, 79, 81–82, 111–12; Darwin and the, 112–15, 119. *See also* design argument; natural theology
Thoreau, Henry David, 77
Thorndike, Lynn, 197
Thorton, Alice, 11
Tidyman, Philip, 101
tikkun olam, 172
Tourgée, Albion Winegar, 129
Towner-Sterling Bill, 179
tracheotomies, 151–52, 159
transcendentalism, 77, 82, 124, 161, 193
transmutation. *See* evolution
Treatise on the Physiological and Moral Management of Infancy, 66–67, 134
Treaty of Point Elliott, 214, 215, 219, 221

tuberculosis, 3, 75–76, 155, 165, 212–14, 217–18, 219–20
Tulalip Indian Boarding School, 213, 215
Tulalip Reservation, 213
Turner, Frank, 133
Turner, James, 41, 43, 44, 58
Turner, Thomas Wyatt, 192
Twain, Mark. *See* Clemens, Samuel
Types of Mankind, 101
typhoid, 2, 144, 196

Uncle Tom's Cabin, 99, 113
Unitarianism, 62, 73–74, 75, 84, 86, 139, 210
United States Children's Bureau, 179, 180

vaccination, 32, 60, 62, 143, 160, 170, 184; criticism of, 175–76, 200; promised by treaties, 214; twentieth-century developments in, 223, 226
Vesalius, Andreas, 27, 29, 143
Vestiges of the Natural History of Creation, 79–82, 84, 142
Vindication of the Rights of Women, 48
vis medicatrix naturae, 62, 83, 154
vivisection, 152, 156–57, 159
Voltaire, 42

Wallace, Alfred Russel, 114, 149
Walsh, James Joseph, 146–47, 182, 191
war, first world, 194, 197–98, 220
Warner, Margaret Humphrey, 15
Waterhouse, Benjamin, 53, 60, 62, 66
Welch, William, 158
Wells, H. G., 169
Wesley, John, 13, 28–29
Westminster Confession, 18
Wharton, Thomas, 27
Wheatley, Phillis, 2, 48–52, 61
White, Andrew Dickson, 2, 138–50, 164, 169, 170, 179, 186, 197
White, Frederick, 144, 145
Whitman, Walt, 124
Whole Duty of a Woman, The, 48

whooping cough, 2, 53, 62, 94, 151, 170
Wiggam, Albert, 188–89, 191, 196
Willis, Thomas, 27
Wilson, John B., 82
Wisdom of God, The, 23–25, 26
Wister, Lowry, 45, 46
witchcraft, 35, 182

Wollstonecraft, Mary, 48
Woman's Bible, The, 134
World Health Organization, 226

yellow fever, 2, 101
Youmans, Edward L., 125, 198